Andrey O. Matveev

Farey Sequences

Also of interest

Pattern Recognition on Oriented Matroids
Andrey O. Matveev, 2017
ISBN 978-3-11-053071-1, e-ISBN (PDF) 978-3-11-053114-5,
e-ISBN (EPUB) 978-3-11-053084-1

Graphs for Pattern Recognition. Infeasible Systems of Linear Inequalities
Damir Gainanov, 2016
ISBN 978-3-11-048013-9, e-ISBN (PDF) 978-3-11-048106-8,
e-ISBN (EPUB) 978-3-11-048030-6

Compressive Sensing. Applications to Sensor Systems and Image Processing
Joachim Ender, 2017
ISBN 978-3-11-033531-6, e-ISBN (PDF) 978-3-11-033539-2,
e-ISBN (EBUB) 978-3-11-039027-8

Regular Graphs. A Spectral Approach
Zoran Stanic, 2017
ISBN 978-3-11-035128-6, e-ISBN (PDF) 978-3-11-035134-7,
e-ISBN (EBUB) 978-3-11-038336-2

Discrete Algebraic Methods. Arithmetic, Cryptography, Automata and Groups
Volker Diekert, Manfred Kufleitner, Gerhard Rosenberger, Ulrich Hertrampf, 2016
ISBN 978-3-11-041332-8, e-ISBN (PDF) 978-3-11-041333-5,
e-ISBN (EBUB) 978-3-11-041632-9

Andrey O. Matveev

Farey Sequences

Duality and Maps Between Subsequences

DE GRUYTER

Mathematics Subject Classification 2010
Primary: 11B57; secondary: 05-01, 11-01

Author
Dr. Andrey O. Matveev
Ekaterinburg
Russia
andrey.o.matveev@gmail.com

ISBN 978-3-11-054662-0
e-ISBN (PDF) 978-3-11-054766-5
e-ISBN (EPUB) 978-3-11-054665-1
Set-ISBN 978-3-11-054767-2

Library of Congress Cataloging-in-Publication Data
A CIP catalog record for this book has been applied for at the Library of Congress.

Bibliographic information published by the Deutsche Nationalbibliothek
The Deutsche Nationalbibliothek lists this publication in the Deutsche Nationalbibliografie; detailed bibliographic data are available on the Internet at http://dnb.dnb.de.

Cover image: A Perspective on Farey Duality, 2017, by Andrey L. Kopyrin, Ekaterinburg, Russia
Typesetting: Dimler & Albroscheit, Müncheberg
Printing and binding: CPI books GmbH, Leck
♾ Printed on acid-free paper
Printed in Germany

www.degruyter.com

To the memory of my mother

Irina M. Matveeva

1934–2014

To my father

Oleg S. Matveev

Preface

The *Farey sequence* $\mathcal{F}_n$ of order n is defined to be the increasing sequence of irreducible fractions $\frac{h}{k}$ such that $\frac{0}{1} \le \frac{h}{k} \le \frac{1}{1}$ and $k \le n$. It is one of the most attractive and popular constructs in number theory.

This book is about the standard sequences $\mathcal{F}_n$ and their fundamental *Farey subsequences* of a number-theoretic and combinatorial nature.

We discuss in detail basic components of these sequences and related topics such as *pairs* and *triples of consecutive fractions*, the *number of fractions*, and the *position of a fraction* in the sequences.

A finite n-set with its distinguished proper m-subset is the source of a sequence of fractions, of a combinatorial nature, whose *half sequences* are in *one-to-one correspondence*, called *Farey duality*, with purely number-theoretic Farey (sub)sequences.

We investigate Farey duality in the cases where $n = 2m$ or $n \neq 2m$, and we extend results on Farey (sub)sequences by presenting their useful dual formulations.

The nice inner structure of the Farey sequences can be easily described in the language of *monotone maps between subsequences*. In order to characterize these maps, we use simple manipulations with 2×2 integer-valued matrices.

In Chapter 1 we discuss various subsequences of the Farey sequence $\mathcal{F}_n$ that can be regarded as true Farey subsequences, since they are all characterized by identical number-theoretic and enumerative properties.

The Farey subsequences $\mathcal{F}(\mathbb{B}(n), m)$, $\mathcal{F}^{\le\frac{1}{2}}(\mathbb{B}(n), m)$ and $\mathcal{F}^{\ge\frac{1}{2}}(\mathbb{B}(n), m)$, closely related to *finite sets*, *Boolean lattices* and finite-dimensional *vector spaces*, are defined in Chapter 1 by

$$\begin{aligned}
\mathcal{F}(\mathbb{B}(n), m) &:= \left(\tfrac{h}{k} \in \mathcal{F}_n : m + k - n \le h \le m\right), \\
\mathcal{F}^{\le\frac{1}{2}}(\mathbb{B}(n), m) &:= \left(\tfrac{h}{k} \in \mathcal{F}(\mathbb{B}(n), m) : \tfrac{h}{k} \le \tfrac{1}{2}\right), \\
\mathcal{F}^{\ge\frac{1}{2}}(\mathbb{B}(n), m) &:= \left(\tfrac{h}{k} \in \mathcal{F}(\mathbb{B}(n), m) : \tfrac{h}{k} \ge \tfrac{1}{2}\right),
\end{aligned}$$

where $n > 1$ and $0 < m < n$.

We mention the significance of the *order-reversing* and *bijective mapping* $\frac{h}{k} \mapsto \frac{k-h}{k}$, describe *pairs* and *triples of consecutive fractions* in Farey (sub)sequences, count *the number of fractions*, determine their *positions*, discuss the *rank problem*, and conclude the chapter with a *generating function* of the sequence $\mathcal{F}_n$.

By *Farey duality*, discussed in Chapter 2, we mean the existence of *monotone* and *bijective* maps

$$\begin{aligned}
\mathcal{F}^{\le\frac{1}{2}}(\mathbb{B}(n), m) &\leftrightarrow \mathcal{F}^{m}_{n-m} := \left(\tfrac{h}{k} \in \mathcal{F}_{n-m} : h \le m\right) \qquad \text{(order-preserving)}, \\
\mathcal{F}^{\ge\frac{1}{2}}(\mathbb{B}(n), m) &\leftrightarrow \mathcal{F}^{n-m}_{m} := \left(\tfrac{h}{k} \in \mathcal{F}_{m} : h \le n - m\right) \quad \text{(order-reversing)}
\end{aligned}$$

and, in particular, the existence of *monotone* and *bijective* maps

$$\mathcal{F}^{\le\frac{1}{2}}(\mathbb{B}(2m), m) \leftrightarrow \mathcal{F}_m \leftrightarrow \mathcal{F}^{\ge\frac{1}{2}}(\mathbb{B}(2m), m).$$

DOI 10.1515/9783110547665-202

In Chapter 2 we first discuss *duality properties* of Farey (sub)sequences, describe their connection with the *Farey map*, and then show that almost all results of Chapter 1 have useful dual formulations.

In Chapter 3 Farey duality serves as an auxiliary technique for establishing *monotone bijective maps* $\mathcal{C}' \leftrightarrow \mathcal{C}''$ between subsequences $\mathcal{C}'$ and $\mathcal{C}''$ of the same Farey sequence $\mathcal{F}_n$ of large order n.

If $\mathcal{F}_m$ is a Farey sequence of small order m, then easily computed *matrix products* $\mathbf{M}$ and $\mathbf{N}$, of equal length and with the factors $\left[\begin{smallmatrix}1&0\\1&1\end{smallmatrix}\right]$ or $\left[\begin{smallmatrix}0&1\\-1&2\end{smallmatrix}\right]$, enable us to give concise descriptions of the *bijections*

$$\begin{aligned}&\mathcal{F}_n \supset \mathbf{M}\cdot\mathcal{F}_m =: \mathcal{C}' \to \mathcal{C}'' := \mathbf{N}\cdot\mathcal{F}_m \subset \mathcal{F}_n,\\ &\left[\begin{smallmatrix}h\\k\end{smallmatrix}\right] \mapsto \mathbf{N}\mathbf{M}^{-1}\cdot\left[\begin{smallmatrix}h\\k\end{smallmatrix}\right] \quad \text{(order-preserving)}\end{aligned}$$

and

$$\begin{aligned}&\mathcal{F}_n \supset \mathcal{C}' \to \mathcal{C}'' \subset \mathcal{F}_n,\\ &\left[\begin{smallmatrix}h\\k\end{smallmatrix}\right] \mapsto \mathbf{N}\left[\begin{smallmatrix}-1&1\\0&1\end{smallmatrix}\right]\mathbf{M}^{-1}\cdot\left[\begin{smallmatrix}h\\k\end{smallmatrix}\right] \quad \textit{(order-reversing)},\end{aligned}$$

between subsequences $\mathbf{M}\cdot\mathcal{F}_m := (\mathbf{M}\cdot\left[\begin{smallmatrix}h\\k\end{smallmatrix}\right] : \left[\begin{smallmatrix}h\\k\end{smallmatrix}\right] \in \mathcal{F}_m)$ and $\mathbf{N}\cdot\mathcal{F}_m$ of the sequence $\mathcal{F}_n$, whose fractions are written in vector form.

The book is intended as a detailed exposition of the basic properties of Farey (sub)sequences, and as a handbook of formulas.

I hope that the text will be of interest to researchers and students, to engineers and software developers.

November 2017

Andrey O. Matveev
Ekaterinburg

Contents

DOI 10.1515/9783110547665-203

List of Tables

DOI 10.1515/9783110547665-204

Farey sequences and collective decision making

The Farey sequence $\mathcal{F}_m$ of order m, defined to be the increasing sequence of irreducible fractions $\frac{h}{k}$ such that $\frac{0}{1} \le \frac{h}{k} \le \frac{1}{1}$ and $k \le m$, has a wide area of applications in mathematics, computer science and physics [45].

The Farey sequences $\mathcal{F}_{2m}$ of order $2m$ contain the subsequences

$$\mathcal{F}(\mathbb{B}(2m), m) := \left(\tfrac{h}{k} \in \mathcal{F}_{2m} : k - m \le h \le m\right),$$

which can be useful for an analysis of collective decision-making procedures. One such procedure is *pattern recognition* by means of *committee decision rules*.

The groundwork for the modern *committee theory* was laid by the short notes [1, 2]; see also, e.g., [62, 90, 112, 114, 115].

A finite collection $\mathcal{H}$ of pairwise distinct hyperplanes in the *feature space* $\mathbb{R}^n$ is said to be a *training set* if it is partitioned into two nonempty subsets $\mathcal{A}$ and $\mathcal{B}$; see Figure 1 (a).

A codimension one subspace $\boldsymbol{H} := \{\mathbf{x} \in \mathbb{R}^n : \langle \boldsymbol{p}, \mathbf{x} \rangle := \sum_{j=1}^{n} p_j x_j = 0\}$ of the *hyperplane arrangement* $\mathcal{H}$ is defined by its normal vector $\boldsymbol{p} \in \mathbb{R}^n$, and this hyperplane is oriented: a vector $\boldsymbol{v}$ lies on the *positive side* of $\boldsymbol{H}$ if $\langle \boldsymbol{h}, \boldsymbol{v} \rangle > 0$, where, by convention, $\boldsymbol{h} := -\boldsymbol{p}$ if $\boldsymbol{H} \in \mathcal{A}$, and $\boldsymbol{h} := \boldsymbol{p}$ if $\boldsymbol{H} \in \mathcal{B}$; see Figure 1 (b). In a similar manner, a *region* $\boldsymbol{T}$ of the hyperplane arrangement $\mathcal{H}$, that is, a connected component of the *complement* $\mathcal{T} := \mathbb{R}^n - \mathcal{H}$, lies on the positive side of the hyperplane $\boldsymbol{H}$ if $\langle \boldsymbol{h}, \boldsymbol{v} \rangle > 0$ for some vector $\boldsymbol{v} \in \boldsymbol{T}$. Let $\mathcal{T}_{\boldsymbol{H}}^{+}$ denote the set of all regions lying on the positive side of $\boldsymbol{H}$; see Figure 2.

The oriented hyperplanes of the arrangement $\mathcal{H}$ are called the *training patterns*. The *training samples* $\mathcal{A}$ and $\mathcal{B}$ provide a partial description of two disjoint *classes* **A** and **B**, respectively; a priori, we have $\mathbf{A} \supseteq \mathcal{A}$ and $\mathbf{B} \supseteq \mathcal{B}$.

We call a subset $\mathcal{K}^* \subset \mathcal{T}$ a *committee of regions* for the arrangement $\mathcal{H}$ if $|\mathcal{K}^* \cap \mathcal{T}_{\boldsymbol{H}}^{+}| > \frac{1}{2}|\mathcal{K}^*|$ for each hyperplane $\boldsymbol{H} \in \mathcal{H}$; see Figure 3 (a).

Consider a new pattern $\boldsymbol{G} := \{\mathbf{x} \in \mathbb{R}^n : \langle \boldsymbol{g}, \mathbf{x} \rangle = 0\} \notin \mathcal{H}$ determined by its normal vector $\boldsymbol{g} \in \mathbb{R}^n$. If a system of distinct representatives $\boldsymbol{W} := \{\boldsymbol{w} \in \boldsymbol{K} : \boldsymbol{K} \in \mathcal{K}^*\}$, of cardinality $|\mathcal{K}^*|$, for the committee of regions $\mathcal{K}^*$ is fixed, then the corresponding *committee decision rule* recognizes the pattern $\boldsymbol{G}$ as an element of the class **A** if $|\{\boldsymbol{w} \in \boldsymbol{W} : \langle \boldsymbol{g}, \boldsymbol{w} \rangle > 0\}| < \frac{1}{2}|\boldsymbol{W}|$; the pattern $\boldsymbol{G}$ is recognized as an element of the class **B** if $|\{\boldsymbol{w} \in \boldsymbol{W} : \langle \boldsymbol{g}, \boldsymbol{w} \rangle > 0\}| > \frac{1}{2}|\boldsymbol{W}|$; see Figure 3 (b).

For any hyperplane $\boldsymbol{H}$ of the arrangement $\mathcal{H}$, the increasing collection of irreducible fractions

$$\left(\frac{|\mathcal{R} \cap \mathcal{T}_{\boldsymbol{H}}^{+}|}{\gcd(|\mathcal{R} \cap \mathcal{T}_{\boldsymbol{H}}^{+}|, |\mathcal{R}|)} \Big/ \frac{|\mathcal{R}|}{\gcd(|\mathcal{R} \cap \mathcal{T}_{\boldsymbol{H}}^{+}|, |\mathcal{R}|)} : \mathcal{R} \subseteq \mathcal{T},\ |\mathcal{R}| > 0\right)$$

is the Farey subsequence $\mathcal{F}(\mathbb{B}(|\mathcal{T}|), \frac{1}{2}|\mathcal{T}|)$ whose half sequences are, in a sense, *duals* of the Farey sequence $\mathcal{F}_{|\mathcal{T}|/2}$.

DOI 10.1515/9783110547665-001

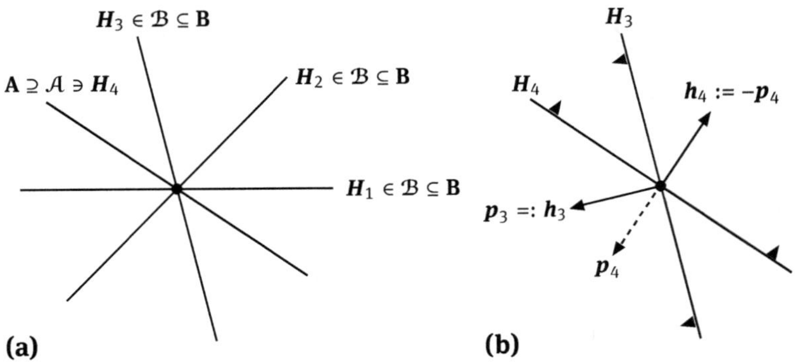

Fig. 1. (a) A training set $\mathcal{H} := \{H_1, H_2, H_3, H_4\}$ in the feature space $\mathbb{R}^2$ partitioned into training samples $\mathcal{A} := \{H_4\}$ and $\mathcal{B} := \{H_1, H_2, H_3\}$, which provide a partial description $\mathbf{A} \supseteq \mathcal{A}$ and $\mathbf{B} \supseteq \mathcal{B}$ of classes **A** and **B**.

(b) The training patterns H_3 and H_4 are defined by $H_3 := \{\mathbf{x} \in \mathbb{R}^2 : \langle p_3, \mathbf{x}\rangle = 0\}$ and $H_4 := \{\mathbf{x} \in \mathbb{R}^2 : \langle p_4, \mathbf{x}\rangle = 0\}$, where p_3 and p_4 are their normal vectors. By convention, the positive sides (marked by triangles) of the oriented hyperplanes $H_3 \in \mathcal{B}$ and $H_4 \in \mathcal{A}$ are determined by the vectors $h_3 := p_3$ and $h_4 := -p_4$.

Indeed, the *left* and *right half sequences*

$$\mathcal{F}^{\le\frac{1}{2}}(\mathbb{B}(2m), m) := \left(\tfrac{h}{k} \in \mathcal{F}(\mathbb{B}(2m), m) \colon \tfrac{h}{k} \le \tfrac{1}{2}\right)$$

and

$$\mathcal{F}^{\ge\frac{1}{2}}(\mathbb{B}(2m), m) := \left(\tfrac{h}{k} \in \mathcal{F}(\mathbb{B}(2m), m) \colon \tfrac{h}{k} \ge \tfrac{1}{2}\right)$$

of the sequence $\mathcal{F}(\mathbb{B}(2m), m)$ and the Farey sequence $\mathcal{F}_m$ all have the same number of fractions; for example,

$$\begin{aligned}
\mathcal{F}_4 &:= \left(\tfrac{0}{1} < \tfrac{1}{4} < \tfrac{1}{3} < \tfrac{1}{2} < \tfrac{2}{3} < \tfrac{3}{4} < \tfrac{1}{1}\right),\\
\mathcal{F}^{\le\frac{1}{2}}(\mathbb{B}(2\cdot 4), 4) &:= \left(\tfrac{0}{1} < \tfrac{1}{5} < \tfrac{1}{4} < \tfrac{1}{3} < \tfrac{2}{5} < \tfrac{3}{7} < \tfrac{1}{2}\right),\\
\mathcal{F}^{\ge\frac{1}{2}}(\mathbb{B}(2\cdot 4), 4) &:= \left(\tfrac{1}{2} < \tfrac{4}{7} < \tfrac{3}{5} < \tfrac{2}{3} < \tfrac{3}{4} < \tfrac{4}{5} < \tfrac{1}{1}\right).
\end{aligned}$$

Moreover, the *numerators* of the fractions of $\mathcal{F}_m$ are the *numerators* of the fractions of the sequence $\mathcal{F}^{\le\frac{1}{2}}(\mathbb{B}(2m), m)$, and the *denominators* of the fractions of $\mathcal{F}_m$ are the *numerators* of the fractions of the sequence $\mathcal{F}^{\ge\frac{1}{2}}(\mathbb{B}(2m), m)$.

A neighborhood of the critical value $\frac{1}{2}$ in $\mathcal{F}(\mathbb{B}(|\mathcal{T}|), \frac{1}{2}|\mathcal{T}|)$ has the simple structure, since the fractions

$$\tfrac{\lceil m/2\rceil - 1}{2\lceil m/2\rceil - 1} < \cdots < \tfrac{m-3}{2m-5} < \tfrac{m-2}{2m-3} < \tfrac{m-1}{2m-1} < \tfrac{1}{2} < \tfrac{m}{2m-1} < \tfrac{m-1}{2m-3} < \tfrac{m-2}{2m-5} < \cdots < \tfrac{\lceil m/2\rceil}{2\lceil m/2\rceil - 1}$$

are known to be *consecutive* in the sequences $\mathcal{F}(\mathbb{B}(2m), m)$.

Note also that the map

$$\mathcal{F}(\mathbb{B}(2m), m) \to \mathcal{F}(\mathbb{B}(2m), m), \quad \tfrac{h}{k} \mapsto \tfrac{k-h}{k},$$

is *order-reversing* and *bijective*.

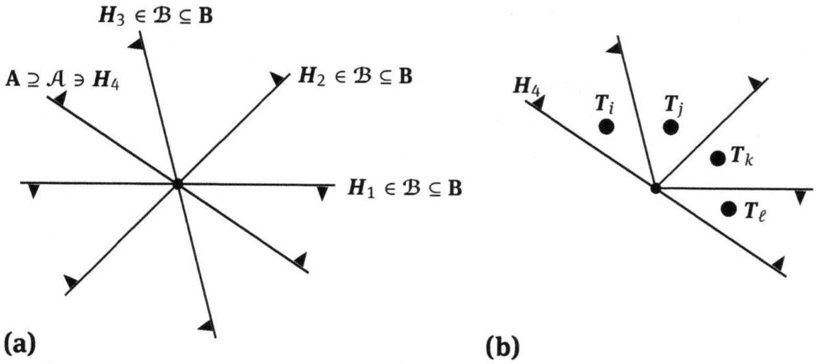

Fig. 2. (a) An arrangement of oriented linear hyperplanes $\mathcal{H}$ in $\mathbb{R}^2$ associated with the training set of Figure 1 (a).

(b) $\mathcal{T}^+_{H_4} = \{T_i, T_j, T_k, T_\ell\}$ is the set of all regions (marked by discs) lying on the positive side of the oriented hyperplane H_4.

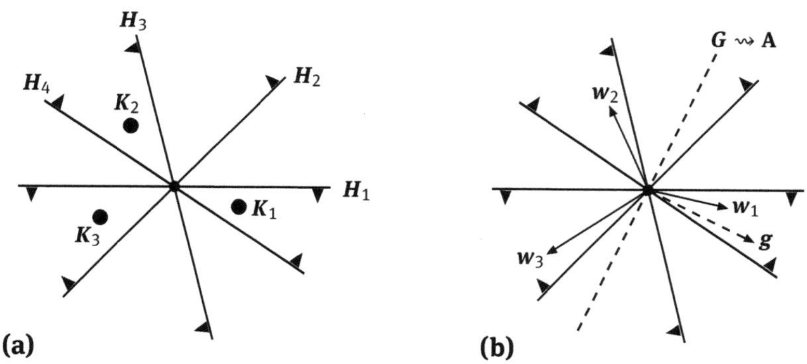

Fig. 3. (a) The set of regions (marked by discs) $\mathcal{K}^* := \{K_1, K_2, K_3\}$ is a committee of regions for the arrangement of oriented hyperplanes $\mathcal{H}$ of Figure 2 (a) associated with the training set of Figure 1 (a).

(b) A new pattern $G := \{\mathbf{x} \in \mathbb{R}^2 : \langle g, \mathbf{x}\rangle = 0\}$ is determined by its normal vector g. A set of vectors $W := \{w_1, w_2, w_3\} = \{w \in K : K \in \mathcal{K}^*\}$ is a system of distinct representatives for the committee of regions $\mathcal{K}^*$. Since $|\{w \in W : \langle g, w\rangle > 0\}| = 1 < \frac{1}{2}|W| = \frac{1}{2} \cdot 3$, the corresponding committee decision rule recognizes the pattern G as an element of the class $\mathbf{A}$.

We have mentioned just a few basic properties of the sequences $\mathcal{F}(\mathbb{B}(2m), m)$ that are discussed in the book.

From the number-theoretic point of view, a subset of regions $\mathcal{K}^* \subset \mathcal{T}$ is a committee for the arrangement of oriented hyperplanes $\mathcal{H}$ if and only if for each hyperplane $H \in \mathcal{H}$, we have

$$\frac{|\mathcal{K}^* \cap \mathcal{T}^+_H|}{\gcd(|\mathcal{K}^* \cap \mathcal{T}^+_H|, |\mathcal{K}^*|)} \Big/ \frac{|\mathcal{K}^*|}{\gcd(|\mathcal{K}^* \cap \mathcal{T}^+_H|, |\mathcal{K}^*|)} \in \mathcal{F}^{\geq \frac{1}{2}}(\mathbb{B}(|\mathcal{T}|), \tfrac{1}{2}|\mathcal{T}|) - \{\tfrac{1}{2}\};$$

thus, the study of the structure of the family of all committees of regions for the hyperplane arrangement $\mathcal{H}$ might involve the sequence $\mathcal{F}^{\geq\frac{1}{2}}(\mathbb{B}(|\mathcal{T}|), \frac{1}{2}|\mathcal{T}|)$. Such an attempt is made in [112, Chapter 7] within the bounds of *oriented matroid* theory [21, 31].

The Farey subsequences $\mathcal{F}(\mathbb{B}(2m), m)$ are closely related to the systems of constraints, given by the arrangements of oriented linear hyperplanes, in view of their central symmetry with respect to the origin. General models of voting and of collective decision making can be accompanied in a natural way by the Farey subsequences

$$\mathcal{F}(\mathbb{B}(n), m) := \left(\tfrac{h}{k} \in \mathcal{F}_n : m + k - n \leq h \leq m\right), \quad n > 1,\ 0 < m < n,$$

extensively studied in the book. In analogy to the particular case of the sequences $\mathcal{F}(\mathbb{B}(2m), m)$, the left and right half sequences of the sequences $\mathcal{F}(\mathbb{B}(n), m)$ will also be shown, via the so-called *Farey duality*, to be *duals* of other Farey (sub)sequences.

As a matter of fact, the subject of the present book, as noted in the Preface, is much wider. The text provides a very detailed and readily accessible exposition of various basic structural and enumerative properties of four fundamental Farey (sub)sequences.

1 Basic properties of Farey sequences

The *Farey sequence* $\mathcal{F}_n$ of order n is defined to be the increasing sequence of irreducible fractions $\frac{h}{k}$ such that $\frac{0}{1} \le \frac{h}{k} \le \frac{1}{1}$ and $k \le n$.

There are many subsequences of $\mathcal{F}_n$ that can be regarded as true *Farey subsequences*, since they are all characterized by identical number-theoretic and enumerative properties.

In this chapter we show that Farey subsequences appear naturally in the context of *finite sets*, *Boolean lattices* and finite-dimensional *vector spaces*.

We mention the significance of the *order-reversing* and *bijective mapping* $\frac{h}{k} \mapsto \frac{k-h}{k}$, describe *pairs* and *triples of consecutive fractions* in Farey (sub)sequences, count the *number of fractions*, determine their *positions*, discuss the *rank problem*, and conclude the chapter with a *generating function* of the sequence $\mathcal{F}_n$.

1.1 Finite sets, Boolean lattices, finite-dimensional vector spaces, and Farey sequences $\mathcal{F}_n$

Let S be a finite set, $|S| > 1$. Given a proper subset $A \subset S$, let us find the ratio of the number of elements in the intersection $B \cap A$, of the set A with a nonempty subset $B \subseteq S$, to the number of elements in B. The ratio of $|B \cap A|$ to $|B|$, written in the lowest integer terms, is just the irreducible fraction

$$\frac{|B \cap A|}{\gcd(|B \cap A|, |B|)} \Big/ \frac{|B|}{\gcd(|B \cap A|, |B|)}.$$

This rational number is $\frac{0}{1} = 0$ if the sets B and A do not intersect, and it is $\frac{1}{1} = 1$ if B is contained in A.

Let us now consider all nonempty subsets $B \subseteq S$, and arrange the corresponding irreducible fractions, without repetition, in increasing order. If $|S| = n$ and $|A| = m$, then we denote the resulting sequence by $\mathcal{F}(\mathbb{B}(n), m)$:

$$\mathcal{F}(\mathbb{B}(n), m) := \left(\frac{|B \cap A|}{\gcd(|B \cap A|, |B|)} \Big/ \frac{|B|}{\gcd(|B \cap A|, |B|)} : B \subseteq S,\ |B| > 0\right), \tag{1.1}$$

where the notation $\mathbb{B}(n)$ is reserved for the Boolean lattice of rank n. Note that the sequence is determined only by the cardinalities of the sets S and A.

For instance, if S is the set $[1, 12] := \{1, 2, \dots, 12\}$ of twelve consecutive positive integers, and A is its subset $[1, 6] := \{1, 2, \dots, 6\} \subset [1, 12]$, then the sequence of fractions $\mathcal{F}(\mathbb{B}(12), 6)$ is as follows:

$$\begin{aligned}\mathcal{F}(\mathbb{B}(12), 6) = \Big(&\tfrac{0}{1} < \tfrac{1}{7} < \tfrac{1}{6} < \tfrac{1}{5} < \tfrac{1}{4} < \tfrac{2}{7} < \tfrac{1}{3} < \tfrac{3}{8} < \tfrac{2}{5} < \tfrac{3}{7} < \tfrac{4}{9} < \tfrac{5}{11} \\ &< \tfrac{1}{2} < \tfrac{6}{11} < \tfrac{5}{9} < \tfrac{4}{7} < \tfrac{3}{5} < \tfrac{5}{8} < \tfrac{2}{3} < \tfrac{5}{7} < \tfrac{3}{4} < \tfrac{4}{5} < \tfrac{5}{6} < \tfrac{6}{7} < \tfrac{1}{1}\Big).\end{aligned}$$

DOI 10.1515/9783110547665-002

Throughout the book, the terms '*precedes*' and '*succeeds*' related to pairs of fractions always mean immediate consecution.

The features of the sequence $\mathcal{F}(\mathbb{B}(12), 6)$ look suspiciously like those of the Farey sequences, and this similarity is no accident. The left and right half sequences of the sequence $\mathcal{F}(\mathbb{B}(12), 6)$ are, in a sense, certain duals of the Farey sequence of order 6:

$$\mathcal{F}_6 = (\tfrac{0}{1} < \tfrac{1}{6} < \tfrac{1}{5} < \tfrac{1}{4} < \tfrac{1}{3} < \tfrac{2}{5} < \tfrac{1}{2} < \tfrac{3}{5} < \tfrac{2}{3} < \tfrac{3}{4} < \tfrac{4}{5} < \tfrac{5}{6} < \tfrac{1}{1}).$$

Recall that the *Farey sequence* $\mathcal{F}_n$ of order n is defined to be the increasing sequence of irreducible fractions $\frac{h}{k} \in \mathbb{Q}$ such that $\frac{0}{1} \le \frac{h}{k} \le \frac{1}{1}$ and $k \le n$. In other words,

$$\mathcal{F}_n := \{\tfrac{0}{1}\} \dot\cup (\tfrac{h}{k} : h, k \in \mathbb{P},\ h < k \le n,\ \gcd(h, k) = 1) \dot\cup \{\tfrac{1}{1}\}. \tag{1.2}$$

Throughout this book we will often write fractions $\frac{h}{k} \in \mathbb{Q}$ in vector form $\left[\begin{smallmatrix} h \\ k \end{smallmatrix}\right] \in \mathbb{Z}^2$.

The subsequence $\mathcal{F}(\mathbb{B}(n), m) \subset \mathcal{F}_n$, where $n > 1$ and $0 < m < n$, can be regarded as the sequence of the form

$$\mathcal{F}(\mathbb{B}(n), m) := (\tfrac{h}{k} \in \mathcal{F}_n : m + k - n \le h \le m); \tag{1.3}$$

in particular,

$$\mathcal{F}(\mathbb{B}(2m), m) := (\tfrac{h}{k} \in \mathcal{F}_{2m} : k - m \le h \le m).$$

The restrictions imposed on the fractions from $\mathcal{F}_n$ in the number-theoretic definition (1.3) are so simple that the sequence $\mathcal{F}(\mathbb{B}(n), m)$ can be redefined in certain natural poset-theoretic and linear algebraic contexts.

Let $\mathbb{B}(n)$ be the Boolean lattice of rank $n > 1$, say the lattice of all subsets of our finite set S of cardinality n, with the meet operation denoted $\wedge$. If a is an element of $\mathbb{B}(n)$ of rank $\rho(a) = m$, where $0 < m < n$, then

$$\mathcal{F}(\mathbb{B}(n), m) := \Big(\frac{\rho(b \wedge a)}{\gcd(\rho(b \wedge a), \rho(b))} \Big/ \frac{\rho(b)}{\gcd(\rho(b \wedge a), \rho(b))} : b \in \mathbb{B}(n),\ \rho(b) > 0\Big).$$

Now let $\mathbb{V}(n)$ be an n-dimensional vector space, where $n > 1$. If $\mathbf{A}$ is a proper m-dimensional subspace of $\mathbb{V}(n)$, then

$$\mathcal{F}(\mathbb{B}(n), m) := \Big(\frac{\dim(\mathbf{B} \cap \mathbf{A})}{\gcd(\dim(\mathbf{B} \cap \mathbf{A}), \dim \mathbf{B})} \Big/ \frac{\dim \mathbf{B}}{\gcd(\dim(\mathbf{B} \cap \mathbf{A}), \dim \mathbf{B})} : \mathbf{B} \text{ subspace of } \mathbb{V}(n),\ \dim \mathbf{B} > 0\Big). \tag{1.4}$$

Why should the sequence $\mathcal{F}(\mathbb{B}(12), 6)$ be nominated for a true Farey subsequence? The answer is that, for example, for the neighboring fractions $\frac{2}{5}$ and $\frac{3}{7}$, we have $-\det\left[\begin{smallmatrix} 2 & 3 \\ 5 & 7 \end{smallmatrix}\right] = 5 \cdot 3 - 2 \cdot 7 = 1$, and the fraction $\frac{3}{7}$ is the mediant $\frac{2}{5} \oplus \frac{4}{9}$ of its neighbors $\frac{2}{5}$ and $\frac{4}{9}$, that is, $\frac{3}{7} = \frac{2+4}{\gcd(2+4,5+9)} \big/ \frac{5+9}{\gcd(2+4,5+9)}$. These observations illustrate simple but important properties, which *all* pairs and *all* triples of consecutive fractions in the sequence $\mathcal{F}(\mathbb{B}(n), m)$ have in common with the Farey sequence $\mathcal{F}_n$: the determinant of every 2×2 matrix, whose columns are neighboring fractions written in vector form, is -1; in any triple of consecutive fractions, the central fraction is the mediant of its neighbors.

Somewhat surprisingly, the three sequences

$$\mathcal{F}^{\leq\frac{1}{2}}(\mathbb{B}(2\cdot 6), 6) := \left(\tfrac{h}{k} \in \mathcal{F}(\mathbb{B}(12), 6) : \tfrac{h}{k} \leq \tfrac{1}{2}\right),$$
$$\mathcal{F}^{\geq\frac{1}{2}}(\mathbb{B}(2\cdot 6), 6) := \left(\tfrac{h}{k} \in \mathcal{F}(\mathbb{B}(12), 6) : \tfrac{h}{k} \geq \tfrac{1}{2}\right)$$

and $\mathcal{F}_6$ are all of the same length. Moreover, the numerators of fractions in $\mathcal{F}_6$ are the numerators of fractions in $\mathcal{F}^{\leq\frac{1}{2}}(\mathbb{B}(2\cdot 6), 6)$, and the denominators of fractions in $\mathcal{F}_6$ are the numerators of fractions in $\mathcal{F}^{\geq\frac{1}{2}}(\mathbb{B}(2\cdot 6), 6)$. Indeed, one can check that the bijections

$$\mathcal{F}_6 \to \mathcal{F}^{\leq\frac{1}{2}}(\mathbb{B}(2\cdot 6), 6), \quad \tfrac{h}{k} \mapsto \tfrac{h}{h+k},$$

and

$$\mathcal{F}_6 \to \mathcal{F}^{\geq\frac{1}{2}}(\mathbb{B}(2\cdot 6), 6), \quad \tfrac{h}{k} \mapsto \tfrac{k}{2k-h},$$

are order-preserving maps, and we will see in Section 2.1.3 that for any m the mappings $\mathcal{F}_m \ni \frac{h}{k} \mapsto \frac{h}{h+k} \in \mathcal{F}^{\leq\frac{1}{2}}(\mathbb{B}(2m), m)$ and $\mathcal{F}_m \ni \frac{h}{k} \mapsto \frac{k}{2k-h} \in \mathcal{F}^{\geq\frac{1}{2}}(\mathbb{B}(2m), m)$ establish order-preserving bijections.

1.2 Farey subsequences

In this section we consider certain subsequences of the Farey sequence whose pairs and triples of consecutive fractions all share the same properties. See Table 1.1 for definitions.

1.2.1 The sequences $\mathcal{F}_n^m$, $\mathcal{G}_n^m$ and $\mathcal{F}(\mathbb{B}(n), m)$

The subsequence of fractions $\mathcal{F}(\mathbb{B}(n), m) \subset \mathcal{F}_n$ defined by (1.3) can be regarded as the intersection

$$\mathcal{F}_n^m \cap \mathcal{G}_n^m =: \mathcal{F}(\mathbb{B}(n), m)$$

of the sequences

$$\mathcal{F}_n^m := \left(\tfrac{h}{k} \in \mathcal{F}_n : h \leq m\right) \tag{1.5}$$

and

$$\mathcal{G}_n^m := \left(\tfrac{h}{k} \in \mathcal{F}_n : m + k - n \leq h\right); \tag{1.6}$$

see Example 1.2. It follows from the definitions (1.5) and (1.6) that

$$\mathcal{F}_n^m \cup \mathcal{G}_n^m = \mathcal{F}_n. \tag{1.7}$$

Note that for any $n \in \mathbb{P}$, we have

$$\mathcal{F}_n^n = \mathcal{F}_n = \mathcal{G}_n^0,$$

and for any $n > 1$, we have

$$\mathcal{F}_n^{n-1} = \mathcal{F}_n = \mathcal{G}_n^1.$$

Table 1.1. Farey (sub)sequences.

Sequence	Definition/description
$\mathcal{F}_n$, $n > 0$	$\{\frac{0}{1}\} \dot\cup (\frac{h}{k} : h, k \in \mathbb{P},\ h < k \le n,\ \gcd(h,k) = 1) \dot\cup \{\frac{1}{1}\}$
$\mathcal{F}_n^m$, $m > 0$	$(\frac{h}{k} \in \mathcal{F}_n : h \le m)$
$\mathcal{G}_n^m$, $m < n$	$(\frac{h}{k} \in \mathcal{F}_n : m + k - n \le h)$
$\mathcal{F}(\mathbb{B}(n), m)$, $n > 1$, $0 < m < n$	$\mathcal{F}_n^m \cap \mathcal{G}_n^m = (\frac{h}{k} \in \mathcal{F}_n : m + k - n \le h \le m)$
$\mathcal{F}(\mathbb{B}(2m), m)$	$(\frac{h}{k} \in \mathcal{F}_{2m} : k - m \le h \le m)$
$\mathcal{F}^{\le\frac{1}{2}}(\mathbb{B}(n), m)$	$(\frac{h}{k} \in \mathcal{F}(\mathbb{B}(n), m) : \frac{h}{k} \le \frac{1}{2})$
$\mathcal{F}^{\ge\frac{1}{2}}(\mathbb{B}(n), m)$	$(\frac{h}{k} \in \mathcal{F}(\mathbb{B}(n), m) : \frac{h}{k} \ge \frac{1}{2})$
$\mathcal{F}(\mathbb{B}(n), m)^{\ell}$, $0 < \ell \le m$	$(\frac{h}{k} \in \mathcal{F}(\mathbb{B}(n), m) : h \le \ell) = \mathcal{F}(\mathbb{B}(n - m + \ell), \ell)$
$\mathcal{F}(\mathbb{B}(2m), m)^{\ell}$	$\mathcal{F}(\mathbb{B}(m + \ell), \ell)$
$\mathcal{G}(\mathbb{B}(n), m)^{\ell}$, $m \le \ell < n$	$(\frac{h}{k} \in \mathcal{F}(\mathbb{B}(n), m) : \ell + k - n \le h) = \mathcal{F}(\mathbb{B}(n + m - \ell), m)$
$\mathcal{G}(\mathbb{B}(2m), m)^{\ell}$	$\mathcal{F}(\mathbb{B}(3m - \ell), m)$

1.2.2 Well-structured Farey subsequences

It follows directly from the definitions (1.5), (1.6) and (1.3), that the following Farey subsequences have the very simple structure:

$$\begin{aligned}
\mathcal{F}_n^1 &= (\tfrac{0}{1} < \tfrac{1}{n} < \tfrac{1}{n-1} < \cdots < \tfrac{1}{3} < \tfrac{1}{2} < \tfrac{1}{1}),\\
\mathcal{G}_n^{n-1} &= (\tfrac{0}{1} < \tfrac{1}{2} < \tfrac{2}{3} < \cdots < \tfrac{n-2}{n-1} < \tfrac{n-1}{n} < \tfrac{1}{1}),\\
\mathcal{F}(\mathbb{B}(n), 1) &= \mathcal{F}_n^1,\\
\mathcal{F}(\mathbb{B}(n), n-1) &= \mathcal{G}_n^{n-1}.
\end{aligned}$$

Additional observations in this direction will be made in Section 2.7.

1.2.3 The sequences $\mathcal{F}^{\le\frac{1}{2}}(\mathbb{B}(n), m)$ and $\mathcal{F}^{\ge\frac{1}{2}}(\mathbb{B}(n), m)$

Define the subsequences $\mathcal{F}^{\le\frac{1}{2}}(\mathbb{B}(n), m)$ and $\mathcal{F}^{\ge\frac{1}{2}}(\mathbb{B}(n), m)$ of the sequence $\mathcal{F}(\mathbb{B}(n), m)$ by

$$\mathcal{F}^{\le\frac{1}{2}}(\mathbb{B}(n), m) := (\tfrac{h}{k} \in \mathcal{F}(\mathbb{B}(n), m) : \tfrac{h}{k} \le \tfrac{1}{2})$$

and

$$\mathcal{F}^{\ge\frac{1}{2}}(\mathbb{B}(n), m) := (\tfrac{h}{k} \in \mathcal{F}(\mathbb{B}(n), m) : \tfrac{h}{k} \ge \tfrac{1}{2}).$$

These sequences play a key role in the Farey duality results discussed in Chapter 2.

1.2.4 The sequences $\mathcal{F}(\mathbb{B}(n), m)^{\ell}$ and $\mathcal{G}(\mathbb{B}(n), m)^{\ell}$

In analogy to the sequences $\mathcal{F}_n^m$ and $\mathcal{G}_n^m$ defined by (1.5) and (1.6), we introduce similar subsequences of the sequence $\mathcal{F}(\mathbb{B}(n), m)$ as follows: given an integer ℓ,

$$\mathcal{F}(\mathbb{B}(n), m)^{\ell} := \left(\tfrac{h}{k} \in \mathcal{F}(\mathbb{B}(n), m) : h \le \ell\right) \quad \text{for } 0 < \ell \le m,$$

and

$$\mathcal{G}(\mathbb{B}(n), m)^{\ell} := \left(\tfrac{h}{k} \in \mathcal{F}(\mathbb{B}(n), m) : \ell + k - n \le h\right) \quad \text{for } m \le \ell < n.$$

Thus, $\mathcal{F}(\mathbb{B}(n), m)^m = \mathcal{G}(\mathbb{B}(n), m)^m = \mathcal{F}(\mathbb{B}(n), m)$. Moreover, if λ is an integer such that $0 < \lambda \le n - m$, then

$$\mathcal{F}(\mathbb{B}(n), m)^{\ell} \cap \mathcal{G}(\mathbb{B}(n), m)^{n-\lambda} = \mathcal{F}(\mathbb{B}(\ell + \lambda), \ell).$$

In fact, as noted below in Remark 1.1 (i), the sequences $\mathcal{F}(\mathbb{B}(n), m)^{\ell}$ and $\mathcal{G}(\mathbb{B}(n), m)^{\ell}$ are all sequences of the form $\mathcal{F}(\mathbb{B}(\,\cdot\,), \cdot\,)$ for any allowed ℓ. Indeed, they can be redefined as follows:

$$\mathcal{F}(\mathbb{B}(n), m)^{\ell} := \mathcal{F}^{\ell}_{n-m+\ell} \cap \mathcal{G}^{\ell}_{n-m+\ell}, \qquad \mathcal{G}(\mathbb{B}(n), m)^{\ell} := \mathcal{F}^{m}_{n+m-\ell} \cap \mathcal{G}^{m}_{n+m-\ell};$$

in particular,

$$\mathcal{F}(\mathbb{B}(2m), m)^{\ell} := \mathcal{F}^{\ell}_{m+\ell} \cap \mathcal{G}^{\ell}_{m+\ell}, \qquad \mathcal{G}(\mathbb{B}(2m), m)^{\ell} := \mathcal{F}^{m}_{3m-\ell} \cap \mathcal{G}^{m}_{3m-\ell}.$$

Remark 1.1. (i) If $0 < \ell \le m$, then

$$\mathcal{F}(\mathbb{B}(n), m)^{\ell} = \mathcal{F}(\mathbb{B}(n - m + \ell), \ell), \qquad \mathcal{F}(\mathbb{B}(2m), m)^{\ell} = \mathcal{F}(\mathbb{B}(m + \ell), \ell).$$

(ii) If $m \le \ell < n$, then

$$\mathcal{G}(\mathbb{B}(n), m)^{\ell} = \mathcal{F}(\mathbb{B}(n + m - \ell), m), \qquad \mathcal{G}(\mathbb{B}(2m), m)^{\ell} = \mathcal{F}(\mathbb{B}(3m - \ell), m).$$

Example 1.2 (see also Example 1.5). Suppose $n := 6$, $m := 4$, $\ell' := 3$ and $\ell'' := 5$. Then we have

$$\begin{aligned}
\mathcal{F}_n &= \left(\tfrac{0}{1} < \tfrac{1}{6} < \tfrac{1}{5} < \tfrac{1}{4} < \tfrac{1}{3} < \tfrac{2}{5} < \tfrac{1}{2} < \tfrac{3}{5} < \tfrac{2}{3} < \tfrac{3}{4} < \tfrac{4}{5} < \tfrac{5}{6} < \tfrac{1}{1}\right),\\
\mathcal{F}_n^m &= \left(\tfrac{0}{1} < \tfrac{1}{6} < \tfrac{1}{5} < \tfrac{1}{4} < \tfrac{1}{3} < \tfrac{2}{5} < \tfrac{1}{2} < \tfrac{3}{5} < \tfrac{2}{3} < \tfrac{3}{4} < \tfrac{4}{5} < \quad < \tfrac{1}{1}\right),\\
\mathcal{G}_n^m &= \left(\tfrac{0}{1} < \quad < \quad < \quad < \tfrac{1}{3} < \quad < \tfrac{1}{2} < \tfrac{3}{5} < \tfrac{2}{3} < \tfrac{3}{4} < \tfrac{4}{5} < \tfrac{5}{6} < \tfrac{1}{1}\right),\\
\mathcal{F}(\mathbb{B}(n), m) &= \left(\tfrac{0}{1} < \quad < \quad < \quad < \tfrac{1}{3} < \quad < \tfrac{1}{2} < \tfrac{3}{5} < \tfrac{2}{3} < \tfrac{3}{4} < \tfrac{4}{5} < \quad < \tfrac{1}{1}\right),\\
\mathcal{F}(\mathbb{B}(n), m)^{\ell'} &= \mathcal{F}(\mathbb{B}(n - m + \ell'), \ell')\\
&= \left(\tfrac{0}{1} < \quad < \quad < \quad < \tfrac{1}{3} < \quad < \tfrac{1}{2} < \tfrac{3}{5} < \tfrac{2}{3} < \tfrac{3}{4} < \quad < \quad < \tfrac{1}{1}\right),\\
\mathcal{G}(\mathbb{B}(n), m)^{\ell''} &= \mathcal{F}(\mathbb{B}(n + m - \ell''), m)\\
&= \left(\tfrac{0}{1} < \quad < \quad < \quad < \quad < \quad < \tfrac{1}{2} < \quad < \tfrac{2}{3} < \tfrac{3}{4} < \tfrac{4}{5} < \quad < \tfrac{1}{1}\right).
\end{aligned}$$

Table 1.2. Order-reversing and bijective mapping $\frac{h}{k} \mapsto \frac{k-h}{k}$; see also Section 3.6.

Mapping $\left[\begin{smallmatrix} h \\ k \end{smallmatrix}\right] \mapsto \left[\begin{smallmatrix} -1 & 1 \\ 0 & 1 \end{smallmatrix}\right] \cdot \left[\begin{smallmatrix} h \\ k \end{smallmatrix}\right]$	**See**
$\mathcal{F}_n \to \mathcal{F}_n$ $\mathcal{F}_n^m \to \mathcal{G}_n^{n-m}$ $\mathcal{G}_n^m \to \mathcal{F}_n^{n-m}$ $\mathcal{F}_{2m}^m \to \mathcal{G}_{2m}^m$ $\mathcal{G}_{2m}^m \to \mathcal{F}_{2m}^m$	Remark 1.3
$\mathcal{F}(\mathbb{B}(n), m) \to \mathcal{F}(\mathbb{B}(n), n-m)$ $\mathcal{F}(\mathbb{B}(2m), m) \to \mathcal{F}(\mathbb{B}(2m), m)$	Remark 1.3
$\mathcal{F}(\mathbb{B}(n), m)^\ell \to \mathcal{F}(\mathbb{B}(n), n-\ell)^{n-m}$ $\mathcal{G}(\mathbb{B}(n), m)^\ell \to \mathcal{G}(\mathbb{B}(n), n-\ell)^{n-m}$ $\mathcal{F}(\mathbb{B}(2m), m)^\ell \to \mathcal{F}(\mathbb{B}(2m), 2m-\ell)^m$ $\mathcal{G}(\mathbb{B}(2m), m)^\ell \to \mathcal{G}(\mathbb{B}(2m), 2m-\ell)^m$	Proposition 1.4

1.3 Order-reversing and bijective mapping $\frac{h}{k} \mapsto \frac{k-h}{k}$. I

In the context of Farey (sub)sequences, the mapping

$$\frac{h}{k} \mapsto \frac{k-h}{k}, \quad \left[\begin{smallmatrix} h \\ k \end{smallmatrix}\right] \mapsto \left[\begin{smallmatrix} -1 & 1 \\ 0 & 1 \end{smallmatrix}\right] \cdot \left[\begin{smallmatrix} h \\ k \end{smallmatrix}\right], \tag{1.8}$$

determined by the *involutory* matrix $\left[\begin{smallmatrix} -1 & 1 \\ 0 & 1 \end{smallmatrix}\right]$, is ubiquitous; see Table 1.2. We will deal with this mapping throughout this book. Section 3.6 is also concerned with this topic.

Remark 1.3. It follows directly from the definitions (1.5), (1.6) and (1.2) of the Farey (sub)sequences $\mathcal{F}_n^m$, $\mathcal{G}_n^m$ and $\mathcal{F}_n$ that the maps

$$\mathcal{F}_n^m \to \mathcal{G}_n^{n-m}, \quad \mathcal{F}_{2m}^m \to \mathcal{G}_{2m}^m, \quad \mathcal{G}_n^m \to \mathcal{F}_n^{n-m}, \quad \mathcal{G}_{2m}^m \to \mathcal{F}_{2m}^m$$

and, in particular, the map $\mathcal{F}_n \to \mathcal{F}_n$, defined in (1.8), are all order-reversing and bijective for any allowed m. As a consequence, the maps

$$\mathcal{F}(\mathbb{B}(n), m) \to \mathcal{F}(\mathbb{B}(n), n-m), \quad \mathcal{F}(\mathbb{B}(2m), m) \to \mathcal{F}(\mathbb{B}(2m), m),$$

defined in (1.8), are order-reversing and bijective for any m, $0 < m < n$.

Proposition 1.4. *The maps*

$$\mathcal{F}(\mathbb{B}(n), m)^\ell \to \mathcal{F}(\mathbb{B}(n), n-\ell)^{n-m}, \tag{1.9}$$
$$\mathcal{F}(\mathbb{B}(2m), m)^\ell \to \mathcal{F}(\mathbb{B}(2m), 2m-\ell)^m,$$
$$\mathcal{G}(\mathbb{B}(n), m)^\ell \to \mathcal{G}(\mathbb{B}(n), n-\ell)^{n-m}, \tag{1.10}$$
$$\mathcal{G}(\mathbb{B}(2m), m)^\ell \to \mathcal{G}(\mathbb{B}(2m), 2m-\ell)^m,$$

defined in (1.8), *are order-reversing and bijective.*

Proof. In (1.9) we have $\mathcal{F}(\mathbb{B}(n), n-\ell)^{n-m} = \mathcal{F}(\mathbb{B}(n-m+\ell), n-m)$, by Remark 1.1 (i). Now we can apply the order-reversing and bijective mapping given by (1.8) to the sequences $\mathcal{F}(\mathbb{B}(n-m+\ell), \ell) = \mathcal{F}(\mathbb{B}(n), m)^{\ell}$ and $\mathcal{F}(\mathbb{B}(n-m+\ell), n-m)$.

In (1.10) we have $\mathcal{G}(\mathbb{B}(n), n-\ell)^{n-m} = \mathcal{F}(\mathbb{B}(n+m-\ell), n-\ell)$, by Remark 1.1 (ii), and the mapping given by (1.8) provides the order-reversing bijections between the sequences $\mathcal{F}(\mathbb{B}(n+m-\ell), m) = \mathcal{G}(\mathbb{B}(n), m)^{\ell}$ and $\mathcal{F}(\mathbb{B}(n+m-\ell), n-\ell)$. □

Example 1.5. Suppose $n := 6$, $m := 4$, $\ell' := 3$ and $\ell'' := 5$. Then we have

$$\begin{aligned}
\mathcal{F}_n &= \left(\tfrac{0}{1} < \tfrac{1}{6} < \tfrac{1}{5} < \tfrac{1}{4} < \tfrac{1}{3} < \tfrac{2}{5} < \tfrac{1}{2} < \tfrac{3}{5} < \tfrac{2}{3} < \tfrac{3}{4} < \tfrac{4}{5} < \tfrac{5}{6} < \tfrac{1}{1}\right),\\
\mathcal{F}(\mathbb{B}(n), m)^{\ell'} &= \mathcal{F}(\mathbb{B}(n-m+\ell'), \ell')\\
&= \left(\tfrac{0}{1} < \quad < \quad < \quad < \tfrac{1}{3} < \quad < \tfrac{1}{2} < \tfrac{3}{5} < \tfrac{2}{3} < \tfrac{3}{4} < \quad < \quad < \tfrac{1}{1}\right),\\
\mathcal{F}(\mathbb{B}(n), n-\ell')^{n-m} &= \mathcal{F}(\mathbb{B}(n-m+\ell'), n-m)\\
&= \left(\tfrac{0}{1} < \quad < \quad < \tfrac{1}{4} < \tfrac{1}{3} < \tfrac{2}{5} < \tfrac{1}{2} < \quad < \tfrac{2}{3} < \quad < \quad < \quad < \tfrac{1}{1}\right),\\
\mathcal{G}(\mathbb{B}(n), m)^{\ell''} &= \mathcal{F}(\mathbb{B}(n+m-\ell''), m)\\
&= \left(\tfrac{0}{1} < \quad < \quad < \quad < \quad < \quad < \tfrac{1}{2} < \quad < \tfrac{2}{3} < \tfrac{3}{4} < \tfrac{4}{5} < \quad < \tfrac{1}{1}\right),\\
\mathcal{G}(\mathbb{B}(n), n-\ell'')^{n-m} &= \mathcal{F}(\mathbb{B}(n+m-\ell''), n-\ell'')\\
&= \left(\tfrac{0}{1} < \quad < \tfrac{1}{5} < \tfrac{1}{4} < \tfrac{1}{3} < \quad < \tfrac{1}{2} < \quad < \quad < \quad < \quad < \quad < \tfrac{1}{1}\right).
\end{aligned}$$

Proposition 1.4 guarantees that the maps $\mathcal{F}(\mathbb{B}(n), m)^{\ell'} \leftrightarrow \mathcal{F}(\mathbb{B}(n), n-\ell')^{n-m}$ and $\mathcal{G}(\mathbb{B}(n), m)^{\ell''} \leftrightarrow \mathcal{G}(\mathbb{B}(n), n-\ell'')^{n-m}$ defined by (1.8) are order-reversing and bijective.

1.4 Pairs of neighboring fractions. I

In this section we consider several ways of finding the neighbors of an arbitrary fraction in a Farey (sub)sequence; see Tables 1.3, 1.4 and 1.5 for formulas. The results of the section will be revisited in Section 2.2 and surveyed in Tables 2.1 and 2.2.

The properties of triples $\frac{h_j}{k_j} < \frac{h_{j+1}}{k_{j+1}} < \frac{h_{j+2}}{k_{j+2}}$ of consecutive fractions in Farey (sub)sequences, discussed in Sections 1.5 and 2.3, allow us to easily find the fraction $\frac{h_{j+2}}{k_{j+2}}$ if the pair of neighboring fractions $\frac{h_j}{k_j}$ and $\frac{h_{j+1}}{k_{j+1}}$ is already known. Similarly, the fraction $\frac{h_j}{k_j}$ can be easily found if the neighboring fractions $\frac{h_{j+1}}{k_{j+1}}$ and $\frac{h_{j+2}}{k_{j+2}}$ are known.

1.4.1 Neighboring fractions in $\mathcal{F}_n$

In this subsection we find the neighbors of fractions in the Farey sequence $\mathcal{F}_n$. The computations presented here will be continued in Section 2.2.1.

Remark 1.6. It is clear that $\frac{n-1}{n}$ precedes $\frac{1}{1}$ in $\mathcal{F}_n$ and the fraction $\frac{1}{n}$ succeeds $\frac{0}{1}$ in $\mathcal{F}_n$; see also Table 1.5.

Table 1.3. The fractions that precede fractions $\frac{h}{k}$ in Farey (sub)sequences, $\frac{0}{1} < \frac{h}{k}$. Here x_0 and y_0 are two integers such that $kx_0 \equiv -1 \pmod h$ and $hy_0 \equiv 1 \pmod k$. See Tables 2.1 and 2.2 on simplified expressions.

$(x_0 + ht^*)/(\frac{kx_0+1}{h} + kt^*)$ precedes $\frac{h}{k}$; cf. Table 2.1

Sequence	t^*		**See**
$\mathcal{F}_n$	$\lfloor \frac{hn-kx_0-1}{hk} \rfloor$		Proposition 1.7 (i) (a)
$\mathcal{F}_n^m$, $0 < m < n$	$\lfloor \frac{m-x_0}{h} \rfloor$	if $hn - km \geq 1$	Corollary 1.11 (i) (a)
	$\lfloor \frac{hn-kx_0-1}{hk} \rfloor$	if $hn - km \leq 1$	Corollary 1.11 (i) (b)
$\mathcal{G}_n^m$, $0 < m < n$	$\lfloor \frac{hn-kx_0-1}{hk} \rfloor$	if $hn - km \geq 1$	Corollary 1.15 (i) (a)
	$\lfloor \frac{h(n-m)-(k-h)x_0-1}{h(k-h)} \rfloor$	if $hn - km \leq 1$	Corollary 1.15 (i) (b)
$\mathcal{F}(\mathbb{B}(n), m)$	$\lfloor \frac{m-x_0}{h} \rfloor$	if $hn - km \geq 1$	Theorem 1.20 (i)
	$\lfloor \frac{h(n-m)-(k-h)x_0-1}{h(k-h)} \rfloor$	if $hn - km \leq 1$	

$(\frac{hy_0-1}{k} + ht^*)/(y_0 + kt^*)$ precedes $\frac{h}{k}$; cf. Table 2.2

Sequence	t^*		**See**
$\mathcal{F}_n$	$\lfloor \frac{n-y_0}{k} \rfloor$		Proposition 1.7 (i) (b)
$\mathcal{F}_n^m$, $0 < m < n$	$\lfloor \frac{km-hy_0+1}{hk} \rfloor$	if $hn - km \geq 1$	Corollary 1.11 (i) (a)
	$\lfloor \frac{n-y_0}{k} \rfloor$	if $hn - km \leq 1$	Corollary 1.11 (i) (b)
$\mathcal{G}_n^m$, $0 < m < n$	$\lfloor \frac{n-y_0}{k} \rfloor$	if $hn - km \geq 1$	Corollary 1.15 (i) (a)
	$\lfloor \frac{k(n-m)-(k-h)y_0-1}{(k-h)k} \rfloor$	if $hn - km \leq 1$	Corollary 1.15 (i) (b)
$\mathcal{F}(\mathbb{B}(n), m)$	$\lfloor \frac{km-hy_0+1}{hk} \rfloor$	if $hn - km \geq 1$	Theorem 1.20 (i)
	$\lfloor \frac{k(n-m)-(k-h)y_0-1}{(k-h)k} \rfloor$	if $hn - km \leq 1$	

Proposition 1.7. (i) *Let $\frac{h}{k} \in \mathcal{F}_n - \{\frac{0}{1}\}$.*

(a) *Let x_0 be an integer such that $kx_0 \equiv -1 \pmod h$. Define two integers y_0 and t^* by $y_0 := \frac{kx_0+1}{h}$ and $t^* := \lfloor \frac{n-y_0}{k} \rfloor$. Then the fraction*

$$\frac{x_0 + ht^*}{y_0 + kt^*} := \frac{x_0 + h\lfloor \frac{hn-kx_0-1}{hk} \rfloor}{\frac{kx_0+1}{h} + k\lfloor \frac{hn-kx_0-1}{hk} \rfloor} \tag{1.11}$$

precedes the fraction $\frac{h}{k}$ in $\mathcal{F}_n$.

(b) *In other words, let y_0 be an integer such that $hy_0 \equiv 1 \pmod k$. Define two integers x_0 and t^* by $x_0 := \frac{hy_0-1}{k}$ and $t^* := \lfloor \frac{n-y_0}{k} \rfloor$. Then the fraction*

$$\frac{x_0 + ht^*}{y_0 + kt^*} := \frac{\frac{hy_0-1}{k} + h\lfloor \frac{n-y_0}{k} \rfloor}{y_0 + k\lfloor \frac{n-y_0}{k} \rfloor},$$

which coincides with the fraction defined by (1.11), precedes $\frac{h}{k}$ in $\mathcal{F}_n$.

Table 1.4. The fractions that succeed fractions $\frac{h}{k}$ in Farey (sub)sequences, $\frac{0}{1} < \frac{h}{k} < \frac{1}{1}$. Here x_0 and y_0 are two integers such that $kx_0 \equiv 1 \pmod{h}$ and $hy_0 \equiv -1 \pmod{k}$. See Tables 2.3 and 2.4 on simplified expressions.

$(x_0 + ht^*)/(\frac{kx_0-1}{h} + kt^*)$ succeeds $\frac{h}{k}$; cf. Table 2.3			
Sequence	t^*		**See**
$\mathcal{F}_n$	$\lfloor\frac{hn-kx_0+1}{hk}\rfloor$		Proposition 1.7 (ii) (a)
$\mathcal{F}_n^m$, $0 < m < n$	$\lfloor\frac{hn-kx_0+1}{hk}\rfloor$	if $km - hn \ge 1$	Corollary 1.11 (ii) (a)
	$\lfloor\frac{m-x_0}{h}\rfloor$	if $km - hn \le 1$	Corollary 1.11 (ii) (b)
$\mathcal{G}_n^m$, $0 < m < n$	$\lfloor\frac{h(n-m)-(k-h)x_0+1}{h(k-h)}\rfloor$	if $km - hn \ge 1$	Corollary 1.15 (ii) (a)
	$\lfloor\frac{hn-kx_0+1}{hk}\rfloor$	if $km - hn \le 1$	Corollary 1.15 (ii) (b)
$\mathcal{F}(\mathbb{B}(n), m)$	$\lfloor\frac{h(n-m)-(k-h)x_0+1}{h(k-h)}\rfloor$	if $km - hn \ge 1$	Theorem 1.20 (ii)
	$\lfloor\frac{m-x_0}{h}\rfloor$	if $km - hn \le 1$	

$(\frac{hy_0+1}{k} + ht^*)/(y_0 + kt^*)$ succeeds $\frac{h}{k}$; cf. Table 2.4			
Sequence	t^*		**See**
$\mathcal{F}_n$	$\lfloor\frac{n-y_0}{k}\rfloor$		Proposition 1.7 (ii) (b)
$\mathcal{F}_n^m$, $0 < m < n$	$\lfloor\frac{n-y_0}{k}\rfloor$	if $km - hn \ge 1$	Corollary 1.11 (ii) (a)
	$\lfloor\frac{km-hy_0-1}{hk}\rfloor$	if $km - hn \le 1$	Corollary 1.11 (ii) (b)
$\mathcal{G}_n^m$, $0 < m < n$	$\lfloor\frac{k(n-m)-(k-h)y_0+1}{(k-h)k}\rfloor$	if $km - hn \ge 1$	Corollary 1.15 (ii) (a)
	$\lfloor\frac{n-y_0}{k}\rfloor$	if $km - hn \le 1$	Corollary 1.15 (ii) (b)
$\mathcal{F}(\mathbb{B}(n), m)$	$\lfloor\frac{k(n-m)-(k-h)y_0+1}{(k-h)k}\rfloor$	if $km - hn \ge 1$	Theorem 1.20 (ii)
	$\lfloor\frac{km-hy_0-1}{hk}\rfloor$	if $km - hn \le 1$	

(ii) *Let* $\frac{h}{k} \in \mathcal{F}_n - \{\frac{0}{1}, \frac{1}{1}\}$.

(a) *Let* x_0 *be an integer such that* $kx_0 \equiv 1 \pmod{h}$. *Define two integers* y_0 *and* t^* *by* $y_0 := \frac{kx_0-1}{h}$ *and* $t^* := \lfloor\frac{n-y_0}{k}\rfloor$. *Then the fraction*

$$\frac{x_0 + ht^*}{y_0 + kt^*} := \frac{x_0 + h\lfloor\frac{hn-kx_0+1}{hk}\rfloor}{\frac{kx_0-1}{h} + k\lfloor\frac{hn-kx_0+1}{hk}\rfloor} \tag{1.12}$$

succeeds the fraction $\frac{h}{k}$ *in* $\mathcal{F}_n$.

(b) *In other words, let* y_0 *be an integer such that* $hy_0 \equiv -1 \pmod{k}$. *Define two integers* x_0 *and* t^* *by* $x_0 := \frac{hy_0+1}{k}$ *and* $t^* := \lfloor\frac{n-y_0}{k}\rfloor$. *Then the fraction*

$$\frac{x_0 + ht^*}{y_0 + kt^*} := \frac{\frac{hy_0+1}{k} + h\lfloor\frac{n-y_0}{k}\rfloor}{y_0 + k\lfloor\frac{n-y_0}{k}\rfloor},$$

which coincides with the fraction defined by (1.12), *succeeds* $\frac{h}{k}$ *in* $\mathcal{F}_n$.

Table 1.5. The neighbors of the fractions $\frac{1}{1}$ and $\frac{0}{1}$ in Farey (sub)sequences.

Sequence	**The predecessor of $\frac{1}{1}$**	**The successor of $\frac{0}{1}$**	**See**
$\mathcal{F}_n$	$\frac{n-1}{n}$	$\frac{1}{n}$	Remark 1.6
$\mathcal{F}_n^m$, $0 < m < n$	$\frac{m}{m+1}$	$\frac{1}{n}$	Remark 1.9
$\mathcal{G}_n^m$, $0 < m < n$	$\frac{n-1}{n}$	$\frac{1}{n-m+1}$	Remark 1.13
$\mathcal{F}(\mathbb{B}(n), m)$	$\frac{m}{m+1}$	$\frac{1}{n-m+1}$	Remark 1.17
$\mathcal{F}(\mathbb{B}(2m), m)$	$\frac{m}{m+1}$	$\frac{1}{m+1}$	

Recall that two integers a and b, not both of which are zero, are *relatively prime*, that is, $\gcd(a, b) = 1$, if and only if there exist integers x_0 and y_0 such that

$$ay_0 - bx_0 = 1.$$

Proof. (i) (a) If a pair $(x_0, y_0) \in \mathbb{Z}^2$ is a solution to the linear Diophantine equation

$$-k\mathrm{x} + h\mathrm{y} = 1, \tag{1.13}$$

in the unknowns x and y, then the pair $(x_0 + ht, y_0 + kt)$ is also a solution for any integer t.

Consider the inequalities $0 \le x_0 + h\mathrm{t}$ and $y_0 + k\mathrm{t} \le n$, in the unknown t, or the inequality

$$-\frac{x_0}{h} \le \mathrm{t} \le \frac{n - y_0}{k}. \tag{1.14}$$

Throughout many computations in this book, we will use the identities

$$\left\lceil \frac{a}{b} \right\rceil = \left\lfloor \frac{a+b-1}{b} \right\rfloor = \left\lfloor \frac{a-1}{b} \right\rfloor + 1, \tag{1.15}$$

$$\left\lfloor \frac{a}{b} \right\rfloor = \left\lceil \frac{a-b+1}{b} \right\rceil = \left\lceil \frac{a+1}{b} \right\rceil - 1 \tag{1.16}$$

for $a, b \in \mathbb{Z}$, where $b > 0$.

Inequality (1.14) has at least one integer solution, namely, $\lceil -\frac{x_0}{h} \rceil$, since

$$\begin{aligned}
\frac{n - y_0}{k} - \left\lceil -\frac{x_0}{h} \right\rceil &= \frac{n - y_0}{k} - \left\lfloor \frac{h - x_0 - 1}{h} \right\rfloor \ge \frac{n - y_0}{k} - \frac{h - x_0 - 1}{h} \\
&= \frac{h(n-k) + k - 1}{hk} \ge \frac{h(n-k) + k - n}{hk} \\
&= \frac{(h-1)(n-k)}{hk} \ge 0.
\end{aligned}$$

For any integer solution t, we have

$$(y_0 + kt) - (x_0 + ht) := \Big(\frac{kx_0 + 1}{h} + kt\Big) - (x_0 + ht) = \frac{(k-h)(x_0 + ht) + 1}{h} > 0,$$

that is,

$$x_0 + ht < y_0 + kt.$$

If t' and t'' are two integer solutions to inequality (1.14), with $t' < t''$, that is,

$$\Big\lceil -\frac{x_0}{h} \Big\rceil \le t' < t'' \le \Big\lfloor \frac{n - y_0}{k} \Big\rfloor,$$

then

$$\frac{x_0 + ht''}{y_0 + kt''} - \frac{x_0 + ht'}{y_0 + kt'} = \frac{t'' - t'}{(y_0 + kt')(y_0 + kt'')} > 0;$$

in other words,

$$\frac{x_0 + ht'}{y_0 + kt'} < \frac{x_0 + ht''}{y_0 + kt''}.$$

Define two integers x^* and y^* by

$$x^* := x_0 + ht^* \quad \text{and} \quad y^* := y_0 + kt^*,$$

where $t^* := \lfloor \frac{n-y_0}{k} \rfloor$, and suppose that there exists a fraction $\frac{i}{j} \in \mathcal{F}_n$ such that $\frac{x^*}{y^*} < \frac{i}{j} < \frac{h}{k}$. Then we have $jh - ik \ge 1$ and $iy^* - jx^* \ge 1$. Hence, on the one hand,

$$\frac{h}{k} - \frac{x^*}{y^*} = \Big(\frac{h}{k} - \frac{i}{j}\Big) + \Big(\frac{i}{j} - \frac{x^*}{y^*}\Big) = \frac{jh - ik}{jk} + \frac{iy^* - jx^*}{jy^*} \ge \frac{1}{jk} + \frac{1}{jy^*} = \frac{y^* + k}{jky^*}.$$

On the other hand,

$$\frac{h}{k} - \frac{x^*}{y^*} = \frac{hy^* - kx^*}{ky^*} = \frac{1}{ky^*},$$

which implies $\frac{y^*+k}{jky^*} \le \frac{1}{ky^*}$, that is, $y^* + k \le j$. But $1 < j \le n$, therefore,

$$y^* < n - k + 1. \tag{1.17}$$

For the maximal integer solution $t^* := \lfloor \frac{n-y_0}{k} \rfloor$ to inequality (1.14), we have

$$y^* := y_0 + k\Big\lfloor \frac{n - y_0}{k} \Big\rfloor = y_0 + k\Big\lceil \frac{n - y_0 - k + 1}{k} \Big\rceil \ge y_0 + k\frac{n - y_0 - k + 1}{k},$$

that is, $y^* \ge n - k + 1$, which contradicts (1.17).

Thus, the fraction $\frac{x^*}{y^*}$ precedes the fraction $\frac{h}{k}$ in $\mathcal{F}_n$.

(i) (b) is a restatement of (i) (a).

(ii) (a) If a pair $(x_0, y_0) \in \mathbb{Z}^2$ is a solution to the linear Diophantine equation

$$k\mathrm{x} - h\mathrm{y} = 1, \tag{1.18}$$

in the unknowns x and y, then the pair $(x_0 + ht, y_0 + kt)$ is also a solution for any integer t.

Consider the inequalities $1 \le x_0 + ht$ and $y_0 + kt \le n$, in the unknown t, or the inequality

$$\frac{1-x_0}{h} \le t \le \frac{n-y_0}{k}. \tag{1.19}$$

This inequality has at least one integer solution, namely, $\lceil \frac{1-x_0}{h} \rceil$, since

$$\frac{n-y_0}{k} - \left\lceil \frac{1-x_0}{h} \right\rceil = \frac{n-y_0}{k} - \left\lfloor \frac{h-x_0}{h} \right\rfloor \ge \frac{n-y_0}{k} - \frac{h-x_0}{h} = \frac{h(n-k)+1}{hk} > 0.$$

For any integer solution t, we have

$$(y_0+kt)-(x_0+ht) := \left(\frac{kx_0-1}{h}+kt\right)-(x_0+ht) = \frac{(k-h)(x_0+ht)-1}{h} \ge 0,$$

that is, $x_0 + ht \le y_0 + kt$.

If t' and t'' are two integer solutions to inequality (1.19), with $t' < t''$, that is,

$$\left\lceil \frac{1-x_0}{h} \right\rceil \le t' < t'' \le \left\lfloor \frac{n-y_0}{k} \right\rfloor,$$

then

$$\frac{x_0+ht''}{y_0+kt''} - \frac{x_0+ht'}{y_0+kt'} = \frac{t'-t''}{(y_0+kt')(y_0+kt'')} < 0;$$

in other words,

$$\frac{x_0+ht'}{y_0+kt'} > \frac{x_0+ht''}{y_0+kt''}.$$

Define two integers x^* and y^* by

$$x^* := x_0 + ht^* \quad \text{and} \quad y^* := y_0 + kt^*,$$

where $t^* := \lfloor \frac{n-y_0}{k} \rfloor$, and suppose that there exists a fraction $\frac{i}{j} \in \mathcal{F}_n$ such that $\frac{h}{k} < \frac{i}{j} < \frac{x^*}{y^*}$. Then we have $jx^* - iy^* \ge 1$ and $ik - jh \ge 1$. Hence, on the one hand,

$$\frac{x^*}{y^*} - \frac{h}{k} = \left(\frac{x^*}{y^*} - \frac{i}{j}\right) + \left(\frac{i}{j} - \frac{h}{k}\right) = \frac{jx^*-iy^*}{jy^*} + \frac{ik-jh}{jk} \ge \frac{1}{jy^*} + \frac{1}{jk} = \frac{k+y^*}{jky^*}.$$

On the other hand,

$$\frac{x^*}{y^*} - \frac{h}{k} = \frac{kx^*-hy^*}{ky^*} = \frac{1}{ky^*},$$

which implies $\frac{k+y^*}{jky^*} \le \frac{1}{ky^*}$, that is, $k + y^* \le j$. But $1 < j \le n$, therefore,

$$y^* < n - k + 1. \tag{1.20}$$

For the maximal integer solution $t^* := \lfloor \frac{n-y_0}{k} \rfloor$ to inequality (1.19), we have

$$y^* := y_0 + k\left\lfloor \frac{n-y_0}{k} \right\rfloor = y_0 + k\left\lceil \frac{n-y_0-k+1}{k} \right\rceil \ge y_0 + k\frac{n-y_0-k+1}{k},$$

that is, $y^* \ge n - k + 1$, which contradicts (1.20).

Thus, the fraction $\frac{x^*}{y^*}$ succeeds the fraction $\frac{h}{k}$ in $\mathcal{F}_n$.

(ii) (b) is a restatement of (ii) (a). □

If in Proposition 1.7 we imposed reasonable restrictions on the search intervals for the integers x_0 and y_0, we could then simplify the results; see Lemma 2.9.

Remark 1.8. For any pair $\frac{h_j}{k_j} < \frac{h_{j+1}}{k_{j+1}}$ of neighboring fractions in $\mathcal{F}_n$, we have

$$k_j + k_{j+1} > n.$$

Indeed, this is obvious for the pairs $\frac{n-1}{n} < \frac{1}{1}$ and $\frac{0}{1} < \frac{1}{n}$, mentioned in Remark 1.6. For other pairs, in view of Proposition 1.7 (i) (b), we have $k_j = y_0 + k_{j+1}\lfloor \frac{n-y_0}{k_{j+1}} \rfloor$ for a certain integer y_0, and thus

$$\begin{aligned} k_j + k_{j+1} &= \Big(y_0 + k_{j+1}\Big\lfloor \frac{n-y_0}{k_{j+1}} \Big\rfloor\Big) + k_{j+1} = \Big(y_0 + k_{j+1}\Big\lceil \frac{n-y_0-k_{j+1}+1}{k_{j+1}} \Big\rceil\Big) + k_{j+1} \\ &\geq \Big(y_0 + k_{j+1}\frac{n-y_0-k_{j+1}+1}{k_{j+1}}\Big) + k_{j+1} = n+1. \end{aligned}$$

See also Remarks 1.12, 1.16 and 1.22.

1.4.2 Neighboring fractions in $\mathcal{F}_n^m$

Recall that the Farey subsequence $\mathcal{F}_n^m$ is defined by $\mathcal{F}_n^m := (\frac{h}{k} \in \mathcal{F}_n : h \leq m)$ for any integer $m > 0$. Nevertheless, in this subsection we will impose an additional restriction on the parameter m, namely, $m < n$.

Let us find the neighbors of a fraction in the sequence $\mathcal{F}_n^m$.

Remark 1.9. It is clear (and verified with the help of Proposition 1.10 (i), Corollary 1.11 (i) (a) and especially Lemma 2.13 (i) (a)) that $\frac{m}{m+1}$ precedes $\frac{1}{1}$ in $\mathcal{F}_n^m$, where $0 < m < n$.

According to Remark 1.13, $\frac{n-1}{n}$ precedes $\frac{1}{1}$ in $\mathcal{G}_n^{n-m}$, where $0 < m < n$. Thus, the order-reversing bijections $\mathcal{G}_n^{n-m} \leftrightarrow \mathcal{F}_n^m$ defined in (1.8) guarantee that $\frac{1}{n}$ succeeds $\frac{0}{1}$ in $\mathcal{F}_n^m$. See also Table 1.5.

Proposition 1.10. (i) *Let $\frac{h}{k} \in \mathcal{F}_n^m - \{\frac{0}{1}\}$, where $0 < m < n$.*

(a) *Let x_0 be an integer such that $kx_0 \equiv -1 \pmod h$. Define two integers y_0 and t^* by $y_0 := \frac{kx_0+1}{h}$ and $t^* := \lfloor \min\{\frac{m-x_0}{h}, \frac{n-y_0}{k}\} \rfloor$. Then the fraction*

$$\frac{x_0 + ht^*}{y_0 + kt^*} := \frac{x_0 + h\lfloor \min\{\frac{m-x_0}{h}, \frac{hn-kx_0-1}{hk}\} \rfloor}{\frac{kx_0+1}{h} + k\lfloor \min\{\frac{m-x_0}{h}, \frac{hn-kx_0-1}{hk}\} \rfloor} \tag{1.21}$$

precedes the fraction $\frac{h}{k}$ in $\mathcal{F}_n^m$.

(b) *In other words, let y_0 be an integer such that $hy_0 \equiv 1 \pmod k$. Define two integers x_0 and t^* by $x_0 := \frac{hy_0-1}{k}$ and $t^* := \lfloor \min\{\frac{m-x_0}{h}, \frac{n-y_0}{k}\} \rfloor$. Then the fraction*

$$\frac{x_0 + ht^*}{y_0 + kt^*} := \frac{\frac{hy_0-1}{k} + h\lfloor \min\{\frac{km-hy_0+1}{hk}, \frac{n-y_0}{k}\} \rfloor}{y_0 + k\lfloor \min\{\frac{km-hy_0+1}{hk}, \frac{n-y_0}{k}\} \rfloor},$$

which coincides with the fraction defined by (1.21), precedes $\frac{h}{k}$ in $\mathcal{F}_n^m$.

(ii) *Let* $\frac{h}{k} \in \mathcal{F}_n^m - \{\frac{0}{1}, \frac{1}{1}\}$, *where* $0 < m < n$.

(a) *Let* x_0 *be an integer such that* $kx_0 \equiv 1 \pmod{h}$. *Define two integers* y_0 *and* t^* *by* $y_0 := \frac{kx_0-1}{h}$ *and* $t^* := \lfloor \min\{\frac{m-x_0}{h}, \frac{n-y_0}{k}\}\rfloor$. *Then the fraction*

$$\frac{x_0 + ht^*}{y_0 + kt^*} := \frac{x_0 + h\lfloor \min\{\frac{m-x_0}{h}, \frac{hn-kx_0+1}{hk}\}\rfloor}{\frac{kx_0-1}{h} + k\lfloor \min\{\frac{m-x_0}{h}, \frac{hn-kx_0+1}{hk}\}\rfloor} \tag{1.22}$$

succeeds the fraction $\frac{h}{k}$ *in* $\mathcal{F}_n^m$.

(b) *In other words, let* y_0 *be an integer such that* $hy_0 \equiv -1 \pmod{k}$. *Define two integers* x_0 *and* t^* *by* $x_0 := \frac{hy_0+1}{k}$ *and* $t^* := \lfloor \min\{\frac{m-x_0}{h}, \frac{n-y_0}{k}\}\rfloor$. *Then the fraction*

$$\frac{x_0 + ht^*}{y_0 + kt^*} := \frac{\frac{hy_0+1}{k} + h\lfloor \min\{\frac{km-hy_0-1}{hk}, \frac{n-y_0}{k}\}\rfloor}{y_0 + k\lfloor \min\{\frac{km-hy_0-1}{hk}, \frac{n-y_0}{k}\}\rfloor},$$

which coincides with the fraction defined by (1.22), *succeeds* $\frac{h}{k}$ *in* $\mathcal{F}_n^m$.

Proof. (i) (a) If a pair $(x_0, y_0) \in \mathbb{Z}^2$ is a solution to the linear Diophantine equation

$$-kx + hy = 1,$$

in the unknowns x and y, then the pair $(x_0 + ht, y_0 + kt)$ is also a solution for any integer t.

Consider the inequalities $0 \le x_0 + ht \le m$ and $y_0 + kt \le n$, in the unknown t, or the equivalent system

$$\begin{cases} -\frac{x_0}{h} \le t \le \frac{m-x_0}{h}, \\ t \le \frac{n-y_0}{k}, \end{cases}$$

or the inequality

$$-\frac{x_0}{h} \le t \le \min\left\{\frac{m-x_0}{h}, \frac{n-y_0}{k}\right\}. \tag{1.23}$$

This inequality has at least one integer solution, namely, $\lceil -\frac{x_0}{h}\rceil$, since

$$\frac{m-x_0}{h} - \left\lceil -\frac{x_0}{h}\right\rceil = \frac{m-x_0}{h} - \left\lfloor \frac{h-x_0-1}{h}\right\rfloor \ge \frac{m-x_0}{h} - \frac{h-x_0-1}{h} = \frac{m-h+1}{h} > 0$$

and

$$\begin{aligned}\frac{n-y_0}{k} - \left\lceil -\frac{x_0}{h}\right\rceil &= \frac{n-y_0}{k} - \left\lfloor \frac{h-x_0-1}{h}\right\rfloor \ge \frac{n-y_0}{k} - \frac{h-x_0-1}{h} \\ &= \frac{h(n-k)+k-1}{hk} \ge \frac{h(n-k)+k-n}{hk} = \frac{(h-1)(n-k)}{hk} \ge 0.\end{aligned}$$

For any integer solution t, we have

$$(y_0 + kt) - (x_0 + ht) := \left(\frac{kx_0+1}{h} + kt\right) - (x_0 + ht) = \frac{(k-h)(x_0+ht)+1}{h} > 0,$$

that is,

$$x_0 + ht < y_0 + kt.$$

If t' and t'' are two integer solutions to inequality (1.23), with $t' < t''$, that is,

$$\left\lceil -\frac{x_0}{h} \right\rceil \le t' < t'' \le \left\lfloor \min\left\{\frac{m-x_0}{h}, \frac{n-y_0}{k}\right\} \right\rfloor,$$

then

$$\frac{x_0+ht''}{y_0+kt''} - \frac{x_0+ht'}{y_0+kt'} = \frac{t''-t'}{(y_0+kt')(y_0+kt'')} > 0;$$

in other words,

$$\frac{x_0+ht'}{y_0+kt'} < \frac{x_0+ht''}{y_0+kt''}.$$

Define two integers x^* and y^* by

$$x^* := x_0 + ht^* \quad \text{and} \quad y^* := y_0 + kt^*,$$

where $t^* := \lfloor \min\{\frac{m-x_0}{h}, \frac{n-y_0}{k}\}\rfloor$.

The assertion is proved straightforwardly in the case $\frac{h}{k} := \frac{1}{n}$. Indeed, here we have

$$\frac{m-x_0}{h} - \frac{n-y_0}{k} := \frac{m-x_0}{1} - \frac{n-y_0}{n} = \frac{n(m-1)+1}{n} > 0$$

and, as a consequence, $t^* = \lfloor \frac{n-y_0}{n} \rfloor = \lfloor -x_0 + 1 - \frac{1}{n} \rfloor = -x_0$. Therefore, for any $x_0 \in \mathbb{Z}$, the fraction

$$\frac{x_0+ht^*}{y_0+kt^*} := \frac{x_0+1\cdot t^*}{y_0+n\cdot t^*} = \frac{x_0+1\cdot(-x_0)}{nx_0+1+n\cdot(-x_0)} = \frac{0}{1}$$

precedes the fraction $\frac{1}{n}$ in $\mathfrak{F}_n^m$, in accordance with Remark 1.9.

Now suppose that $\frac{h}{k} > \frac{1}{n}$. If $t^* = \lfloor \frac{n-y_0}{k} \rfloor$, that is, $t^* = \lfloor \frac{hn-kx_0-1}{hk} \rfloor$, then

$$\begin{aligned} x^* := x_0 + ht^* &= x_0 + h\left\lfloor \frac{hn-kx_0-1}{hk} \right\rfloor = x_0 + h\left\lceil \frac{h(n-k)-kx_0}{hk} \right\rceil \\ &\ge x_0 + h\frac{h(n-k)-kx_0}{hk} = \frac{h}{k}(n-k) > \frac{1}{n}(n-k) \ge 0. \end{aligned}$$

If $t^* = \lfloor \frac{m-x_0}{h} \rfloor$, then

$$x^* := x_0 + ht^* = x_0 + h\left\lfloor \frac{m-x_0}{h} \right\rfloor = x_0 + h\left\lceil \frac{m-x_0-h+1}{h} \right\rceil \ge x_0 + h\frac{m-x_0-h+1}{h},$$

that is,

$$x^* \ge m - h + 1 > 0. \tag{1.24}$$

We see that $\frac{1}{n} < \frac{h}{k}$ implies $x^* > 0$.

Suppose that there exists a fraction $\frac{i}{j} \in \mathfrak{F}_n^m$ such that $\frac{x^*}{y^*} < \frac{i}{j} < \frac{h}{k}$. Then we have $jh - ik \ge 1$ and $iy^* - jx^* \ge 1$. Hence, on the one hand,

$$\frac{h}{k} - \frac{x^*}{y^*} = \left(\frac{h}{k} - \frac{i}{j}\right) + \left(\frac{i}{j} - \frac{x^*}{y^*}\right) = \frac{jh-ik}{jk} + \frac{iy^*-jx^*}{jy^*} \ge \frac{1}{jk} + \frac{1}{jy^*} = \frac{y^*+k}{jky^*}.$$

On the other hand,

$$\frac{h}{k} - \frac{x^*}{y^*} = \frac{hy^*-kx^*}{ky^*} = \frac{1}{ky^*},$$

which implies $\frac{y^*+k}{jky^*} \le \frac{1}{ky^*}$, that is, $y^* + k \le j$. But $1 < j \le n$, and therefore

$$y^* < n - k + 1. \tag{1.25}$$

Since $\frac{k}{h} < \frac{j}{i} < \frac{y^*}{x^*}$, we obtain

$$\frac{y^*}{x^*} - \frac{k}{h} = \left(\frac{y^*}{x^*} - \frac{j}{i}\right) + \left(\frac{j}{i} - \frac{k}{h}\right) = \frac{iy^* - jx^*}{ix^*} + \frac{jh - ik}{ih} \geq \frac{1}{ix^*} + \frac{1}{ih} = \frac{h + x^*}{ihx^*}.$$

On the other hand,

$$\frac{y^*}{x^*} - \frac{k}{h} = \frac{hy^* - kx^*}{hx^*} = \frac{1}{hx^*},$$

which implies $\frac{h+x^*}{ihx^*} \leq \frac{1}{hx^*}$, that is, $h + x^* \leq i$. But $1 \leq i \leq m$, and therefore

$$x^* < m - h + 1. \tag{1.26}$$

Suppose that $t^* = \lfloor \frac{n-y_0}{k} \rfloor$. Then

$$y^* := y_0 + k\left\lfloor \frac{n - y_0}{k} \right\rfloor = y_0 + k\left\lceil \frac{n - y_0 - k + 1}{k} \right\rceil \geq y_0 + k\frac{n - y_0 - k + 1}{k},$$

that is, $y^* \geq n - k + 1$, which contradicts (1.25).

Suppose that $t^* = \lfloor \frac{m-x_0}{h} \rfloor$. As noted in (1.24), we have $x^* \geq m - h + 1$, but this contradicts (1.26).

Thus, the fraction $\frac{x^*}{y^*}$ precedes the fraction $\frac{h}{k}$ in $\mathfrak{F}_n^m$.

(i) (b) is a restatement of (i) (a).

(ii) (a) If a pair $(x_0, y_0) \in \mathbb{Z}^2$ is a solution to the linear Diophantine equation

$$k\mathrm{x} - h\mathrm{y} = 1,$$

in the unknowns x and y, then the pair $(x_0 + ht, y_0 + kt)$ is also a solution for any integer t.

Consider the inequalities $1 \leq x_0 + ht \leq m$ and $y_0 + k\mathrm{t} \leq n$, in the unknown t, or the equivalent system

$$\begin{cases} \frac{1-x_0}{h} \leq \mathrm{t} \leq \frac{m-x_0}{h}, \\ \mathrm{t} \leq \frac{n-y_0}{k}, \end{cases}$$

or the inequality

$$\frac{1 - x_0}{h} \leq \mathrm{t} \leq \min\left\{\frac{m - x_0}{h}, \frac{n - y_0}{k}\right\}. \tag{1.27}$$

This inequality has at least one integer solution, namely, $\lceil \frac{1-x_0}{h} \rceil$, since

$$\frac{m - x_0}{h} - \left\lceil \frac{1 - x_0}{h} \right\rceil = \frac{m - x_0}{h} - \left\lfloor \frac{h - x_0}{h} \right\rfloor \geq \frac{m - x_0}{h} - \frac{h - x_0}{h} = \frac{m - h}{h} \geq 0$$

and

$$\frac{n - y_0}{k} - \left\lceil \frac{1 - x_0}{h} \right\rceil = \frac{n - y_0}{k} - \left\lfloor \frac{h - x_0}{h} \right\rfloor \geq \frac{n - y_0}{k} - \frac{h - x_0}{h} = \frac{h(n - k) + 1}{hk} > 0.$$

For any integer solution t, we have $(y_0 + kt) - (x_0 + ht) = \frac{(k-h)(x_0+ht)-1}{h} \geq 0$, that is,

$$x_0 + ht \leq y_0 + kt.$$

If t' and t'' are two integer solutions to inequality (1.27), with $t' < t''$, that is,

$$\left\lceil \frac{1 - x_0}{h} \right\rceil \le t' < t'' \le \left\lfloor \min\left\{ \frac{m - x_0}{h}, \frac{n - y_0}{k} \right\} \right\rfloor,$$

then

$$\frac{x_0 + ht''}{y_0 + kt''} - \frac{x_0 + ht'}{y_0 + kt'} = \frac{t' - t''}{(y_0 + kt')(y_0 + kt'')} < 0;$$

in other words,

$$\frac{x_0 + ht'}{y_0 + kt'} > \frac{x_0 + ht''}{y_0 + kt''}.$$

Define two integers x^* and y^* by

$$x^* := x_0 + ht^* \quad \text{and} \quad y^* := y_0 + kt^*,$$

where $t^* := \lfloor \min\{\frac{m-x_0}{h}, \frac{n-y_0}{k}\} \rfloor$, and recall that $x^* > 0$.

Suppose that there exists a fraction $\frac{i}{j} \in \mathcal{F}_n^m$ such that $\frac{h}{k} < \frac{i}{j} < \frac{x^*}{y^*}$. Then we have $jx^* - iy^* \ge 1$ and $ik - jh \ge 1$. Hence, on the one hand,

$$\frac{x^*}{y^*} - \frac{h}{k} = \left(\frac{x^*}{y^*} - \frac{i}{j}\right) + \left(\frac{i}{j} - \frac{h}{k}\right) = \frac{jx^* - iy^*}{jy^*} + \frac{ik - jh}{jk} \ge \frac{1}{jy^*} + \frac{1}{jk} = \frac{k + y^*}{jky^*}.$$

On the other hand,

$$\frac{x^*}{y^*} - \frac{h}{k} = \frac{kx^* - hy^*}{ky^*} = \frac{1}{ky^*},$$

which implies $\frac{k+y^*}{jky^*} \le \frac{1}{ky^*}$, that is, $k + y^* \le j$. But $1 < j \le n$, and therefore

$$y^* < n - k + 1. \tag{1.28}$$

Since $\frac{y^*}{x^*} < \frac{j}{i} < \frac{k}{h}$, we obtain

$$\frac{k}{h} - \frac{y^*}{x^*} = \left(\frac{k}{h} - \frac{j}{i}\right) + \left(\frac{j}{i} - \frac{y^*}{x^*}\right) = \frac{ik - jh}{ih} + \frac{jx^* - iy^*}{ix^*} \ge \frac{1}{ih} + \frac{1}{ix^*} = \frac{x^* + h}{ihx^*}.$$

On the other hand,

$$\frac{k}{h} - \frac{y^*}{x^*} = \frac{kx^* - hy^*}{hx^*} = \frac{1}{hx^*},$$

which implies $\frac{x^*+h}{ihx^*} \le \frac{1}{hx^*}$, that is, $x^* + h \le i$. But $1 \le i \le m$, and therefore

$$x^* < m - h + 1. \tag{1.29}$$

If $t^* = \lfloor \frac{n-y_0}{k} \rfloor$, then we have

$$y^* := y_0 + k\left\lfloor \frac{n - y_0}{k} \right\rfloor = y_0 + k\left\lceil \frac{n - y_0 - k + 1}{k} \right\rceil \ge y_0 + k\frac{n - y_0 - k + 1}{k},$$

that is, $y^* \ge n - k + 1$, which contradicts (1.28).

If $t^* = \lfloor \frac{m-x_0}{h} \rfloor$, then we have

$$x^* := x_0 + h\left\lfloor \frac{m - x_0}{h} \right\rfloor = x_0 + h\left\lceil \frac{m - x_0 - h + 1}{h} \right\rceil \ge x_0 + h\frac{m - x_0 - h + 1}{h},$$

that is, $x^* \ge m - h + 1$, which contradicts (1.29).

Thus, the fraction $\frac{x^*}{y^*}$ succeeds the fraction $\frac{h}{k}$ in $\mathcal{F}_n^m$.

(ii) (b) is a restatement of (ii) (a). □

The results of Proposition 1.10 will be streamlined in Corollary 1.11 and Lemma 2.13.

Corollary 1.11. (i) *Let* $\frac{h}{k} \in \mathcal{F}_n^m - \{\frac{0}{1}\}$, *where* $0 < m < n$. *Let* x_0 *and* y_0 *be two integers such that* $kx_0 \equiv -1 \pmod{h}$ *and* $hy_0 \equiv 1 \pmod{k}$.

(a) *If* $hn - km \geq 1$, *then the fraction*

$$\frac{x_0 + h\lfloor\frac{m-x_0}{h}\rfloor}{\frac{kx_0+1}{h} + k\lfloor\frac{m-x_0}{h}\rfloor} = \frac{\frac{hy_0-1}{k} + h\lfloor\frac{km-hy_0+1}{hk}\rfloor}{y_0 + k\lfloor\frac{km-hy_0+1}{hk}\rfloor} \tag{1.30}$$

precedes the fraction $\frac{h}{k}$ *in* $\mathcal{F}_n^m$.

(b) *If* $hn - km \leq 1$, *then the fraction*

$$\frac{x_0 + h\lfloor\frac{hn-kx_0-1}{hk}\rfloor}{\frac{kx_0+1}{h} + k\lfloor\frac{hn-kx_0-1}{hk}\rfloor} = \frac{\frac{hy_0-1}{k} + h\lfloor\frac{n-y_0}{k}\rfloor}{y_0 + k\lfloor\frac{n-y_0}{k}\rfloor} \tag{1.31}$$

precedes the fraction $\frac{h}{k}$ *in* $\mathcal{F}_n^m$.

(ii) *Let* $\frac{h}{k} \in \mathcal{F}_n^m - \{\frac{0}{1}, \frac{1}{1}\}$, *where* $0 < m < n$. *Let* x_0 *and* y_0 *be two integers such that* $kx_0 \equiv 1 \pmod{h}$ *and* $hy_0 \equiv -1 \pmod{k}$.

(a) *If* $km - hn \geq 1$, *then the fraction*

$$\frac{x_0 + h\lfloor\frac{hn-kx_0+1}{hk}\rfloor}{\frac{kx_0-1}{h} + k\lfloor\frac{hn-kx_0+1}{hk}\rfloor} = \frac{\frac{hy_0+1}{k} + h\lfloor\frac{n-y_0}{k}\rfloor}{y_0 + k\lfloor\frac{n-y_0}{k}\rfloor} \tag{1.32}$$

succeeds the fraction $\frac{h}{k}$ *in* $\mathcal{F}_n^m$.

(b) *If* $km - hn \leq 1$, *then the fraction*

$$\frac{x_0 + h\lfloor\frac{m-x_0}{h}\rfloor}{\frac{kx_0-1}{h} + k\lfloor\frac{m-x_0}{h}\rfloor} = \frac{\frac{hy_0+1}{k} + h\lfloor\frac{km-hy_0-1}{hk}\rfloor}{y_0 + k\lfloor\frac{km-hy_0-1}{hk}\rfloor} \tag{1.33}$$

succeeds the fraction $\frac{h}{k}$ *in* $\mathcal{F}_n^m$.

Proof. To prove assertion (i) (a), it suffices to note, in view of Proposition 1.10 (i) (a), that in (1.21), we have $\min\{\frac{m-x_0}{h}, \frac{hn-kx_0-1}{hk}\} = \frac{m-x_0}{h}$ if and only if $hn - km \geq 1$, and so on. □

By imposing natural restrictions on the integers x_0 and y_0 in Corollary 1.11, we come to even simpler results in Lemma 2.13.

We conclude the subsection with yet another observation on the neighboring fractions in $\mathcal{F}_n^m$; cf. Remarks 1.8, 1.16 and 1.22.

Remark 1.12. (i) For any pair $\frac{h_j}{k_j} < \frac{h_{j+1}}{k_{j+1}}$ of neighboring fractions in $\mathcal{F}_n^m$, where $0 < m < n$, such that

$$h_{j+1}n - k_{j+1}m \geq 1, \tag{1.34}$$

or

$$\frac{h_j}{k_j} > \frac{0}{1}, \quad k_jm - h_jn \leq 1, \tag{1.35}$$

we have

$$h_j + h_{j+1} > m. \tag{1.36}$$

Indeed, if relation (1.34) holds, then we see, according to Corollary 1.11 (i) (a), that $h_j = x_0 + h_{j+1}\lfloor\frac{m-x_0}{h_{j+1}}\rfloor$ for a certain integer x_0 and, as a consequence, we have

$$h_j+h_{j+1} = x_0+h_{j+1}\left\lceil\frac{m-x_0-h_{j+1}+1}{h_{j+1}}\right\rceil+h_{j+1} \geq x_0+h_{j+1}\frac{m-x_0-h_{j+1}+1}{h_{j+1}}+h_{j+1} = m+1.$$

If relation (1.35) holds, then, according to Corollary 1.11 (ii) (b), we see that $h_{j+1} = x_0 + h_j\lfloor\frac{m-x_0}{h_j}\rfloor$ for a certain integer x_0, and (1.36) follows.

(ii) For any pair $\frac{h_j}{k_j} < \frac{h_{j+1}}{k_{j+1}}$ of neighboring fractions in $\mathcal{F}_n^m$, where $0 < m < n$, such that

$$h_{j+1}n - k_{j+1}m \leq 1, \tag{1.37}$$

or

$$\frac{h_j}{k_j} > \frac{0}{1}, \quad k_jm - h_jn \geq 1, \tag{1.38}$$

we have

$$k_j + k_{j+1} > n. \tag{1.39}$$

Indeed, if relation (1.37) holds, then we see, according to Corollary 1.11 (i) (b), that $k_j = y_0 + k_{j+1}\lfloor\frac{n-y_0}{k_{j+1}}\rfloor$ for a certain integer y_0 and, as a consequence, we have

$$k_j+k_{j+1} = y_0+k_{j+1}\left\lceil\frac{n-y_0-k_{j+1}+1}{k_{j+1}}\right\rceil+k_{j+1} \geq y_0+k_{j+1}\frac{n-y_0-k_{j+1}+1}{k_{j+1}}+k_{j+1} = n+1.$$

If relation (1.38) holds, then, according to Corollary 1.11 (ii) (a), we see that $k_{j+1} = y_0 + k_j\lfloor\frac{n-y_0}{k_j}\rfloor$ for a certain integer y_0, and (1.39) follows.

1.4.3 Neighboring fractions in $\mathcal{G}_n^m$

Recall that the Farey subsequence $\mathcal{G}_n^m$ is defined by $\mathcal{G}_n^m := (\frac{h}{k} \in \mathcal{F}_n : m + k - n \leq h)$ for any integer $m < n$. However, throughout this subsection, we will additionally suppose that $m > 0$.

Let us find the neighbors of a fraction in the sequence $\mathcal{G}_n^m$.

Remark 1.13. It is clear (and verified with the help of Proposition 1.14 (ii), Corollary 1.15 (ii) (b) and especially Lemma 2.15 (ii) (b)) that $\frac{n-1}{n}$ precedes $\frac{1}{1}$ in $\mathcal{G}_n^m$, where $0 < m < n$.

According to Remark 1.9, $\frac{n-m}{n-m+1}$ precedes $\frac{1}{1}$ in $\mathcal{F}_n^{n-m}$, where $0 < m < n$. Thus, the order-reversing bijections $\mathcal{F}_n^{n-m} \leftrightarrow \mathcal{G}_n^m$, defined in (1.8), guarantee that $\frac{1}{n-m+1}$ succeeds $\frac{0}{1}$ in $\mathcal{G}_n^m$. See also Table 1.5.

Proposition 1.14. (i) *Let $\frac{h}{k} \in \mathcal{G}_n^m - \{\frac{0}{1}, \frac{1}{1}\}$, where $0 < m < n$.*

(a) *Let x_0 be an integer such that $kx_0 \equiv -1 \pmod{h}$. Define two integers y_0 and t^* by $y_0 := \frac{kx_0+1}{h}$ and $t^* := \lfloor\min\{\frac{n-y_0}{k}, \frac{(n-m)-(y_0-x_0)}{k-h}\}\rfloor$. Then the fraction*

$$\frac{x_0+ht^*}{y_0+kt^*} := \frac{x_0 + h\lfloor\min\{\frac{hn-kx_0-1}{hk}, \frac{h(n-m)-(k-h)x_0-1}{h(k-h)}\}\rfloor}{\frac{kx_0+1}{h} + k\lfloor\min\{\frac{hn-kx_0-1}{hk}, \frac{h(n-m)-(k-h)x_0-1}{h(k-h)}\}\rfloor} \tag{1.40}$$

precedes the fraction $\frac{h}{k}$ in $\mathcal{G}_n^m$.

(b) *In other words, let* y_0 *be an integer such that* $hy_0 \equiv 1 \pmod{k}$. *Define two integers* x_0 *and* t^* *by* $x_0 := \frac{hy_0-1}{k}$ *and* $t^* := \lfloor\min\{\frac{n-y_0}{k}, \frac{(n-m)-(y_0-x_0)}{k-h}\}\rfloor$. *Then the fraction*

$$\frac{x_0+ht^*}{y_0+kt^*} := \frac{\frac{hy_0-1}{k} + h\lfloor\min\{\frac{n-y_0}{k}, \frac{k(n-m)-(k-h)y_0-1}{(k-h)k}\}\rfloor}{y_0 + k\lfloor\min\{\frac{n-y_0}{k}, \frac{k(n-m)-(k-h)y_0-1}{(k-h)k}\}\rfloor},$$

which coincides with the fraction defined by (1.40), *precedes* $\frac{h}{k}$ *in* $\mathcal{G}_n^m$.

(ii) *Let* $\frac{h}{k} \in \mathcal{G}_n^m - \{\frac{0}{1}, \frac{1}{1}\}$, *where* $0 < m < n$.

(a) *Let* x_0 *be an integer such that* $kx_0 \equiv 1 \pmod{h}$. *Define two integers* y_0 *and* t^* *by* $y_0 := \frac{kx_0-1}{h}$ *and* $t^* := \lfloor\min\{\frac{n-y_0}{k}, \frac{(n-m)-(y_0-x_0)}{k-h}\}\rfloor$. *Then the fraction*

$$\frac{x_0+ht^*}{y_0+kt^*} := \frac{x_0 + h\lfloor\min\{\frac{hn-kx_0+1}{hk}, \frac{h(n-m)-(k-h)x_0+1}{h(k-h)}\}\rfloor}{\frac{kx_0-1}{h} + k\lfloor\min\{\frac{hn-kx_0+1}{hk}, \frac{h(n-m)-(k-h)x_0+1}{h(k-h)}\}\rfloor} \tag{1.41}$$

succeeds the fraction $\frac{h}{k}$ *in* $\mathcal{G}_n^m$.

(b) *In other words, let* y_0 *be an integer such that* $hy_0 \equiv -1 \pmod{k}$. *Define two integers* x_0 *and* t^* *by* $x_0 := \frac{hy_0+1}{k}$ *and* $t^* := \lfloor\min\{\frac{n-y_0}{k}, \frac{(n-m)-(y_0-x_0)}{k-h}\}\rfloor$. *Then the fraction*

$$\frac{x_0+ht^*}{y_0+kt^*} := \frac{\frac{hy_0+1}{k} + h\lfloor\min\{\frac{n-y_0}{k}, \frac{k(n-m)-(k-h)y_0+1}{(k-h)k}\}\rfloor}{y_0 + k\lfloor\min\{\frac{n-y_0}{k}, \frac{k(n-m)-(k-h)y_0+1}{(k-h)k}\}\rfloor},$$

which coincides with the fraction defined by (1.41), *succeeds* $\frac{h}{k}$ *in* $\mathcal{G}_n^m$.

Proof. (i) (a) If a pair $(x_0, y_0) \in \mathbb{Z}^2$ is a solution to the linear Diophantine equation

$$-kx + hy = 1,$$

in the unknowns x and y, then the pair $(x_0 + ht, y_0 + kt)$ is also a solution for any integer t.

Consider the inequalities $0 \le x_0 + ht, y_0 + kt \le n$ and $(y_0 + kt) - (x_0 + ht) \le n - m$, in the unknown t, or the equivalent system

$$\begin{cases} -\frac{x_0}{h} \le t, \\ t \le \frac{n-y_0}{k}, \\ t \le \frac{(n-m)-(y_0-x_0)}{k-h}, \end{cases}$$

or the inequality

$$-\frac{x_0}{h} \le t \le \min\Big\{\frac{n-y_0}{k}, \frac{(n-m)-(y_0-x_0)}{k-h}\Big\}. \tag{1.42}$$

This inequality has at least one integer solution, namely, $\lceil -\frac{x_0}{h} \rceil$, since

$$\frac{n-y_0}{k} - \Big\lceil -\frac{x_0}{h} \Big\rceil = \frac{n-y_0}{k} - \Big\lfloor \frac{h-x_0-1}{h} \Big\rfloor \ge \frac{n-y_0}{k} - \frac{h-x_0-1}{h}$$
$$= \frac{h(n-k)+k-1}{hk} \ge \frac{h(n-k)+k-n}{hk} = \frac{(n-k)(h-1)}{hk} \ge 0$$

and

$$\begin{aligned}\frac{(n-m)-(y_0-x_0)}{k-h} - \left\lceil -\frac{x_0}{h} \right\rceil &= \frac{(n-m)-(y_0-x_0)}{k-h} - \left\lfloor \frac{h-x_0-1}{h} \right\rfloor \\ &\geq \frac{(n-m)-(y_0-x_0)}{k-h} - \frac{h-x_0-1}{h} \\ &= \frac{h((n-m)-(k-h))+(k-h-1)}{h(k-h)} \geq 0.\end{aligned}$$

For any integer solution t, we have

$$(y_0+kt)-(x_0+ht) := \left(\frac{kx_0+1}{h}+kt\right)-(x_0+ht) = \frac{(k-h)(x_0+ht)+1}{h} > 0,$$

that is, $x_0+ht < y_0+kt$.

If t' and t'' are two integer solutions to inequality (1.42), with $t' < t''$, that is,

$$\left\lceil -\frac{x_0}{h} \right\rceil \leq t' < t'' \leq \left\lfloor \min\left\{\frac{n-y_0}{k}, \frac{(n-m)-(y_0-x_0)}{k-h}\right\} \right\rfloor,$$

then

$$\frac{x_0+ht''}{y_0+kt''} - \frac{x_0+ht'}{y_0+kt'} = \frac{t''-t'}{(y_0+kt')(y_0+kt'')} > 0;$$

in other words,

$$\frac{x_0+ht'}{y_0+kt'} < \frac{x_0+ht''}{y_0+kt''}.$$

Define two integers x^* and y^* by

$$x^* := x_0+ht^* \quad \text{and} \quad y^* := y_0+kt^*,$$

where $t^* := \lfloor \min\{\frac{n-y_0}{k}, \frac{(n-m)-(y_0-x_0)}{k-h}\} \rfloor$, and recall that $y^*-x^* > 0$.

Suppose that there exists a fraction $\frac{i}{j} \in \mathcal{G}_n^m$ such that $\frac{x^*}{y^*} < \frac{i}{j} < \frac{h}{k}$. Then we have $jh-ik \geq 1$ and $iy^*-jx^* \geq 1$. Hence, on the one hand,

$$\frac{h}{k}-\frac{x^*}{y^*} = \left(\frac{h}{k}-\frac{i}{j}\right)+\left(\frac{i}{j}-\frac{x^*}{y^*}\right) = \frac{jh-ik}{jk}+\frac{iy^*-jx^*}{jy^*} \geq \frac{1}{jk}+\frac{1}{jy^*} = \frac{y^*+k}{jky^*}.$$

On the other hand,

$$\frac{h}{k}-\frac{x^*}{y^*} = \frac{hy^*-kx^*}{ky^*} = \frac{1}{ky^*},$$

which implies $\frac{y^*+k}{jky^*} \leq \frac{1}{ky^*}$, that is, $y^*+k \leq j$. But $1 < j \leq n$, and therefore

$$y^* < n-k+1. \tag{1.43}$$

Since $\frac{x^*}{y^*-x^*} < \frac{i}{j-i} < \frac{h}{k-h}$, we obtain

$$\begin{aligned}\frac{h}{k-h}-\frac{x^*}{y^*-x^*} &= \left(\frac{h}{k-h}-\frac{i}{j-i}\right)+\left(\frac{i}{j-i}-\frac{x^*}{y^*-x^*}\right) \\ &= \frac{(j-i)h-i(k-h)}{(j-i)(k-h)}+\frac{i(y^*-x^*)-(j-i)x^*}{(j-i)(y^*-x^*)} \\ &\geq \frac{1}{(j-i)(k-h)}+\frac{1}{(j-i)(y^*-x^*)} \\ &= \frac{(y^*-x^*)+(k-h)}{(j-i)(k-h)(y^*-x^*)}.\end{aligned}$$

On the other hand,

$$\frac{h}{k-h} - \frac{x^*}{y^*-x^*} = \frac{hy^* - kx^*}{(k-h)(y^*-x^*)} = \frac{1}{(k-h)(y^*-x^*)},$$

which implies $\frac{(y^*-x^*)+(k-h)}{(j-i)(k-h)(y^*-x^*)} \leq \frac{1}{(k-h)(y^*-x^*)}$, that is, $(y^*-x^*)+(k-h) \leq j-i$. But $j-i \leq n-m$, and therefore

$$y^* - x^* < (n-m) - (k-h) + 1. \tag{1.44}$$

Suppose that $t^* = \lfloor \frac{n-y_0}{k} \rfloor$. Then

$$y^* := y_0 + k\Big\lfloor \frac{n-y_0}{k} \Big\rfloor = y_0 + k\Big\lceil \frac{n-y_0-k+1}{k} \Big\rceil \geq y_0 + k\frac{n-y_0-k+1}{k},$$

that is, $y^* \geq n-k+1$, which contradicts (1.43).

Suppose that $t^* = \lfloor \frac{(n-m)-(y_0-x_0)}{k-h} \rfloor$. Then

$$\begin{aligned} y^* - x^* &:= (y_0 - x_0) + (k-h)\Big\lfloor \frac{(n-m)-(y_0-x_0)}{k-h} \Big\rfloor \\ &= (y_0 - x_0) + (k-h)\Big\lceil \frac{(n-m)-(y_0-x_0)-(k-h)+1}{k-h} \Big\rceil \\ &\geq (y_0 - x_0) + (k-h)\frac{(n-m)-(y_0-x_0)-(k-h)+1}{k-h}, \end{aligned}$$

that is, $y^* - x^* \geq (n-m)-(k-h)+1$, which contradicts (1.44).

Thus, the fraction $\frac{x^*}{y^*}$ precedes the fraction $\frac{h}{k}$ in $\mathcal{G}_n^m$.

(i) (b) is a restatement of (i) (a).

(ii) (a) If a pair $(x_0, y_0) \in \mathbb{Z}^2$ is a solution to the linear Diophantine equation

$$kx - hy = 1,$$

in the unknowns x and y, then the pair $(x_0 + ht, y_0 + kt)$ is also a solution for any integer t.

Consider the inequalities $1 \leq x_0 + ht$, $y_0 + kt \leq n$ and $(y_0 + kt) - (x_0 + ht) \leq n - m$, in the unknown t, or the equivalent system

$$\begin{cases} \frac{1-x_0}{h} \leq t, \\ t \leq \frac{n-y_0}{k}, \\ t \leq \frac{(n-m)-(y_0-x_0)}{k-h}, \end{cases}$$

or the inequality

$$\frac{1-x_0}{h} \leq t \leq \min\Big\{\frac{n-y_0}{k}, \frac{(n-m)-(y_0-x_0)}{k-h}\Big\}. \tag{1.45}$$

This inequality has at least one integer solution, namely, $\lceil \frac{1-x_0}{h} \rceil$, since

$$\frac{n-y_0}{k} - \Big\lceil \frac{1-x_0}{h} \Big\rceil = \frac{n-y_0}{k} - \Big\lfloor \frac{h-x_0}{h} \Big\rfloor \geq \frac{n-y_0}{k} - \frac{h-x_0}{h} = \frac{h(n-k)+1}{k} > 0$$

and

$$\frac{(n-m)-(y_0-x_0)}{k-h} - \left\lceil \frac{1-x_0}{h} \right\rceil = \frac{(n-m)-(y_0-x_0)}{k-h} - \left\lfloor \frac{h-x_0}{h} \right\rfloor \geq \frac{(n-m)-(y_0-x_0)}{k-h} - \frac{h-x_0}{h} = \frac{h((n-m)-(k-h))+1}{h(k-h)} > 0.$$

For any integer solution t, we have

$$(y_0+kt)-(x_0+ht) := \left(\frac{kx_0-1}{h}+kt\right)-(x_0+ht) = \frac{(k-h)(x_0+ht)-1}{h} \geq 0$$

and, as a consequence, $x_0 + ht \leq y_0 + kt$.

If t' and t'' are two integer solutions to inequality (1.45), with $t' < t''$, that is,

$$\left\lceil \frac{1-x_0}{h} \right\rceil \leq t' < t'' \leq \left\lfloor \min\left\{\frac{n-y_0}{k}, \frac{(n-m)-(y_0-x_0)}{k-h}\right\} \right\rfloor,$$

then

$$\frac{x_0+ht''}{y_0+kt''} - \frac{x_0+ht'}{y_0+kt'} = \frac{t'-t''}{(y_0+kt')(y_0+kt'')} < 0;$$

in other words,

$$\frac{x_0+ht'}{y_0+kt'} > \frac{x_0+ht''}{y_0+kt''}.$$

Define two integers x^* and y^* by

$$x^* := x_0 + ht^* \quad \text{and} \quad y^* := y_0 + kt^*,$$

where $t^* := \lfloor \min\{\frac{n-y_0}{k}, \frac{(n-m)-(y_0-x_0)}{k-h}\}\rfloor$.

The assertion is proved straightforwardly in the case $\frac{h}{k} := \frac{n-1}{n}$. Indeed, here we have

$$\frac{(n-m)-(y_0-x_0)}{k-h} - \frac{n-y_0}{k} := \frac{(n-m)-(y_0-x_0)}{n-(n-1)} - \frac{n-y_0}{n} = n-m-1 \geq 0,$$

that is, $t^* = \lfloor \frac{n-y_0}{n} \rfloor = \lfloor -\frac{x_0}{n-1} + 1 + \frac{1}{(n-1)n} \rfloor$. Therefore, for any $x_0 \in \{s(n-1)+1 : s \in \mathbb{Z}\}$, we have $t^* = \lfloor -s + 1 - \frac{1}{n} \rfloor = s$ and, as a consequence, the fraction

$$\frac{x_0+ht^*}{y_0+kt^*} := \frac{x_0+(n-1)t^*}{y_0+nt^*} := \frac{x_0+(n-1)t^*}{\frac{nx_0-1}{n-1}+nt^*} = \frac{s(n-1)+1+(n-1)(-s)}{\frac{n(s(n-1)+1)-1}{n-1}+n(-s)} = \frac{1}{1}$$

succeeds the fraction $\frac{n-1}{n}$ in $\mathcal{G}_n^m$, in accordance with Remark 1.13.

Now suppose that

$$\frac{h}{k} < \frac{n-1}{n}$$

or, equivalently,

$$\frac{k-h}{k} > \frac{1}{n}.$$

If $t^* = \lfloor\frac{(n-m)-(y_0-x_0)}{k-h}\rfloor$, then

$$\begin{aligned}y^*-x^* &:= (y_0-x_0)+(k-h)\Big\lfloor\frac{(n-m)-(y_0-x_0)}{k-h}\Big\rfloor\\ &= (y_0-x_0)+(k-h)\Big\lceil\frac{(n-m)-(y_0-x_0)-(k-h)+1}{k-h}\Big\rceil\\ &\geq (y_0-x_0)+(k-h)\frac{(n-m)-(y_0-x_0)-(k-h)+1}{k-h},\end{aligned}$$

that is,

$$y^*-x^* \geq (n-m)-(k-h)+1 > 0. \tag{1.46}$$

If $t^* = \lfloor\frac{n-y_0}{k}\rfloor$, then

$$\begin{aligned}y^*-x^* &:= (y_0-x_0)+(k-h)\Big\lfloor\frac{n-y_0}{k}\Big\rfloor = (y_0-x_0)+(k-h)\Big\lceil\frac{n-y_0-k+1}{k}\Big\rceil\\ &\geq (y_0-x_0)+(k-h)\frac{n-y_0-k+1}{k} = \frac{(k-h)(n-k+1)-1}{k}\\ &= \frac{k-h}{k}(n-k+1)-\frac{1}{k} > \frac{1}{n}(n-k+1)-\frac{1}{k} = \frac{(n-k)(k-1)}{kn} \geq 0.\end{aligned}$$

We see that $\frac{h}{k} < \frac{n-1}{n}$ implies $y^*-x^* > 0$.

Suppose that there exists a fraction $\frac{i}{j} \in \mathcal{G}_n^m$ such that $\frac{h}{k} < \frac{i}{j} < \frac{x^*}{y^*}$. Then we have $jx^*-iy^* \geq 1$ and $ik-jh \geq 1$. Hence, on the one hand,

$$\frac{x^*}{y^*}-\frac{h}{k} = \Big(\frac{x^*}{y^*}-\frac{i}{j}\Big)+\Big(\frac{i}{j}-\frac{h}{k}\Big) = \frac{jx^*-iy^*}{jy^*}+\frac{ik-jh}{jk} \geq \frac{1}{jy^*}+\frac{1}{jk} = \frac{k+y^*}{jky^*}.$$

On the other hand,

$$\frac{x^*}{y^*}-\frac{h}{k} = \frac{kx^*-hy^*}{ky^*} = \frac{1}{ky^*},$$

which implies $\frac{k+y^*}{jky^*} \leq \frac{1}{ky^*}$, that is, $k+y^* \leq j$. But $1 < j \leq n$, and therefore

$$y^* < n-k+1. \tag{1.47}$$

Since $\frac{h}{k-h} < \frac{i}{j-i} < \frac{x^*}{y^*-x^*}$, we obtain

$$\begin{aligned}\frac{x^*}{y^*-x^*}-\frac{h}{k-h} &= \Big(\frac{x^*}{y^*-x^*}-\frac{i}{j-i}\Big)+\Big(\frac{i}{j-i}-\frac{h}{k-h}\Big)\\ &= \frac{(j-i)x^*-i(y^*-x^*)}{(y^*-x^*)(j-i)}+\frac{i(k-h)-(j-i)h}{(k-h)(j-i)}\\ &\geq \frac{1}{(y^*-x^*)(j-i)}+\frac{1}{(k-h)(j-i)}\\ &= \frac{(k-h)+(y^*-x^*)}{(k-h)(j-i)(y^*-x^*)}.\end{aligned}$$

On the other hand, $\frac{x^*}{y^*-x^*}-\frac{h}{k-h} = \frac{1}{(k-h)(y^*-x^*)}$, which implies

$$\frac{(k-h)+(y^*-x^*)}{(k-h)(j-i)(y^*-x^*)} \leq \frac{1}{(k-h)(y^*-x^*)},$$

that is, $(k-h)+(y^*-x^*) \le j-i$. But $j-i \le n-m$, and therefore

$$y^*-x^* < (n-m)-(k-h)+1. \tag{1.48}$$

Suppose that $t^* = \lfloor \frac{n-y_0}{k} \rfloor$. Then

$$y^* := y_0 + k\Big\lfloor \frac{n-y_0}{k} \Big\rfloor = y_0 + k\Big\lceil \frac{n-y_0-k+1}{k} \Big\rceil \ge y_0 + k\frac{n-y_0-k+1}{k},$$

that is, $y^* \ge n-k+1$, which contradicts (1.47).

Suppose that $t^* = \lfloor \frac{(n-m)-(y_0-x_0)}{k-h} \rfloor$. Then, as noted in (1.46), we obtain

$$y^*-x^* \ge (n-m)-(k-h)+1,$$

but this contradicts (1.48).

Thus, the fraction $\frac{x^*}{y^*}$ succeeds the fraction $\frac{h}{k}$ in $\mathcal{G}_n^m$.

(ii) (b) is a restatement of (ii) (a). □

The results of Proposition 1.14 will be streamlined in Corollary 1.15 and Lemma 2.15.

Corollary 1.15. (i) *Let $\frac{h}{k} \in \mathcal{G}_n^m - \{\frac{0}{1}, \frac{1}{1}\}$, where $0 < m < n$. Let x_0 and y_0 be two integers such that $kx_0 \equiv -1 \pmod{h}$ and $hy_0 \equiv 1 \pmod{k}$.*

(a) *If $hn - km \ge 1$, then the fraction*

$$\frac{x_0 + h\lfloor \frac{hn-kx_0-1}{hk} \rfloor}{\frac{kx_0+1}{h} + k\lfloor \frac{hn-kx_0-1}{hk} \rfloor} = \frac{\frac{hy_0-1}{k} + h\lfloor \frac{n-y_0}{k} \rfloor}{y_0 + k\lfloor \frac{n-y_0}{k} \rfloor} \tag{1.49}$$

precedes the fraction $\frac{h}{k}$ in $\mathcal{G}_n^m$.

(b) *If $hn - km \le 1$, then the fraction*

$$\frac{x_0 + h\lfloor \frac{h(n-m)-(k-h)x_0-1}{h(k-h)} \rfloor}{\frac{kx_0+1}{h} + k\lfloor \frac{h(n-m)-(k-h)x_0-1}{h(k-h)} \rfloor} = \frac{\frac{hy_0-1}{k} + h\lfloor \frac{k(n-m)-(k-h)y_0-1}{(k-h)k} \rfloor}{y_0 + k\lfloor \frac{k(n-m)-(k-h)y_0-1}{(k-h)k} \rfloor} \tag{1.50}$$

precedes the fraction $\frac{h}{k}$ in $\mathcal{G}_n^m$.

(ii) *Let $\frac{h}{k} \in \mathcal{G}_n^m - \{\frac{0}{1}, \frac{1}{1}\}$, where $0 < m < n$. Let x_0 and y_0 be two integers such that $kx_0 \equiv 1 \pmod{h}$ and $hy_0 \equiv -1 \pmod{k}$.*

(a) *If $km - hn \ge 1$, then the fraction*

$$\frac{x_0 + h\lfloor \frac{h(n-m)-(k-h)x_0+1}{h(k-h)} \rfloor}{\frac{kx_0-1}{h} + k\lfloor \frac{h(n-m)-(k-h)x_0+1}{h(k-h)} \rfloor} = \frac{\frac{hy_0+1}{k} + h\lfloor \frac{k(n-m)-(k-h)y_0+1}{(k-h)k} \rfloor}{y_0 + k\lfloor \frac{k(n-m)-(k-h)y_0+1}{(k-h)k} \rfloor} \tag{1.51}$$

succeeds the fraction $\frac{h}{k}$ in $\mathcal{G}_n^m$.

(b) *If $km - hn \le 1$, then the fraction*

$$\frac{x_0 + h\lfloor \frac{hn-kx_0+1}{hk} \rfloor}{\frac{kx_0-1}{h} + k\lfloor \frac{hn-kx_0+1}{hk} \rfloor} = \frac{\frac{hy_0+1}{k} + h\lfloor \frac{n-y_0}{k} \rfloor}{y_0 + k\lfloor \frac{n-y_0}{k} \rfloor} \tag{1.52}$$

succeeds the fraction $\frac{h}{k}$ in $\mathcal{G}_n^m$.

Proof. To prove assertion (i) (a), it suffices to note, in view of Proposition 1.14 (i) (a), that in (1.40), we have

$$\min\left\{\frac{hn-kx_0-1}{hk}, \frac{h(n-m)-(k-h)x_0-1}{h(k-h)}\right\} = \frac{hn-kx_0-1}{hk}$$

if and only if $hn - km \geq 1$, and so on. □

By imposing natural restrictions on the integers x_0 and y_0 in Corollary 1.15, we come to even simpler results in Lemma 2.15.

We conclude the subsection with yet another observation on the neighboring fractions in $\mathcal{G}_n^m$; cf. Remarks 1.8, 1.12 and 1.22.

Remark 1.16. (i) For any pair $\frac{h_j}{k_j} < \frac{h_{j+1}}{k_{j+1}}$ of neighboring fractions in $\mathcal{G}_n^m$, where $0 < m < n$, such that

$$\frac{h_{j+1}}{k_{j+1}} < \frac{1}{1}, \quad h_{j+1}n - k_{j+1}m \geq 1, \tag{1.53}$$

or

$$\frac{h_j}{k_j} > \frac{0}{1}, \quad k_jm - h_jn \leq 1, \tag{1.54}$$

we have

$$k_j + k_{j+1} > n. \tag{1.55}$$

Indeed, if relation (1.53) holds, then we see, according to Corollary 1.15 (i) (a), that $k_j = y_0 + k_{j+1}\lfloor\frac{n-y_0}{k_{j+1}}\rfloor$ for a certain integer y_0 and, as a consequence, we have

$$k_j+k_{j+1} = y_0+k_{j+1}\left\lceil\frac{n-y_0-k_{j+1}+1}{k_{j+1}}\right\rceil+k_{j+1} \geq y_0+k_{j+1}\frac{n-y_0-k_{j+1}+1}{k_{j+1}}+k_{j+1} = n+1.$$

If relation (1.54) holds, then we see, according to Corollary 1.15 (ii) (b), that $k_{j+1} = y_0 + k_j\lfloor\frac{n-y_0}{k_j}\rfloor$ for a certain integer y_0, and (1.55) follows.

(ii) For any pair $\frac{h_j}{k_j} < \frac{h_{j+1}}{k_{j+1}}$ of neighboring fractions in $\mathcal{G}_n^m$, where $0 < m < n$, such that

$$\frac{h_j}{k_j} < \frac{1}{1}, \quad h_{j+1}n - k_{j+1}m \leq 1, \tag{1.56}$$

or

$$\frac{h_j}{k_j} > \frac{0}{1}, \quad k_jm - h_jn \geq 1, \tag{1.57}$$

we have

$$(k_j - h_j) + (k_{j+1} - h_{j+1}) > n - m. \tag{1.58}$$

Indeed, if relation (1.56) holds, then we see, according to Corollary 1.15 (i) (b), that

$$k_j - h_j = \left(y_0 - \frac{h_{j+1}y_0 - 1}{k_{j+1}}\right) + (k_{j+1} - h_{j+1})\left\lfloor\frac{k_{j+1}(n-m) - (k_{j+1} - h_{j+1})y_0 - 1}{(k_{j+1} - h_{j+1})k_{j+1}}\right\rfloor$$

for a certain integer y_0 and, as a consequence, we have

$$\begin{aligned}(k_j - h_j) + (k_{j+1} - h_{j+1}) &= \frac{(k_{j+1} - h_{j+1})y_0 + 1}{k_{j+1}} \\ &\quad + (k_{j+1} - h_{j+1})\left\lceil \frac{k_{j+1}(n-m) - (k_{j+1} - h_{j+1})(k_{j+1} + y_0)}{(k_{j+1} - h_{j+1})k_{j+1}} \right\rceil \\ &\quad + (k_{j+1} - h_{j+1}) \\ &\geq \frac{(k_{j+1} - h_{j+1})y_0 + 1}{k_{j+1}} \\ &\quad + (k_{j+1} - h_{j+1})\frac{k_{j+1}(n-m) - (k_{j+1} - h_{j+1})(k_{j+1} + y_0)}{(k_{j+1} - h_{j+1})k_{j+1}} \\ &\quad + (k_{j+1} - h_{j+1}) \\ &= n - m + \frac{1}{k},\end{aligned}$$

and (1.58) follows.

If relation (1.57) holds, then we see, according to Corollary 1.15 (ii) (a), that

$$k_{j+1} - h_{j+1} = \left(y_0 - \frac{h_j y_0 + 1}{k_j}\right) + (k_j - h_j)\left\lfloor \frac{k_j(n-m) - (k_j - h_j)y_0 + 1}{(k_j - h_j)k_j} \right\rfloor$$

for a certain integer y_0, and (1.58) follows.

1.4.4 Neighboring fractions in $\mathcal{F}(\mathbb{B}(n), m)$

In this subsection we find the neighbors of fractions in the Farey subsequence $\mathcal{F}(\mathbb{B}(n), m)$.

Remark 1.17. It is clear (and verified, e.g., with the help of Corollary 1.19 (ii) (c), Theorem 1.20 (ii) and especially Proposition 1.21 (ii) (b)) that $\frac{m}{m+1}$ precedes $\frac{1}{1}$ in $\mathcal{F}(\mathbb{B}(n), m)$.

Since $\frac{n-m}{n-m+1}$ precedes $\frac{1}{1}$ in $\mathcal{F}(\mathbb{B}(n), n-m)$, we see, thanks to the order-reversing bijections $\mathcal{F}(\mathbb{B}(n), n-m) \leftrightarrow \mathcal{F}(\mathbb{B}(n), m)$, defined in (1.8), that $\frac{1}{n-m+1}$ succeeds $\frac{0}{1}$ in $\mathcal{F}(\mathbb{B}(n), m)$. See also Table 1.5.

We continue with a statement whose lengthy proof, which is a combination of fragments of the proofs of Propositions 1.10 and 1.14, is more important than the statement itself. The reader should be warned that Lemma 1.18 has no applied significance. However, the results of the lemma will be considered, in more detail, in Corollary 1.19, and they will be presented, in a more refined form, in Theorem 1.20.

Lemma 1.18. (i) *Let* $\frac{h}{k} \in \mathcal{F}(\mathbb{B}(n), m) - \{\frac{0}{1}, \frac{1}{1}\}$.

(a) *Let* x_0 *be an integer such that* $kx_0 \equiv -1 \pmod{h}$. *Define two integers* y_0 *and* t^* *by* $y_0 := \frac{kx_0+1}{h}$ *and* $t^* := \lfloor \min\{\frac{m-x_0}{h}, \frac{n-y_0}{k}, \frac{(n-m)-(y_0-x_0)}{k-h}\} \rfloor$. *Then the fraction*

$$\frac{x_0 + ht^*}{y_0 + kt^*} := \frac{x_0 + h\lfloor \min\{\frac{m-x_0}{h}, \frac{hn-kx_0-1}{hk}, \frac{h(n-m)-(k-h)x_0-1}{h(k-h)}\}\rfloor}{\frac{kx_0+1}{h} + k\lfloor \min\{\frac{m-x_0}{h}, \frac{hn-kx_0-1}{hk}, \frac{h(n-m)-(k-h)x_0-1}{h(k-h)}\}\rfloor} \tag{1.59}$$

precedes the fraction $\frac{h}{k}$ *in* $\mathcal{F}(\mathbb{B}(n), m)$.

(b) *In other words, let* y_0 *be an integer such that* $hy_0 \equiv 1 \pmod{k}$. *Define two integers* x_0 *and* t^* *by* $x_0 := \frac{hy_0-1}{k}$ *and* $t^* := \lfloor \min\{\frac{m-x_0}{h}, \frac{n-y_0}{k}, \frac{(n-m)-(y_0-x_0)}{k-h}\}\rfloor$. *Then the fraction*

$$\frac{x_0+ht^*}{y_0+kt^*} := \frac{\frac{hy_0-1}{k} + h\lfloor \min\{\frac{km-hy_0+1}{hk}, \frac{n-y_0}{k}, \frac{k(n-m)-(k-h)y_0-1}{(k-h)k}\}\rfloor}{y_0 + k\lfloor \min\{\frac{km-hy_0+1}{hk}, \frac{n-y_0}{k}, \frac{k(n-m)-(k-h)y_0-1}{(k-h)k}\}\rfloor},$$

which coincides with the fraction defined by (1.59), *precedes* $\frac{h}{k}$ *in* $\mathcal{F}(\mathbb{B}(n), m)$.

(ii) *Let* $\frac{h}{k} \in \mathcal{F}(\mathbb{B}(n), m) - \{\frac{0}{1}, \frac{1}{1}\}$.

(a) *Let* x_0 *be an integer such that* $kx_0 \equiv 1 \pmod{h}$. *Define two integers* y_0 *and* t^* *by* $y_0 := \frac{kx_0-1}{h}$ *and* $t^* := \lfloor \min\{\frac{m-x_0}{h}, \frac{n-y_0}{k}, \frac{(n-m)-(y_0-x_0)}{k-h}\}\rfloor$. *Then the fraction*

$$\frac{x_0+ht^*}{y_0+kt^*} := \frac{x_0 + h\lfloor \min\{\frac{m-x_0}{h}, \frac{hn-kx_0+1}{hk}, \frac{h(n-m)-(k-h)x_0+1}{h(k-h)}\}\rfloor}{\frac{kx_0-1}{h} + k\lfloor \min\{\frac{m-x_0}{h}, \frac{hn-kx_0+1}{hk}, \frac{h(n-m)-(k-h)x_0+1}{h(k-h)}\}\rfloor} \tag{1.60}$$

succeeds the fraction $\frac{h}{k}$ *in* $\mathcal{F}(\mathbb{B}(n), m)$.

(b) *In other words, let* y_0 *be an integer such that* $hy_0 \equiv -1 \pmod{k}$. *Define two integers* x_0 *and* t^* *by* $x_0 := \frac{hy_0+1}{k}$ *and* $t^* := \lfloor \min\{\frac{m-x_0}{h}, \frac{n-y_0}{k}, \frac{(n-m)-(y_0-x_0)}{k-h}\}\rfloor$. *Then the fraction*

$$\frac{x_0+ht^*}{y_0+kt^*} := \frac{\frac{hy_0+1}{k} + h\lfloor \min\{\frac{km-hy_0-1}{hk}, \frac{n-y_0}{k}, \frac{k(n-m)-(k-h)y_0+1}{(k-h)k}\}\rfloor}{y_0 + k\lfloor \min\{\frac{km-hy_0-1}{hk}, \frac{n-y_0}{k}, \frac{k(n-m)-(k-h)y_0+1}{(k-h)k}\}\rfloor},$$

which coincides with the fraction defined by (1.60), *succeeds* $\frac{h}{k}$ *in* $\mathcal{F}(\mathbb{B}(n), m)$.

Proof. (i) (a) If a pair $(x_0, y_0) \in \mathbb{Z}^2$ is a solution to the linear Diophantine equation

$$-kx + hy = 1,$$

in the unknowns x and y, then the pair $(x_0 + ht, y_0 + kt)$ is also a solution for any integer t.

Consider the inequalities $0 \le x_0 + ht \le m$, $y_0 + kt \le n$, $(y_0 + kt) - (x_0 + ht) \le n - m$, in the unknown t, or the equivalent system

$$\begin{cases} -\frac{x_0}{h} \le t \le \frac{m-x_0}{h}, \\ t \le \frac{n-y_0}{k}, \\ t \le \frac{(n-m)-(y_0-x_0)}{k-h}, \end{cases}$$

or the inequality

$$-\frac{x_0}{h} \le t \le \min\left\{\frac{m-x_0}{h}, \frac{n-y_0}{k}, \frac{(n-m)-(y_0-x_0)}{k-h}\right\}. \tag{1.61}$$

This inequality has at least one integer solution, namely, $\lceil -\frac{x_0}{h} \rceil$, since

$$\frac{m-x_0}{h} - \left\lceil -\frac{x_0}{h} \right\rceil = \frac{m-x_0}{h} - \left\lfloor \frac{h-x_0-1}{h} \right\rfloor \geq \frac{m-x_0}{h} - \frac{h-x_0-1}{h} = \frac{m-h+1}{h} > 0,$$

$$\frac{n-y_0}{k} - \left\lceil -\frac{x_0}{h} \right\rceil = \frac{n-y_0}{k} - \left\lfloor \frac{h-x_0-1}{h} \right\rfloor \geq \frac{n-y_0}{k} - \frac{h-x_0-1}{h} = \frac{h(n-k)+k-1}{hk}$$
$$\geq \frac{h(n-k)+k-n}{hk} = \frac{(n-k)(h-1)}{hk} \geq 0$$

and

$$\frac{(n-m)-(y_0-x_0)}{k-h} - \left\lceil -\frac{x_0}{h} \right\rceil = \frac{(n-m)-(y_0-x_0)}{k-h} - \left\lfloor \frac{h-x_0-1}{h} \right\rfloor$$
$$\geq \frac{(n-m)-(y_0-x_0)}{k-h} - \frac{h-x_0-1}{h}$$
$$= \frac{h((n-m)-(k-h))+(k-h-1)}{h(k-h)} \geq 0.$$

For any integer solution t, we have

$$(y_0+kt)-(x_0+ht) := \left(\frac{kx_0+1}{h}+kt\right)-(x_0+ht) = \frac{(k-h)(x_0+ht)+1}{h} > 0$$

and, as a consequence, $x_0+ht < y_0+kt$.

If t' and t'' are two integer solutions to inequality (1.61), with $t' < t''$, that is,

$$\left\lceil -\frac{x_0}{h} \right\rceil \leq t' < t'' \leq \left\lfloor \min\left\{\frac{m-x_0}{h}, \frac{n-y_0}{k}, \frac{(n-m)-(y_0-x_0)}{k-h}\right\} \right\rfloor,$$

then

$$\frac{x_0+ht''}{y_0+kt''} - \frac{x_0+ht'}{y_0+kt'} = \frac{t''-t'}{(y_0+kt')(y_0+kt'')} > 0;$$

in other words,

$$\frac{x_0+ht'}{y_0+kt'} < \frac{x_0+ht''}{y_0+kt''}.$$

We have

$$\frac{n-y_0}{k} - \frac{m-x_0}{h} = \frac{hn-km-1}{hk},$$
$$\frac{(n-m)-(y_0-x_0)}{k-h} - \frac{m-x_0}{h} = \frac{hn-km-1}{h(k-h)},$$
$$\frac{(n-m)-(y_0-x_0)}{k-h} - \frac{n-y_0}{k} = \frac{hn-km-1}{k(k-h)}.$$

Thus,

$$hn-km=1 \implies \frac{m-x_0}{h} = \frac{n-y_0}{k} = \frac{(n-m)-(y_0-x_0)}{k-h},$$
$$hn-km \geq 1 \implies \frac{m-x_0}{h} \leq \frac{n-y_0}{k} \leq \frac{(n-m)-(y_0-x_0)}{k-h},$$
$$hn-km \leq 1 \implies \frac{(n-m)-(y_0-x_0)}{k-h} \leq \frac{n-y_0}{k} \leq \frac{m-x_0}{h}.$$

Define two integers x^* and y^* by

$$x^* := x_0 + ht^* \quad \text{and} \quad y^* := y_0 + kt^*,$$

where

$$\begin{aligned} t^* &:= \left\lfloor \min\left\{\frac{m - x_0}{h}, \frac{n - y_0}{k}, \frac{(n-m) - (y_0 - x_0)}{k-h}\right\}\right\rfloor \\ &= \left\lfloor \min\left\{\frac{m - x_0}{h}, \frac{(n-m) - (y_0 - x_0)}{k-h}\right\}\right\rfloor. \end{aligned}$$

If $t^* = \lfloor \frac{m-x_0}{h} \rfloor$, then

$$x^* := x_0 + ht^* = x_0 + h\left\lfloor \frac{m - x_0}{h} \right\rfloor = x_0 + h\left\lceil \frac{m - x_0 - h + 1}{h} \right\rceil \geq x_0 + h\frac{m - x_0 - h + 1}{h},$$

and we see that

$$hn - km \geq 1 \implies t^* = \left\lfloor \frac{m - x_0}{h} \right\rfloor, \quad x^* \geq m - h + 1 > 0. \tag{1.62}$$

If $t^* = \lfloor \frac{(n-m)-(y_0-x_0)}{k-h} \rfloor$, then

$$\begin{aligned} y^* - x^* &:= (y_0 - x_0) + (k-h)\left\lfloor \frac{(n-m) - (y_0 - x_0)}{k-h} \right\rfloor \\ &= (y_0 - x_0) + (k-h)\left\lceil \frac{(n-m) - (y_0 - x_0) - (k-h) + 1}{k-h} \right\rceil \\ &\geq (y_0 - x_0) + (k-h)\frac{(n-m) - (y_0 - x_0) - (k-h) + 1}{k-h}, \end{aligned}$$

and we see that

$$hn - km \leq 1 \implies \begin{cases} t^* = \left\lfloor \dfrac{(n-m) - (y_0 - x_0)}{k-h} \right\rfloor, \\ y^* - x^* \geq (n-m) - (k-h) + 1 > 0. \end{cases} \tag{1.63}$$

Suppose that there exists a fraction $\frac{i}{j} \in \mathcal{F}(\mathbb{B}(n), m)$ such that $\frac{x^*}{y^*} < \frac{i}{j} < \frac{h}{k}$. Then we have $jh - ik \geq 1$ and $iy^* - jx^* \geq 1$.

In the case where $hn - km \geq 1$, in view of $\frac{k}{h} < \frac{j}{i} < \frac{y^*}{x^*}$, we have

$$\frac{y^*}{x^*} - \frac{k}{h} = \left(\frac{y^*}{x^*} - \frac{j}{i}\right) + \left(\frac{j}{i} - \frac{k}{h}\right) = \frac{iy^* - jx^*}{ix^*} + \frac{jh - ik}{ih} \geq \frac{1}{ix^*} + \frac{1}{ih} = \frac{h + x^*}{ihx^*}.$$

On the other hand,

$$\frac{y^*}{x^*} - \frac{k}{h} = \frac{hy^* - kx^*}{hx^*} = \frac{1}{hx^*},$$

which implies $\frac{h+x^*}{ihx^*} \leq \frac{1}{hx^*}$, that is, $h + x^* \leq i$. But $1 \leq i \leq m$, and therefore

$$x^* < m - h + 1. \tag{1.64}$$

In the case where $hn - km \leq 1$, in view of $\frac{x^*}{y^*-x^*} < \frac{i}{j-i} < \frac{h}{k-h}$, we have

$$\begin{aligned}
\frac{h}{k-h} - \frac{x^*}{y^*-x^*} &= \left(\frac{h}{k-h} - \frac{i}{j-i}\right) + \left(\frac{i}{j-i} - \frac{x^*}{y^*-x^*}\right) \\
&= \frac{(j-i)h - i(k-h)}{(k-h)(j-i)} + \frac{i(y^*-x^*) - (j-i)x^*}{(y^*-x^*)(j-i)} \\
&\geq \frac{1}{(k-h)(j-i)} + \frac{1}{(y^*-x^*)(j-i)} \\
&= \frac{(y^*-x^*) + (k-h)}{(k-h)(j-i)(y^*-x^*)}.
\end{aligned}$$

On the other hand,

$$\frac{h}{k-h} - \frac{x^*}{y^*-x^*} = \frac{1}{(k-h)(y^*-x^*)},$$

which implies

$$\frac{(y^*-x^*) + (k-h)}{(k-h)(j-i)(y^*-x^*)} \leq \frac{1}{(k-h)(y^*-x^*)},$$

that is, $(y^* - x^*) + (k-h) \leq j - i$. But $j - i \leq n - m$, and therefore

$$y^* - x^* < (n-m) - (k-h) + 1. \tag{1.65}$$

Suppose that $t^* = \lfloor \frac{m-x_0}{h} \rfloor$. According to (1.62), we have

$$x^* \geq m - h + 1,$$

but this contradicts (1.64).

Suppose that $t^* = \lfloor \frac{(n-m)-(y_0-x_0)}{k-h} \rfloor$. According to (1.63), we have

$$y^* - x^* \geq (n-m) - (k-h) + 1,$$

but this contradicts (1.65).

Thus, the fraction $\frac{x^*}{y^*}$ precedes the fraction $\frac{h}{k}$ in $\mathcal{F}(\mathbb{B}(n), m)$.

(i) (b) is a restatement of (i) (a).

(ii) (a) If a pair $(x_0, y_0) \in \mathbb{Z}^2$ is a solution to the linear Diophantine equation

$$kx - hy = 1,$$

in the unknowns x and y, then the pair $(x_0 + ht, y_0 + kt)$ is also a solution for any integer t.

Consider the inequalities $1 \leq x_0 + ht \leq m$, $y_0 + kt \leq n$, $(y_0 + kt) - (x_0 + ht) \leq n - m$, in the unknown t, or the equivalent system

$$\begin{cases}
\frac{1-x_0}{h} \leq t \leq \frac{m-x_0}{h}, \\
t \leq \frac{n-y_0}{k}, \\
t \leq \frac{(n-m)-(y_0-x_0)}{k-h},
\end{cases}$$

or the inequality

$$\frac{1-x_0}{h} \le t \le \min\left\{\frac{m-x_0}{h}, \frac{n-y_0}{k}, \frac{(n-m)-(y_0-x_0)}{k-h}\right\}. \tag{1.66}$$

This inequality has at least one integer solution, namely $\lceil \frac{1-x_0}{h} \rceil$, since

$$\frac{m-x_0}{h} - \left\lceil \frac{1-x_0}{h} \right\rceil = \frac{m-x_0}{h} - \left\lfloor \frac{h-x_0}{h} \right\rfloor \ge \frac{m-x_0}{h} - \frac{h-x_0}{h} = \frac{m-h}{h} \ge 0,$$

$$\frac{n-y_0}{k} - \left\lceil \frac{1-x_0}{h} \right\rceil = \frac{n-y_0}{k} - \left\lfloor \frac{h-x_0}{h} \right\rfloor \ge \frac{n-y_0}{k} - \frac{h-x_0}{h} = \frac{h(n-k)+1}{hk} > 0$$

and

$$\begin{aligned}\frac{(n-m)-(y_0-x_0)}{k-h} - \left\lceil \frac{1-x_0}{h} \right\rceil &= \frac{(n-m)-(y_0-x_0)}{k-h} - \left\lfloor \frac{h-x_0}{k} \right\rfloor \\ &\ge \frac{(n-m)-(y_0-x_0)}{k-h} - \frac{h-x_0}{k} \\ &= \frac{h((n-m)-(k-h))+1}{h(k-h)} > 0.\end{aligned}$$

For any integer solution t, we have

$$(y_0+kt)-(x_0+ht) := \left(\frac{kx_0-1}{h} + kt\right) - (x_0+ht) = \frac{(k-h)(x_0+ht)-1}{h} \ge 0$$

and, as a consequence, $x_0+ht \le y_0+kt$.

If t' and t'' are two integer solutions to inequality (1.66), with $t' < t''$, that is,

$$\left\lceil \frac{1-x_0}{h} \right\rceil \le t' < t'' \le \left\lfloor \min\left\{\frac{m-x_0}{h}, \frac{n-y_0}{k}, \frac{(n-m)-(y_0-x_0)}{k-h}\right\}\right\rfloor,$$

then

$$\frac{x_0+ht''}{y_0+kt''} - \frac{x_0+ht'}{y_0+kt'} = \frac{t'-t''}{(y_0+kt')(y_0+kt'')} < 0;$$

in other words,

$$\frac{x_0+ht'}{y_0+kt'} > \frac{x_0+ht''}{y_0+kt''}.$$

We have

$$\begin{aligned}\frac{n-y_0}{k} - \frac{m-x_0}{h} &= \frac{hn-km+1}{hk}, \\ \frac{(n-m)-(y_0-x_0)}{k-h} - \frac{m-x_0}{h} &= \frac{hn-km+1}{h(k-h)}, \\ \frac{(n-m)-(y_0-x_0)}{k-h} - \frac{n-y_0}{k} &= \frac{hn-km+1}{k(k-h)}.\end{aligned}$$

Thus,

$$\begin{aligned}km-hn=1 &\implies \frac{m-x_0}{h} = \frac{n-y_0}{k} = \frac{(n-m)-(y_0-x_0)}{k-h}, \\ km-hn \ge 1 &\implies \frac{(n-m)-(y_0-x_0)}{k-h} \le \frac{n-y_0}{k} \le \frac{m-x_0}{h}, \\ km-hn \le 1 &\implies \frac{m-x_0}{h} \le \frac{n-y_0}{k} \le \frac{(n-m)-(y_0-x_0)}{k-h}.\end{aligned}$$

Define two integers x^* and y^* by

$$x^* := x_0 + ht^* \quad \text{and} \quad y^* := y_0 + kt^*,$$

where

$$\begin{aligned} t^* &:= \left\lfloor \min\left\{ \frac{m - x_0}{h}, \frac{n - y_0}{k}, \frac{(n-m) - (y_0 - x_0)}{k-h} \right\} \right\rfloor \\ &= \left\lfloor \min\left\{ \frac{m - x_0}{h}, \frac{(n-m) - (y_0 - x_0)}{k-h} \right\} \right\rfloor. \end{aligned}$$

If $t^* = \lfloor \frac{(n-m)-(y_0-x_0)}{k-h} \rfloor$, then

$$\begin{aligned} y^* - x^* &:= (y_0 - x_0) + (k-h)\left\lfloor \frac{(n-m) - (y_0 - x_0)}{k-h} \right\rfloor \\ &= (y_0 - x_0) + (k-h)\left\lceil \frac{(n-m) - (y_0 - x_0) - (k-h) + 1}{k-h} \right\rceil \\ &\geq (y_0 - x_0) + (k-h)\frac{(n-m) - (y_0 - x_0) - (k-h) + 1}{k-h}, \end{aligned}$$

and we see that

$$km - hn \geq 1 \implies \begin{cases} t^* = \left\lfloor \dfrac{(n-m) - (y_0 - x_0)}{k-h} \right\rfloor, \\ y^* - x^* \geq (n-m) - (k-h) + 1 > 0. \end{cases} \tag{1.67}$$

If $t^* = \lfloor \frac{m-x_0}{h} \rfloor$, then

$$x^* := x_0 + ht^* = x_0 + h\left\lfloor \frac{m - x_0}{h} \right\rfloor = x_0 + h\left\lceil \frac{m - x_0 - h + 1}{h} \right\rceil \geq x_0 + h\frac{m - x_0 - h + 1}{h},$$

and we see that

$$km - hn \leq 1 \implies t^* = \left\lfloor \frac{m - x_0}{h} \right\rfloor, \quad x^* \geq m - h + 1 > 0. \tag{1.68}$$

Suppose that there exists a fraction $\frac{i}{j} \in \mathcal{F}(\mathbb{B}(n), m)$ such that $\frac{h}{k} < \frac{i}{j} < \frac{x^*}{y^*}$. Then we have $jx^* - iy^* \geq 1$ and $ik - jh \geq 1$.

In the case where $km - hn \geq 1$, in view of $\frac{h}{k-h} < \frac{i}{j-i} < \frac{x^*}{y^*-x^*}$, we have

$$\begin{aligned} \frac{x^*}{y^* - x^*} - \frac{h}{k-h} &= \left(\frac{x^*}{y^* - x^*} - \frac{i}{j-i} \right) + \left(\frac{i}{j-i} - \frac{h}{k-h} \right) \\ &= \frac{(j-i)x^* - i(y^* - x^*)}{(y^* - x^*)(j-i)} + \frac{i(k-h) - (j-i)h}{(k-h)(j-i)} \\ &\geq \frac{1}{(y^* - x^*)(j-i)} + \frac{1}{(k-h)(j-i)} = \frac{(k-h) + (y^* - x^*)}{(k-h)(j-i)(y^* - x^*)}. \end{aligned}$$

On the other hand,

$$\frac{x^*}{y^* - x^*} - \frac{h}{k-h} = \frac{1}{(k-h)(y^* - x^*)},$$

which implies

$$\frac{(k-h)+(y^*-x^*)}{(k-h)(j-i)(y^*-x^*)} \le \frac{1}{(k-h)(y^*-x^*)},$$

that is, $(k-h)+(y^*-x^*) \le j-i$. But $j-i \le n-m$, and therefore

$$y^* - x^* < (n-m)-(k-h)+1. \tag{1.69}$$

In the case where $km - hn \le 1$, in view of $\frac{y^*}{x^*} < \frac{j}{i} < \frac{k}{h}$, we have

$$\frac{k}{h} - \frac{y^*}{x^*} = \left(\frac{k}{h} - \frac{j}{i}\right) + \left(\frac{j}{i} - \frac{y^*}{x^*}\right) = \frac{ik-jh}{ih} + \frac{jx^*-iy^*}{ix^*} \ge \frac{1}{ih} + \frac{1}{ix^*} = \frac{x^*+h}{ihx^*}.$$

On the other hand,

$$\frac{k}{h} - \frac{y^*}{x^*} = \frac{kx^*-hy^*}{hx^*} = \frac{1}{hx^*},$$

which implies $\frac{x^*+h}{ihx^*} \le \frac{1}{hx^*}$, that is, $x^*+h \le i$. But $1 \le i \le m$, therefore,

$$x^* < m-h+1. \tag{1.70}$$

Suppose that $t^* = \lfloor \frac{(n-m)-(y_0-x_0)}{k-h} \rfloor$. According to (1.67), we have

$$y^* - x^* \ge (n-m)-(k-h)+1,$$

but this contradicts (1.69).

Suppose that $t^* = \lfloor \frac{m-x_0}{h} \rfloor$. According to (1.68), we have $x^* \ge m-h+1$, but this contradicts (1.70).

Thus, the fraction $\frac{x^*}{y^*}$ succeeds the fraction $\frac{h}{k}$ in $\mathcal{F}(\mathbb{B}(n), m)$.

(ii) (b) is a restatement of (ii) (a). □

Corollary 1.19. (i) *Let* $\frac{h}{k} \in \mathcal{F}(\mathbb{B}(n), m) - \{\frac{0}{1}, \frac{1}{1}\}$. *Let* x_0 *and* y_0 *be two integers such that* $kx_0 \equiv -1 \pmod{h}$ *and* $hy_0 \equiv 1 \pmod{k}$.

(a) *If* $hn - km \ge 1$, *then the fraction*

$$\frac{x_0 + h\lfloor \frac{m-x_0}{h} \rfloor}{\frac{kx_0+1}{h} + k\lfloor \frac{m-x_0}{h} \rfloor} = \frac{\frac{hy_0-1}{k} + h\lfloor \frac{km-hy_0+1}{hk} \rfloor}{y_0 + k\lfloor \frac{km-hy_0+1}{hk} \rfloor}$$

precedes the fraction $\frac{h}{k}$ *in* $\mathcal{F}(\mathbb{B}(n), m)$.

(b) *If* $hn - km = 1$, *then the fraction*

$$\frac{x_0 + h\lfloor \frac{m-x_0}{h} \rfloor}{\frac{kx_0+1}{h} + k\lfloor \frac{m-x_0}{h} \rfloor} = \frac{\frac{hy_0-1}{k} + h\lfloor \frac{n-y_0}{k} \rfloor}{y_0 + k\lfloor \frac{n-y_0}{k} \rfloor}$$

precedes the fraction $\frac{h}{k}$ *in* $\mathcal{F}(\mathbb{B}(n), m)$.

(c) *If* $hn - km \le 1$, *then the fraction*

$$\frac{x_0 + h\lfloor \frac{h(n-m)-(k-h)x_0-1}{h(k-h)} \rfloor}{\frac{kx_0+1}{h} + k\lfloor \frac{h(n-m)-(k-h)x_0-1}{h(k-h)} \rfloor} = \frac{\frac{hy_0-1}{k} + h\lfloor \frac{k(n-m)-(k-h)y_0-1}{(k-h)k} \rfloor}{y_0 + k\lfloor \frac{k(n-m)-(k-h)y_0-1}{(k-h)k} \rfloor}$$

precedes the fraction $\frac{h}{k}$ *in* $\mathcal{F}(\mathbb{B}(n), m)$.

(ii) *Let* $\frac{h}{k} \in \mathcal{F}(\mathbb{B}(n), m) - \{\frac{0}{1}, \frac{1}{1}\}$. *Let* x_0 *and* y_0 *be two integers such that* $kx_0 \equiv 1 \pmod{h}$ *and* $hy_0 \equiv -1 \pmod{k}$.

(a) *If* $km - hn \geq 1$, *then the fraction*

$$\frac{x_0 + h\lfloor\frac{h(n-m)-(k-h)x_0+1}{h(k-h)}\rfloor}{\frac{kx_0-1}{h} + k\lfloor\frac{h(n-m)-(k-h)x_0+1}{h(k-h)}\rfloor} = \frac{\frac{hy_0+1}{k} + h\lfloor\frac{k(n-m)-(k-h)y_0+1}{(k-h)k}\rfloor}{y_0 + \lfloor\frac{k(n-m)-(k-h)y_0+1}{(k-h)k}\rfloor}$$

succeeds the fraction $\frac{h}{k}$ *in* $\mathcal{F}(\mathbb{B}(n), m)$.

(b) *If* $km - hn = 1$, *then the fraction*

$$\frac{x_0 + h\lfloor\frac{m-x_0}{h}\rfloor}{\frac{kx_0-1}{h} + k\lfloor\frac{m-x_0}{h}\rfloor} = \frac{\frac{hy_0+1}{k} + h\lfloor\frac{n-y_0}{k}\rfloor}{y_0 + k\lfloor\frac{n-y_0}{k}\rfloor}$$

succeeds the fraction $\frac{h}{k}$ *in* $\mathcal{F}(\mathbb{B}(n), m)$.

(c) *If* $km - hn \leq 1$, *then the fraction*

$$\frac{x_0 + h\lfloor\frac{m-x_0}{h}\rfloor}{\frac{kx_0-1}{h} + k\lfloor\frac{m-x_0}{h}\rfloor} = \frac{\frac{hy_0+1}{k} + h\lfloor\frac{km-hy_0-1}{hk}\rfloor}{y_0 + k\lfloor\frac{km-hy_0-1}{hk}\rfloor}$$

succeeds the fraction $\frac{h}{k}$ *in* $\mathcal{F}(\mathbb{B}(n), m)$.

Proof. To prove assertion (i) (a), it suffices to note that in (1.59) of Lemma 1.18 (i) (a), we have

$$\min\left\{\frac{m - x_0}{h}, \frac{hn - kx_0 - 1}{hk}, \frac{h(n-m) - (k-h)x_0 - 1}{h(k-h)}\right\} = \frac{m - x_0}{h}$$

if and only if $hn - km \geq 1$, and so on. □

Corollary 1.19 allows us to refine Lemma 1.18 as follows.

Theorem 1.20.

(i) *Let* $\frac{h}{k} \in \mathcal{F}(\mathbb{B}(n), m) - \{\frac{0}{1}, \frac{1}{1}\}$. *Let* x_0 *be an integer such that* $kx_0 \equiv -1 \pmod{h}$. *Define two integers* y_0 *and* t^* *by* $y_0 := \frac{kx_0+1}{h}$ *and* $t^* := \lfloor\min\{\frac{m-x_0}{h}, \frac{(n-m)-(y_0-x_0)}{k-h}\}\rfloor$. *Then the fraction*

$$\frac{x_0 + ht^*}{y_0 + kt^*} := \frac{x_0 + h\lfloor\min\{\frac{m-x_0}{h}, \frac{h(n-m)-(k-h)x_0-1}{h(k-h)}\}\rfloor}{\frac{kx_0+1}{h} + k\lfloor\min\{\frac{m-x_0}{h}, \frac{h(n-m)-(k-h)x_0-1}{h(k-h)}\}\rfloor} \tag{1.71}$$

precedes the fraction $\frac{h}{k}$ *in* $\mathcal{F}(\mathbb{B}(n), m)$.

In other words, let y_0 *be an integer such that* $hy_0 \equiv 1 \pmod{k}$. *Define two integers* x_0 *and* t^* *by* $x_0 := \frac{hy_0-1}{k}$ *and* $t^* := \lfloor\min\{\frac{m-x_0}{h}, \frac{(n-m)-(y_0-x_0)}{k-h}\}\rfloor$. *Then the fraction*

$$\frac{x_0 + ht^*}{y_0 + kt^*} := \frac{\frac{hy_0-1}{k} + h\lfloor\min\{\frac{km-hy_0+1}{hk}, \frac{k(n-m)-(k-h)y_0-1}{(k-h)k}\}\rfloor}{y_0 + k\lfloor\min\{\frac{km-hy_0+1}{hk}, \frac{k(n-m)-(k-h)y_0-1}{(k-h)k}\}\rfloor},$$

which coincides with the fraction defined by (1.71), *precedes* $\frac{h}{k}$ *in* $\mathcal{F}(\mathbb{B}(n), m)$.

More precisely, if $hn - km \geq 1$, *then the fraction*

$$\frac{x_0 + h\lfloor\frac{m-x_0}{h}\rfloor}{\frac{kx_0+1}{h} + k\lfloor\frac{m-x_0}{h}\rfloor} = \frac{\frac{hy_0-1}{k} + h\lfloor\frac{km-hy_0+1}{hk}\rfloor}{y_0 + k\lfloor\frac{km-hy_0+1}{hk}\rfloor} \tag{1.72}$$

precedes the fraction $\frac{h}{k}$ *in* $\mathcal{F}(\mathbb{B}(n), m)$. *If* $hn - km \leq 1$, *then the fraction*

$$\frac{x_0 + h\lfloor\frac{h(n-m)-(k-h)x_0-1}{h(k-h)}\rfloor}{\frac{kx_0+1}{h} + k\lfloor\frac{h(n-m)-(k-h)x_0-1}{h(k-h)}\rfloor} = \frac{\frac{hy_0-1}{k} + h\lfloor\frac{k(n-m)-(k-h)y_0-1}{(k-h)k}\rfloor}{y_0 + k\lfloor\frac{k(n-m)-(k-h)y_0-1}{(k-h)k}\rfloor} \tag{1.73}$$

precedes $\frac{h}{k}$ *in* $\mathcal{F}(\mathbb{B}(n), m)$.

(ii) *Let* $\frac{h}{k} \in \mathcal{F}(\mathbb{B}(n), m) - \{\frac{0}{1}, \frac{1}{1}\}$. *Let* x_0 *be an integer such that* $kx_0 \equiv 1 \pmod{h}$. *Define two integers* y_0 *and* t^* *by* $y_0 := \frac{kx_0-1}{h}$ *and* $t^* := \lfloor\min\{\frac{m-x_0}{h}, \frac{(n-m)-(y_0-x_0)}{k-h}\}\rfloor$. *Then the fraction*

$$\frac{x_0 + ht^*}{y_0 + kt^*} := \frac{x_0 + h\lfloor\min\{\frac{m-x_0}{h}, \frac{h(n-m)-(k-h)x_0+1}{h(k-h)}\}\rfloor}{\frac{kx_0-1}{h} + k\lfloor\min\{\frac{m-x_0}{h}, \frac{h(n-m)-(k-h)x_0+1}{h(k-h)}\}\rfloor} \tag{1.74}$$

succeeds the fraction $\frac{h}{k}$ *in* $\mathcal{F}(\mathbb{B}(n), m)$.

In other words, let y_0 *be an integer such that* $hy_0 \equiv -1 \pmod{k}$. *Define two integers* x_0 *and* t^* *by* $x_0 := \frac{hy_0+1}{k}$ *and* $t^* := \lfloor\min\{\frac{m-x_0}{h}, \frac{(n-m)-(y_0-x_0)}{k-h}\}\rfloor$. *Then the fraction*

$$\frac{x_0 + ht^*}{y_0 + kt^*} := \frac{\frac{hy_0+1}{k} + h\lfloor\min\{\frac{km-hy_0-1}{hk}, \frac{k(n-m)-(k-h)y_0+1}{(k-h)k}\}\rfloor}{y_0 + k\lfloor\min\{\frac{km-hy_0-1}{hk}, \frac{k(n-m)-(k-h)y_0+1}{(k-h)k}\}\rfloor},$$

which coincides with the fraction defined by (1.74), *succeeds* $\frac{h}{k}$ *in* $\mathcal{F}(\mathbb{B}(n), m)$.

More precisely, if $km - hn \geq 1$, *then the fraction*

$$\frac{x_0 + h\lfloor\frac{h(n-m)-(k-h)x_0+1}{h(k-h)}\rfloor}{\frac{kx_0-1}{h} + k\lfloor\frac{h(n-m)-(k-h)x_0+1}{h(k-h)}\rfloor} = \frac{\frac{hy_0+1}{k} + h\lfloor\frac{k(n-m)-(k-h)y_0+1}{(k-h)k}\rfloor}{y_0 + k\lfloor\frac{k(n-m)-(k-h)y_0+1}{(k-h)k}\rfloor} \tag{1.75}$$

succeeds the fraction $\frac{h}{k}$ *in* $\mathcal{F}(\mathbb{B}(n), m)$. *If* $km - hn \leq 1$, *then the fraction*

$$\frac{x_0 + h\lfloor\frac{m-x_0}{h}\rfloor}{\frac{kx_0-1}{h} + k\lfloor\frac{m-x_0}{h}\rfloor} = \frac{\frac{hy_0+1}{k} + h\lfloor\frac{km-hy_0-1}{hk}\rfloor}{y_0 + k\lfloor\frac{km-hy_0-1}{hk}\rfloor} \tag{1.76}$$

succeeds $\frac{h}{k}$ *in* $\mathcal{F}(\mathbb{B}(n), m)$.

The following statement, in the case of the sequences $\mathcal{F}(\mathbb{B}(n), m)$, with $n \neq 2m$, will be revisited in Propositions 2.18 and 2.19, by means of the Farey duality discussed in Chapter 2.

Proposition 1.21. (i) *Let* $\frac{h}{k} \in \mathcal{F}(\mathbb{B}(n), m) - \{\frac{0}{1}, \frac{1}{1}\}$.

(a) *Suppose that* $hn - km \geq 1$. *Let* a *and* b *be two integers such that*

$$ka \equiv -1 \pmod{h}, \quad m - h + 1 \leq a \leq m, \tag{1.77}$$

$$hb \equiv 1 \pmod{k}, \quad \left\lceil \frac{km+2}{h} \right\rceil - k \leq b \leq \left\lceil \frac{km+2}{h} \right\rceil - 1. \tag{1.78}$$

Then the fraction $a/\frac{ka+1}{h} = \frac{hb-1}{k}/b$ *precedes the fraction* $\frac{h}{k}$ *in* $\mathcal{F}(\mathbb{B}(n), m)$.

(b) *Suppose that* $hn - km \leq 1$. *Let* a *and* b *be two integers such that*

$$ka \equiv -1 \pmod{h}, \quad \left\lceil \frac{h(n-m)}{k-h} \right\rceil - h \leq a \leq \left\lceil \frac{h(n-m)}{k-h} \right\rceil - 1, \tag{1.79}$$

$$hb \equiv 1 \pmod{k}, \quad \left\lceil \frac{k(n-m)}{k-h} \right\rceil - k \leq b \leq \left\lceil \frac{k(n-m)}{k-h} \right\rceil - 1. \tag{1.80}$$

Then the fraction $a/\frac{ka+1}{h} = \frac{hb-1}{k}/b$ *precedes the fraction* $\frac{h}{k}$ *in* $\mathcal{F}(\mathbb{B}(n), m)$.

(ii) *Let* $\frac{h}{k} \in \mathcal{F}(\mathbb{B}(n), m) - \{\frac{0}{1}, \frac{1}{1}\}$.

(a) *Suppose that* $km - hn \geq 1$. *Let* a *and* b *be two integers such that*

$$ka \equiv 1 \pmod{h}, \quad \left\lceil \frac{h(n-m)+2}{k-h} \right\rceil - h \leq a \leq \left\lceil \frac{h(n-m)+2}{k-h} \right\rceil - 1, \tag{1.81}$$

$$hb \equiv -1 \pmod{k}, \quad \left\lceil \frac{k(n-m)+2}{k-h} \right\rceil - k \leq b \leq \left\lceil \frac{k(n-m)+2}{k-h} \right\rceil - 1. \tag{1.82}$$

Then the fraction $a/\frac{ka-1}{h} = \frac{hb+1}{k}/b$ *succeeds the fraction* $\frac{h}{k}$ *in* $\mathcal{F}(\mathbb{B}(n), m)$.

(b) *Suppose that* $km - hn \leq 1$. *Let* a *and* b *be two integers such that*

$$ka \equiv 1 \pmod{h}, \quad m - h + 1 \leq a \leq m, \tag{1.83}$$

$$hb \equiv -1 \pmod{k}, \quad \left\lceil \frac{km}{h} \right\rceil - k \leq b \leq \left\lceil \frac{km}{h} \right\rceil - 1. \tag{1.84}$$

Then the fraction $a/\frac{ka-1}{h} = \frac{hb+1}{k}/b$ *succeeds the fraction* $\frac{h}{k}$ *in* $\mathcal{F}(\mathbb{B}(n), m)$.

Proof. To prove assertion (i) (a), it suffices to establish those search intervals for the integers x_0 and y_0 in Theorem 1.20 (i), for which the multipliers of the form $\lfloor \cdot \rfloor$ in (1.72) vanish. These search intervals are precisely the intervals for the integers a and b in (1.77) and (1.78).

Assertion (i) (b) also follows from Theorem 1.20 (i); it suffices to apply the search intervals for the integers a and b in (1.79) and (1.80) to the integers x_0 and y_0 in (1.73), respectively.

Similarly, assertion (ii) follows from Theorem 1.20 (ii). Indeed, the search intervals for the integers a and b in (1.81) and (1.82), applied to the integers x_0 and y_0 in (1.75), respectively, turn the multipliers of the form $\lfloor \cdot \rfloor$ in (1.75) into zero.

The search intervals for a and b in (1.83) and (1.84), applied to x_0 and y_0 in (1.76), respectively, also turn the multipliers of the form $\lfloor \cdot \rfloor$ in (1.76) into zero. □

We conclude the subsection with a statement which is a combination of fragments of Remarks 1.12 and 1.16; see also Remark 1.8.

Remark 1.22. (i) For any pair $\frac{h_j}{k_j} < \frac{h_{j+1}}{k_{j+1}}$ of neighboring fractions in $\mathcal{F}(\mathbb{B}(n), m)$, such that

$$\frac{h_j}{k_j} < \frac{1}{1}, \quad h_{j+1}n - k_{j+1}m \geq 1, \tag{1.85}$$

or

$$\frac{h_j}{k_j} > \frac{0}{1}, \quad k_j m - h_j n \leq 1, \tag{1.86}$$

we have

$$h_j + h_{j+1} > m. \tag{1.87}$$

Indeed, if relation (1.85) holds, then we see, according to Theorem 1.20 (i), that $h_j = x_0 + h_{j+1}\lfloor \frac{m-x_0}{h_{j+1}} \rfloor$ for a certain integer x_0 and, as a consequence, we have

$$h_j + h_{j+1} = x_0 + h_{j+1}\left\lceil \frac{m - x_0 - h_{j+1} + 1}{h_{j+1}} \right\rceil + h_{j+1} \geq x_0 + h_{j+1}\frac{m - x_0 - h_{j+1} + 1}{h_{j+1}} + h_{j+1} = m+1.$$

On the other hand, if relation (1.86) holds, then we see, according to Theorem 1.20 (ii), that $h_{j+1} = x_0 + h_j\lfloor \frac{m-x_0}{h_j} \rfloor$ for a certain integer x_0, and (1.87) follows.

(ii) For any pair $\frac{h_j}{k_j} < \frac{h_{j+1}}{k_{j+1}}$ of neighboring fractions in $\mathcal{F}(\mathbb{B}(n), m)$ such that

$$\frac{h_j}{k_j} < \frac{1}{1}, \quad h_{j+1}n - k_{j+1}m \leq 1, \tag{1.88}$$

or

$$\frac{h_j}{k_j} > \frac{0}{1}, \quad k_j m - h_j n \geq 1, \tag{1.89}$$

we have

$$(k_j - h_j) + (k_{j+1} - h_{j+1}) > n - m. \tag{1.90}$$

Indeed, if relation (1.88) holds, then we see, according to Theorem 1.20 (i), that

$$k_j - h_j = \left(y_0 - \frac{h_{j+1}y_0 - 1}{k_{j+1}}\right) + (k_{j+1} - h_{j+1})\left\lfloor \frac{k_{j+1}(n-m) - (k_{j+1} - h_{j+1})y_0 - 1}{(k_{j+1} - h_{j+1})k_{j+1}} \right\rfloor$$

for a certain integer y_0 and, as a consequence, we have

$$\begin{aligned}
(k_j - h_j) + (k_{j+1} - h_{j+1}) &= \frac{(k_{j+1} - h_{j+1})y_0 + 1}{k_{j+1}} \\
&\quad + (k_{j+1} - h_{j+1})\left\lceil \frac{k_{j+1}(n-m) - (k_{j+1} - h_{j+1})(k_{j+1} + y_0)}{(k_{j+1} - h_{j+1})k_{j+1}} \right\rceil \\
&\quad + (k_{j+1} - h_{j+1}) \\
&\geq \frac{(k_{j+1} - h_{j+1})y_0 + 1}{k_{j+1}} \\
&\quad + (k_{j+1} - h_{j+1})\frac{k_{j+1}(n-m) - (k_{j+1} - h_{j+1})(k_{j+1} + y_0)}{(k_{j+1} - h_{j+1})k_{j+1}} \\
&\quad + (k_{j+1} - h_{j+1}) \\
&= n - m + \frac{1}{k},
\end{aligned}$$

and (1.90) follows.

If relation (1.89) holds, then we see, according to Theorem 1.20 (ii), that

$$k_{j+1} - h_{j+1} = \Big(y_0 - \frac{h_j y_0 + 1}{k_j}\Big) + (k_j - h_j)\Big\lfloor \frac{k_j(n-m) - (k_j - h_j)y_0 + 1}{(k_j - h_j)k_j} \Big\rfloor$$

for a certain integer y_0, and (1.90) follows.

1.4.5 The det = −1 property

Let us consider the simplest Farey sequence $\mathcal{F}_1 := (\frac{0}{1} < \frac{1}{1})$, and regard its fractions as the consecutive columns of the 2×2 matrix $\left[\begin{smallmatrix} 0 & 1 \\ 1 & 1 \end{smallmatrix}\right]$; the determinant of the matrix is -1, and this observation is also true for any pair of neighboring fractions in Farey (sub)sequences.

Proposition 1.23. *If $\frac{h_j}{k_j} < \frac{h_{j+1}}{k_{j+1}}$ are two consecutive fractions in a Farey (sub)sequence $(\frac{h_0}{k_0} < \frac{h_1}{k_1} < \cdots < \frac{h_\ell}{k_\ell})$ of the form $\mathcal{F}_n$, $\mathcal{F}_n^m$, $\mathcal{G}_n^m$ or $\mathcal{F}(\mathbb{B}(n), m)$, then*

$$k_j h_{j+1} - h_j k_{j+1} = 1 \tag{1.91}$$

or, in other words, $\det\left[\begin{smallmatrix} h_j & h_{j+1} \\ k_j & k_{j+1} \end{smallmatrix}\right] = -1$. *As a consequence,*

$$\frac{h_{j+1}}{k_{j+1}} - \frac{h_j}{k_j} = \frac{1}{k_j k_{j+1}}, \quad \sum_{j=0}^{\ell-1} \frac{1}{k_j k_{j+1}} = 1.$$

Proof. Let us turn to assertions (i) (a) of Propositions 1.7, 1.10, 1.14 and Lemma 1.18, for the Farey (sub)sequences $\mathcal{F}_n$, $\mathcal{F}_n^m$, $\mathcal{G}_n^m$ and $\mathcal{F}(\mathbb{B}(n), m)$, respectively. In terms of these statements, we have $h_j = x_0 + t^* h_{j+1}$ and $k_j = y_0 + t^* k_{j+1}$ for certain integers x_0, y_0 and t^*, and we obtain

$$\begin{aligned} k_j h_{j+1} - k_{j+1} h_j &= (y_0 + t^* k_{j+1}) h_{j+1} - k_{j+1}(x_0 + t^* h_{j+1}) = h_{j+1} y_0 - k_{j+1} x_0 \\ &= \frac{k_{j+1} x_0 + 1}{h_{j+1}} h_{j+1} - k_{j+1} x_0 = 1. \end{aligned}$$

Similarly, (1.91) follows from assertions (ii) (a) of Propositions 1.7, 1.10, 1.14 and Lemma 1.18. Indeed, we have $h_{j+1} = x_0 + t^* h_j$ and $k_{j+1} = y_0 + t^* k_j$ for certain integers x_0, y_0 and t^*, and we obtain

$$k_j h_{j+1} - k_{j+1} h_j = k_j (x_0 + t^* h_j) - (y_0 + t^* k_j) h_j = k_j x_0 - h_j y_0 = k_j x_0 - \frac{k_j x_0 - 1}{h_j} h_j = 1.$$

The proof is complete. □

1.5 Triples of consecutive fractions. I

If a pair of neighboring fractions in a Farey (sub)sequence is known (the previous Section 1.4, as well as Section 2.2, are devoted to such pairs), then the fractions that precede and succeed the pair can be found with the help of straightforward procedures discussed in this section (see Table 1.6) and in Section 2.3.

1.5.1 The mediant property

Recall that all pairs of consecutive fractions in any Farey (sub)sequence $\mathcal{F}_n$, $\mathcal{F}_n^m$, $\mathcal{G}_n^m$ or $\mathcal{F}(\mathbb{B}(n), m)$ share an interesting property mentioned in Section 1.4.5.

Another interesting property, the *mediant property* of the triples of consecutive fractions in Farey (sub)sequences, is as follows.

Proposition 1.24. *If* $\frac{h_j}{k_j} < \frac{h_{j+1}}{k_{j+1}} < \frac{h_{j+2}}{k_{j+2}}$ *are three consecutive fractions in a Farey (sub)sequence of the form* $\mathcal{F}_n$, $\mathcal{F}_n^m$, $\mathcal{G}_n^m$ *or* $\mathcal{F}(\mathbb{B}(n), m)$, *then*

$$\frac{h_{j+1}}{k_{j+1}} = \frac{h_j + h_{j+2}}{\gcd(h_j + h_{j+2}, k_j + k_{j+2})} \Big/ \frac{k_j + k_{j+2}}{\gcd(h_j + h_{j+2}, k_j + k_{j+2})}. \tag{1.92}$$

Proof. Note that $k_j h_{j+1} - h_j k_{j+1} = 1$ and $k_{j+1} h_{j+2} - h_{j+1} k_{j+2} = 1$, by Proposition 1.23. Therefore, $h_j = \frac{k_j h_{j+1} - 1}{k_{j+1}}$ and $h_{j+2} = \frac{h_{j+1} k_{j+2} + 1}{k_{j+1}}$, and as a consequence, we have $h_j + h_{j+2} = \frac{h_{j+1}}{k_{j+1}}(k_j + k_{j+2})$. Hence, (1.92) follows. □

1.5.2 Triples of consecutive fractions in $\mathcal{F}_n$

Let us consider recurrent expressions that allow us to determine the fractions preceding or succeeding any pair of consecutive fractions in the Farey sequence $\mathcal{F}_n$.

Proposition 1.25. *If* $\frac{h_j}{k_j} < \frac{h_{j+1}}{k_{j+1}} < \frac{h_{j+2}}{k_{j+2}}$ *are three consecutive fractions in* $\mathcal{F}_n$, *then*

$$h_j = \left\lfloor \frac{n + k_{j+2}}{k_{j+1}} \right\rfloor h_{j+1} - h_{j+2}, \tag{1.93}$$

$$k_j = \left\lfloor \frac{n + k_{j+2}}{k_{j+1}} \right\rfloor k_{j+1} - k_{j+2}, \tag{1.94}$$

$$h_{j+2} = \left\lfloor \frac{n + k_j}{k_{j+1}} \right\rfloor h_{j+1} - h_j, \tag{1.95}$$

$$k_{j+2} = \left\lfloor \frac{n + k_j}{k_{j+1}} \right\rfloor k_{j+1} - k_j. \tag{1.96}$$

Proof. In view of Proposition 1.24, we have

$$\gcd(h_j + h_{j+2}, k_j + k_{j+2}) \cdot h_{j+1} = h_j + h_{j+2} \le (n-1) + h_{j+2},$$
$$\gcd(h_j + h_{j+2}, k_j + k_{j+2}) \cdot k_{j+1} = k_j + k_{j+2} \le n + k_{j+2}.$$

Table 1.6. Triples of consecutive fractions $\frac{h_j}{k_j} < \frac{h_{j+1}}{k_{j+1}} < \frac{h_{j+2}}{k_{j+2}}$ in Farey (sub)sequences; see also Table 2.9.

Sequence and Formulas		**See**
$\mathcal{F}_n$		
$h_j = \lfloor\frac{n+k_{j+2}}{k_{j+1}}\rfloor h_{j+1} - h_{j+2}$, $k_j = \lfloor\frac{n+k_{j+2}}{k_{j+1}}\rfloor k_{j+1} - k_{j+2}$ $h_{j+2} = \lfloor\frac{n+k_j}{k_{j+1}}\rfloor h_{j+1} - h_j$, $k_{j+2} = \lfloor\frac{n+k_j}{k_{j+1}}\rfloor k_{j+1} - k_j$		Proposition 1.25
$\mathcal{F}_n^m$, $0 < m < n$		
$h_j = \lfloor\frac{m+h_{j+2}}{h_{j+1}}\rfloor h_{j+1} - h_{j+2}$, $k_j = \lfloor\frac{m+h_{j+2}}{h_{j+1}}\rfloor k_{j+1} - k_{j+2}$	if $h_{j+1}n - k_{j+1}m \geq 1$	Proposition 1.26 (ii) (a)
$h_j = \lfloor\frac{n+k_{j+2}}{k_{j+1}}\rfloor h_{j+1} - h_{j+2}$, $k_j = \lfloor\frac{n+k_{j+2}}{k_{j+1}}\rfloor k_{j+1} - k_{j+2}$	if $h_{j+1}n - k_{j+1}m \leq 1$	
$h_{j+2} = \lfloor\frac{n+k_j}{k_{j+1}}\rfloor h_{j+1} - h_j$, $k_{j+2} = \lfloor\frac{n+k_j}{k_{j+1}}\rfloor k_{j+1} - k_j$	if $k_{j+1}m - h_{j+1}n \geq 1$	Proposition 1.26 (ii) (b)
$h_{j+2} = \lfloor\frac{m+h_j}{h_{j+1}}\rfloor h_{j+1} - h_j$, $k_{j+2} = \lfloor\frac{m+h_j}{h_{j+1}}\rfloor k_{j+1} - k_j$	if $k_{j+1}m - h_{j+1}n \leq 1$	
$\mathcal{G}_n^m$, $0 < m < n$		
$h_j = \lfloor\frac{n+k_{j+2}}{k_{j+1}}\rfloor h_{j+1} - h_{j+2}$, $k_j = \lfloor\frac{n+k_{j+2}}{k_{j+1}}\rfloor k_{j+1} - k_{j+2}$	if $h_{j+1}n - k_{j+1}m \geq 1$	Proposition 1.27 (ii) (a)
$h_j = \lfloor\frac{(n-m)+(k_{j+2}-h_{j+2})}{k_{j+1}-h_{j+1}}\rfloor h_{j+1} - h_{j+2}$, $k_j = \lfloor\frac{(n-m)+(k_{j+2}-h_{j+2})}{k_{j+1}-h_{j+1}}\rfloor k_{j+1} - k_{j+2}$	if $h_{j+1}n - k_{j+1}m \leq 1$	
$h_{j+2} = \lfloor\frac{(n-m)+(k_j-h_j)}{k_{j+1}-h_{j+1}}\rfloor h_{j+1} - h_j$, $k_{j+2} = \lfloor\frac{(n-m)+(k_j-h_j)}{k_{j+1}-h_{j+1}}\rfloor k_{j+1} - k_j$	if $k_{j+1}m - h_{j+1}n \geq 1$	Proposition 1.27 (ii) (b)
$h_{j+2} = \lfloor\frac{n+k_j}{k_{j+1}}\rfloor h_{j+1} - h_j$, $k_{j+2} = \lfloor\frac{n+k_j}{k_{j+1}}\rfloor k_{j+1} - k_j$	if $k_{j+1}m - h_{j+1}n \leq 1$	
$\mathcal{F}(\mathbb{B}(n), m)$		
$h_j = \lfloor\frac{m+h_{j+2}}{h_{j+1}}\rfloor h_{j+1} - h_{j+2}$, $k_j = \lfloor\frac{m+h_{j+2}}{h_{j+1}}\rfloor k_{j+1} - k_{j+2}$	if $h_{j+1}n - k_{j+1}m \geq 1$	Proposition 1.28 (ii) (a)
$h_j = \lfloor\frac{(n-m)+(k_{j+2}-h_{j+2})}{k_{j+1}-h_{j+1}}\rfloor h_{j+1} - h_{j+2}$, $k_j = \lfloor\frac{(n-m)+(k_{j+2}-h_{j+2})}{k_{j+1}-h_{j+1}}\rfloor k_{j+1} - k_{j+2}$	if $h_{j+1}n - k_{j+1}m \leq 1$	
$h_{j+2} = \lfloor\frac{(n-m)+(k_j-h_j)}{k_{j+1}-h_{j+1}}\rfloor h_{j+1} - h_j$, $k_{j+2} = \lfloor\frac{(n-m)+(k_j-h_j)}{k_{j+1}-h_{j+1}}\rfloor k_{j+1} - k_j$	if $k_{j+1}m - h_{j+1}n \geq 1$	Proposition 1.28 (ii) (b)
$h_{j+2} = \lfloor\frac{m+h_j}{h_{j+1}}\rfloor h_{j+1} - h_j$, $k_{j+2} = \lfloor\frac{m+h_j}{h_{j+1}}\rfloor k_{j+1} - k_j$	if $k_{j+1}m - h_{j+1}n \leq 1$	

Since $k_{j+1}h_{j+2} - h_{j+1}k_{j+2} = 1$, by Proposition 1.23, we have

$$\begin{aligned}\frac{(n-1)+h_{j+2}}{h_{j+1}} - \frac{n+k_{j+2}}{k_{j+1}} &= \frac{n(k_{j+1}-h_{j+1})-(k_{j+1}-1)}{h_{j+1}k_{j+1}} > \frac{n(k_{j+1}-h_{j+1})-(k_{j+1}+1)}{h_{j+1}k_{j+1}}\\ &\geq \frac{(k_{j+1}+1)(k_{j+1}-h_{j+1})-(k_{j+1}+1)}{h_{j+1}k_{j+1}}\\ &= \frac{(k_{j+1}+1)(k_{j+1}-h_{j+1}-1)}{h_{j+1}k_{j+1}} \geq 0,\end{aligned}$$

that is, $\min\{\frac{(n-1)+h_{j+2}}{h_{j+1}}, \frac{n+k_{j+2}}{k_{j+1}}\} = \frac{n+k_{j+2}}{k_{j+1}}$.

Further, we have

$$\left\lfloor \frac{n+k_{j+2}}{k_{j+1}} \right\rfloor = \left\lfloor \frac{k_j+k_{j+2}+(n-k_j)}{k_{j+1}} \right\rfloor = \frac{k_j+k_{j+2}}{k_{j+1}} + \left\lfloor \frac{n-k_j}{k_{j+1}} \right\rfloor.$$

But the summand $\lfloor \frac{n-k_j}{k_{j+1}} \rfloor$ vanishes, in view of Remark 1.8, since $k_j + k_{j+1} > n$. Thus, we obtain

$$\frac{k_j+k_{j+2}}{k_{j+1}} =: \gcd(h_j+h_{j+2}, k_j+k_{j+2}) = \left\lfloor \frac{n+k_{j+2}}{k_{j+1}} \right\rfloor. \tag{1.97}$$

Hence, (1.93) and (1.94) follow.

Similarly, we have

$$\begin{aligned}\gcd(h_j+h_{j+2}, k_j+k_{j+2}) \cdot h_{j+1} &= h_j + h_{j+2} \leq h_j + (n-1),\\ \gcd(h_j+h_{j+2}, k_j+k_{j+2}) \cdot k_{j+1} &= k_j + k_{j+2} \leq k_j + n,\end{aligned}$$

and as a consequence, we have

$$\gcd(h_j+h_{j+2}, k_j+k_{j+2}) = \left\lfloor \frac{n+k_j}{k_{j+1}} \right\rfloor. \tag{1.98}$$

Hence, (1.95) and (1.96) follow. □

1.5.3 Triples of consecutive fractions in $\mathfrak{F}_n^m$

In this subsection we find recurrent expressions that describe triples of consecutive fractions in the Farey subsequence $\mathfrak{F}_n^m$.

Proposition 1.26. *Let $\frac{h_j}{k_j} < \frac{h_{j+1}}{k_{j+1}} < \frac{h_{j+2}}{k_{j+2}}$ be three consecutive fractions in $\mathfrak{F}_n^m$, $0 < m < n$.*

(i) *We have*

$$h_j = \left\lfloor \min\left\{\frac{m+h_{j+2}}{h_{j+1}}, \frac{n+k_{j+2}}{k_{j+1}}\right\} \right\rfloor h_{j+1} - h_{j+2}, \tag{1.99}$$

$$k_j = \left\lfloor \min\left\{\frac{m+h_{j+2}}{h_{j+1}}, \frac{n+k_{j+2}}{k_{j+1}}\right\} \right\rfloor k_{j+1} - k_{j+2}, \tag{1.100}$$

$$h_{j+2} = \left\lfloor \min\left\{\frac{m+h_j}{h_{j+1}}, \frac{n+k_j}{k_{j+1}}\right\} \right\rfloor h_{j+1} - h_j, \tag{1.101}$$

$$k_{j+2} = \left\lfloor \min\left\{\frac{m+h_j}{h_{j+1}}, \frac{n+k_j}{k_{j+1}}\right\} \right\rfloor k_{j+1} - k_j. \tag{1.102}$$

(ii) *More precisely:*
(a) *If* $h_{j+1}n - k_{j+1}m \geq 1$, *then*

$$h_j = \left\lfloor \frac{m + h_{j+2}}{h_{j+1}} \right\rfloor h_{j+1} - h_{j+2}, \quad k_j = \left\lfloor \frac{m + h_{j+2}}{h_{j+1}} \right\rfloor k_{j+1} - k_{j+2}.$$

If $h_{j+1}n - k_{j+1}m \leq 1$, *then*

$$h_j = \left\lfloor \frac{n + k_{j+2}}{k_{j+1}} \right\rfloor h_{j+1} - h_{j+2}, \quad k_j = \left\lfloor \frac{n + k_{j+2}}{k_{j+1}} \right\rfloor k_{j+1} - k_{j+2}.$$

(b) *If* $k_{j+1}m - h_{j+1}n \geq 1$, *then*

$$h_{j+2} = \left\lfloor \frac{n + k_j}{k_{j+1}} \right\rfloor h_{j+1} - h_j, \quad k_{j+2} = \left\lfloor \frac{n + k_j}{k_{j+1}} \right\rfloor k_{j+1} - k_j.$$

If $k_{j+1}m - h_{j+1}n \leq 1$, *then*

$$h_{j+2} = \left\lfloor \frac{m + h_j}{h_{j+1}} \right\rfloor h_{j+1} - h_j, \quad k_{j+2} = \left\lfloor \frac{m + h_j}{h_{j+1}} \right\rfloor k_{j+1} - k_j.$$

Proof. (i) In view of Proposition 1.24, we have

$$\gcd(h_j + h_{j+2}, k_j + k_{j+2}) \cdot h_{j+1} = h_j + h_{j+2} \leq m + h_{j+2},$$
$$\gcd(h_j + h_{j+2}, k_j + k_{j+2}) \cdot k_{j+1} = k_j + k_{j+2} \leq n + k_{j+2}.$$

On the one hand, if $\min\{\frac{m+h_{j+2}}{h_{j+1}}, \frac{n+k_{j+2}}{k_{j+1}}\} = \frac{m+h_{j+2}}{h_{j+1}}$, that is, $h_{j+1}n - k_{j+1}m \geq 1$, then we have

$$\left\lfloor \frac{m + h_{j+2}}{h_{j+1}} \right\rfloor = \left\lfloor \frac{h_j + h_{j+2} + (m - h_j)}{h_{j+1}} \right\rfloor = \frac{h_j + h_{j+2}}{h_{j+1}} + \left\lfloor \frac{m - h_j}{h_{j+1}} \right\rfloor.$$

But the summand $\lfloor \frac{m-h_j}{h_{j+1}} \rfloor$ vanishes, since $h_j + h_{j+1} > m$, in view of Remark 1.12 (i) and, as a consequence, we have

$$\frac{h_j + h_{j+2}}{h_{j+1}} = \left\lfloor \frac{m + h_{j+2}}{h_{j+1}} \right\rfloor.$$

On the other hand, if $\min\{\frac{m+h_{j+2}}{h_{j+1}}, \frac{n+k_{j+2}}{k_{j+1}}\} = \frac{n+k_{j+2}}{k_{j+1}}$, that is, $h_{j+1}n - k_{j+1}m \leq 1$, then we have

$$\left\lfloor \frac{n + k_{j+2}}{k_{j+1}} \right\rfloor = \left\lfloor \frac{k_j + k_{j+2} + (n - k_j)}{k_{j+1}} \right\rfloor = \frac{k_j + k_{j+2}}{k_{j+1}} + \left\lfloor \frac{n - k_j}{k_{j+1}} \right\rfloor.$$

But $\lfloor \frac{n-k_j}{k_{j+1}} \rfloor = 0$, since $k_j + k_{j+1} > n$, in view of Remark 1.12 (ii) and, as a consequence, we have

$$\frac{k_j + k_{j+2}}{k_{j+1}} = \left\lfloor \frac{n + k_{j+2}}{k_{j+1}} \right\rfloor.$$

Thus, we have

$$\begin{aligned}\frac{h_j + h_{j+2}}{h_{j+1}} = \frac{k_j + k_{j+2}}{k_{j+1}} &=: \gcd(h_j + h_{j+2}, k_j + k_{j+2}) \\ &= \left\lfloor \min\left\{\frac{m + h_{j+2}}{h_{j+1}}, \frac{n + k_{j+2}}{k_{j+1}}\right\}\right\rfloor,\end{aligned} \tag{1.103}$$

hence (1.99) and (1.100) follow.

Similarly, we have

$$\gcd(h_j + h_{j+2}, k_j + k_{j+2}) \cdot h_{j+1} = h_j + h_{j+2} \le h_j + m,$$
$$\gcd(h_j + h_{j+2}, k_j + k_{j+2}) \cdot k_{j+1} = k_j + k_{j+2} \le k_j + n,$$

that is,

$$\gcd(h_j + h_{j+2}, k_j + k_{j+2}) = \left\lfloor \min\left\{\frac{h_j + m}{h_{j+1}}, \frac{k_j + n}{k_{j+1}}\right\}\right\rfloor, \tag{1.104}$$

hence (1.101) and (1.102) follow.

(ii) Since $k_{j+1}h_{j+2} - h_{j+1}k_{j+2} = 1$ and $k_jh_{j+1} - h_jk_{j+1} = 1$, by Proposition 1.23, we have

$$\frac{n + k_{j+2}}{k_{j+1}} - \frac{m + h_{j+2}}{h_{j+1}} = \frac{h_{j+1}n - k_{j+1}m - 1}{h_{j+1}k_{j+1}} \quad\text{and}\quad \frac{h_j + m}{h_{j+1}} - \frac{k_j + n}{k_{j+1}} = \frac{k_{j+1}m - h_{j+1}n - 1}{h_{j+1}k_{j+1}}.$$

Therefore, we have $\min\{\frac{m+h_{j+2}}{h_{j+1}}, \frac{n+k_{j+2}}{k_{j+1}}\} = \frac{m+h_{j+2}}{h_{j+1}}$ in (1.99) and (1.100) if and only if $h_{j+1}n - k_{j+1}m \ge 1$, and so on. □

1.5.4 Triples of consecutive fractions in $\mathcal{G}_n^m$

We continue to establish recurrent expressions that describe triples of consecutive fractions in Farey subsequences, and this subsection is devoted to such triples in the sequence $\mathcal{G}_n^m$.

Proposition 1.27. *Let $\frac{h_j}{k_j} < \frac{h_{j+1}}{k_{j+1}} < \frac{h_{j+2}}{k_{j+2}}$ be three consecutive fractions in $\mathcal{G}_n^m$, $0 < m < n$.*
(i) *We have*

$$h_j = \left\lfloor \min\left\{\frac{n + k_{j+2}}{k_{j+1}}, \frac{(n-m) + (k_{j+2} - h_{j+2})}{k_{j+1} - h_{j+1}}\right\}\right\rfloor h_{j+1} - h_{j+2}, \tag{1.105}$$

$$k_j = \left\lfloor \min\left\{\frac{n + k_{j+2}}{k_{j+1}}, \frac{(n-m) + (k_{j+2} - h_{j+2})}{k_{j+1} - h_{j+1}}\right\}\right\rfloor k_{j+1} - k_{j+2}, \tag{1.106}$$

$$h_{j+2} = \left\lfloor \min\left\{\frac{n + k_j}{k_{j+1}}, \frac{(n-m) + (k_j - h_j)}{k_{j+1} - h_{j+1}}\right\}\right\rfloor h_{j+1} - h_j, \tag{1.107}$$

$$k_{j+2} = \left\lfloor \min\left\{\frac{n + k_j}{k_{j+1}}, \frac{(n-m) + (k_j - h_j)}{k_{j+1} - h_{j+1}}\right\}\right\rfloor k_{j+1} - k_j. \tag{1.108}$$

(ii) *More precisely:*

(a) *If $h_{j+1}n - k_{j+1}m \ge 1$, then*

$$h_j = \left\lfloor \frac{n + k_{j+2}}{k_{j+1}} \right\rfloor h_{j+1} - h_{j+2}, \quad k_j = \left\lfloor \frac{n + k_{j+2}}{k_{j+1}} \right\rfloor k_{j+1} - k_{j+2}.$$

If $h_{j+1}n - k_{j+1}m \le 1$, then

$$h_j = \left\lfloor \frac{(n-m) + (k_{j+2} - h_{j+2})}{k_{j+1} - h_{j+1}} \right\rfloor h_{j+1} - h_{j+2},$$

$$k_j = \left\lfloor \frac{(n-m) + (k_{j+2} - h_{j+2})}{k_{j+1} - h_{j+1}} \right\rfloor k_{j+1} - k_{j+2}.$$

(b) *If* $k_{j+1}m - h_{j+1}n \geq 1$, *then*

$$h_{j+2} = \left\lfloor \frac{(n-m) + (k_j - h_j)}{k_{j+1} - h_{j+1}} \right\rfloor h_{j+1} - h_j\,,$$
$$k_{j+2} = \left\lfloor \frac{(n-m) + (k_j - h_j)}{k_{j+1} - h_{j+1}} \right\rfloor k_{j+1} - k_j\,.$$

If $k_{j+1}m - h_{j+1}n \leq 1$, *then*

$$h_{j+2} = \left\lfloor \frac{n + k_j}{k_{j+1}} \right\rfloor h_{j+1} - h_j, \quad k_{j+2} = \left\lfloor \frac{n + k_j}{k_{j+1}} \right\rfloor k_{j+1} - k_j.$$

Proof. (i) In view of Proposition 1.24, we have

$$\gcd(h_j + h_{j+2}, k_j + k_{j+2}) \cdot k_{j+1} = k_j + k_{j+2} \leq n + k_{j+2},$$
$$\gcd(h_j + h_{j+2}, k_j + k_{j+2}) \cdot (k_{j+1} - h_{j+1}) = (k_j - h_j) + (k_{j+2} - h_{j+2})$$
$$\leq (n-m) + (k_{j+2} - h_{j+2}).$$

On the one hand, if $\min\{\frac{n+k_{j+2}}{k_{j+1}}, \frac{(n-m)+(k_{j+2}-h_{j+2})}{k_{j+1}-h_{j+1}}\} = \frac{n+k_{j+2}}{k_{j+1}}$, that is, $h_{j+1}n - k_{j+1}m \geq 1$, then we have

$$\left\lfloor \frac{n + k_{j+2}}{k_{j+1}} \right\rfloor = \left\lfloor \frac{k_j + k_{j+2} + (n - k_j)}{k_{j+1}} \right\rfloor = \frac{k_j + k_{j+2}}{k_{j+1}} + \left\lfloor \frac{n - k_j}{k_{j+1}} \right\rfloor.$$

But the summand $\lfloor \frac{n-k_j}{k_{j+1}} \rfloor$ vanishes, since $k_j + k_{j+1} > n$, in view of Remark 1.16 (i), and as a consequence, we have

$$\frac{k_j + k_{j+2}}{k_{j+1}} = \left\lfloor \frac{n + k_{j+2}}{k_{j+1}} \right\rfloor.$$

On the other hand, if

$$\min\left\{\frac{n + k_{j+2}}{k_{j+1}}, \frac{(n-m) + (k_{j+2} - h_{j+2})}{k_{j+1} - h_{j+1}}\right\} = \frac{(n-m) + (k_{j+2} - h_{j+2})}{k_{j+1} - h_{j+1}},$$

that is, $h_{j+1}n - k_{j+1}m \leq 1$, then

$$\left\lfloor \frac{(n-m) + (k_{j+2} - h_{j+2})}{k_{j+1} - h_{j+1}} \right\rfloor = \left\lfloor \frac{(k_j - h_j) + (k_{j+2} - h_{j+2}) + ((n-m) - (k_j - h_j))}{k_{j+1} - h_{j+1}} \right\rfloor$$
$$= \frac{(k_j - h_j) + (k_{j+2} - h_{j+2})}{k_{j+1} - h_{j+1}} + \left\lfloor \frac{(n-m) - (k_j - h_j)}{k_{j+1} - h_{j+1}} \right\rfloor.$$

But $\lfloor \frac{(n-m)-(k_j-h_j)}{k_{j+1}-h_{j+1}} \rfloor = 0$, since $(k_j - h_j) + (k_{j+1} - h_{j+1}) > n - m$, in view of Remark 1.16 (ii). As a consequence, we have

$$\frac{(k_j - h_j) + (k_{j+2} - h_{j+2})}{k_{j+1} - h_{j+1}} = \left\lfloor \frac{(n-m) + (k_{j+2} - h_{j+2})}{k_{j+1} - h_{j+1}} \right\rfloor.$$

Thus, we have

$$\begin{aligned}\frac{k_j + k_{j+2}}{k_{j+1}} &= \frac{(k_j - h_j) + (k_{j+2} - h_{j+2})}{k_{j+1} - h_{j+1}} \\ &=: \gcd(h_j + h_{j+2}, k_j + k_{j+2}) \\ &= \left\lfloor \min\left\{\frac{n + k_{j+2}}{k_{j+1}}, \frac{(n-m) + (k_{j+2} - h_{j+2})}{k_{j+1} - h_{j+1}}\right\}\right\rfloor. \end{aligned} \tag{1.109}$$

Hence, (1.105) and (1.106) follow.

Similarly, we have

$$\begin{aligned}&\gcd(h_j + h_{j+2}, k_j + k_{j+2}) \cdot k_{j+1} = k_j + k_{j+2} \le k_j + n, \\ &\gcd(h_j + h_{j+2}, k_j + k_{j+2}) \cdot (k_{j+1} - h_{j+1}) = (k_j - h_j) + (k_{j+2} - h_{j+2}) \le (k_j - h_j) + (n - m),\end{aligned}$$

that is,

$$\gcd(h_j + h_{j+2}, k_j + k_{j+2}) = \left\lfloor \min\left\{\frac{k_j + n}{k_{j+1}}, \frac{(k_j - h_j) + (n-m)}{k_{j+1} - h_{j+1}}\right\}\right\rfloor. \tag{1.110}$$

Hence, (1.107) and (1.108) follow.

(ii) Since $k_{j+1}h_{j+2} - h_{j+1}k_{j+2} = 1$ and $k_jh_{j+1} - h_jk_{j+1} = 1$, by Proposition 1.23, we have

$$\frac{(n-m) + (k_{j+2} - h_{j+2})}{k_{j+1} - h_{j+1}} - \frac{n + k_{j+2}}{k_{j+1}} = \frac{h_{j+1}n - k_{j+1}m - 1}{(k_{j+1} - h_{j+1})k_{j+1}}$$

and

$$\frac{(n-m) + (k_j - h_j)}{k_{j+1} - h_{j+1}} - \frac{n + k_j}{k_{j+1}} = \frac{h_{j+1}n - k_{j+1}m - 1}{(k_{j+1} - h_{j+1})k_{j+1}}.$$

Thus, we have

$$\min\left\{\frac{n + k_{j+2}}{k_{j+1}}, \frac{(n-m) + (k_{j+2} - h_{j+2})}{k_{j+1} - h_{j+1}}\right\} = \frac{n + k_{j+2}}{k_{j+1}}$$

in (1.105) and (1.106) if and only if $h_{j+1}n - k_{j+1}m \ge 1$, and so on. □

1.5.5 Triples of consecutive fractions in $\mathcal{F}(\mathbb{B}(n), m)$

In this subsection we consider the Farey subsequence $\mathcal{F}(\mathbb{B}(n), m)$ and establish, as in the three previous subsections, recurrent expressions for the triples of consecutive fractions. The following statement will be revisited, via the Farey duality, several times in Section 2.3.

Proposition 1.28. *Let* $\frac{h_j}{k_j} < \frac{h_{j+1}}{k_{j+1}} < \frac{h_{j+2}}{k_{j+2}}$ *be three consecutive fractions in* $\mathcal{F}(\mathbb{B}(n), m)$.

(i) *We have*

$$h_j = \left\lfloor \min\left\{\frac{m + h_{j+2}}{h_{j+1}}, \frac{(n-m) + (k_{j+2} - h_{j+2})}{k_{j+1} - h_{j+1}}\right\}\right\rfloor h_{j+1} - h_{j+2}, \tag{1.111}$$

$$k_j = \left\lfloor \min\left\{\frac{m + h_{j+2}}{h_{j+1}}, \frac{(n-m) + (k_{j+2} - h_{j+2})}{k_{j+1} - h_{j+1}}\right\}\right\rfloor k_{j+1} - k_{j+2}, \tag{1.112}$$

$$h_{j+2} = \left\lfloor \min\left\{\frac{m+h_j}{h_{j+1}}, \frac{(n-m)+(k_j-h_j)}{k_{j+1}-h_{j+1}}\right\}\right\rfloor h_{j+1} - h_j, \tag{1.113}$$

$$k_{j+2} = \left\lfloor \min\left\{\frac{m+h_j}{h_{j+1}}, \frac{(n-m)+(k_j-h_j)}{k_{j+1}-h_{j+1}}\right\}\right\rfloor k_{j+1} - k_j. \tag{1.114}$$

(ii) *More precisely:*

(a) *If* $h_{j+1}n - k_{j+1}m \geq 1$, *then*

$$h_j = \left\lfloor \frac{m+h_{j+2}}{h_{j+1}} \right\rfloor h_{j+1} - h_{j+2}, \quad k_j = \left\lfloor \frac{m+h_{j+2}}{h_{j+1}} \right\rfloor k_{j+1} - k_{j+2}.$$

If $h_{j+1}n - k_{j+1}m \leq 1$, *then*

$$h_j = \left\lfloor \frac{(n-m)+(k_{j+2}-h_{j+2})}{k_{j+1}-h_{j+1}} \right\rfloor h_{j+1} - h_{j+2},$$

$$k_j = \left\lfloor \frac{(n-m)+(k_{j+2}-h_{j+2})}{k_{j+1}-h_{j+1}} \right\rfloor k_{j+1} - k_{j+2}.$$

(b) *If* $k_{j+1}m - h_{j+1}n \geq 1$, *then*

$$h_{j+2} = \left\lfloor \frac{(n-m)+(k_j-h_j)}{k_{j+1}-h_{j+1}} \right\rfloor h_{j+1} - h_j,$$

$$k_{j+2} = \left\lfloor \frac{(n-m)+(k_j-h_j)}{k_{j+1}-h_{j+1}} \right\rfloor k_{j+1} - k_j.$$

If $k_{j+1}m - h_{j+1}n \leq 1$, *then*

$$h_{j+2} = \left\lfloor \frac{m+h_j}{h_{j+1}} \right\rfloor h_{j+1} - h_j, \quad k_{j+2} = \left\lfloor \frac{m+h_j}{h_{j+1}} \right\rfloor k_{j+1} - k_j.$$

Proof. (i) In view of Proposition 1.24, we have

$$\begin{aligned}
\gcd(h_j+h_{j+2}, k_j+k_{j+2}) \cdot h_{j+1} &= h_j + h_{j+2} \leq m + h_{j+2}, \\
\gcd(h_j+h_{j+2}, k_j+k_{j+2}) \cdot k_{j+1} &= k_j + k_{j+2} \leq n + k_{j+2}, \\
\gcd(h_j+h_{j+2}, k_j+k_{j+2}) \cdot (k_{j+1}-h_{j+1}) &= (k_j - h_j) + (k_{j+2} - h_{j+2}) \\
&\leq (n-m) + (k_{j+2} - h_{j+2}).
\end{aligned}$$

Recall that $k_{j+1}h_{j+2} - h_{j+1}k_{j+2} = 1$ and $k_jh_{j+1} - h_jk_{j+1} = 1$, by Proposition 1.23. We have

$$\frac{n+k_{j+2}}{k_{j+1}} - \frac{m+h_{j+2}}{h_{j+1}} = \frac{h_{j+1}n - k_{j+1}m - 1}{h_{j+1}k_{j+1}}$$

and

$$\frac{(n-m)+(k_{j+2}-h_{j+2})}{k_{j+1}-h_{j+1}} - \frac{m+h_{j+2}}{h_{j+1}} = \frac{h_{j+1}n - k_{j+1}m - 1}{(k_{j+1}-h_{j+1})h_{j+1}},$$

that is,

$$\min\Big\{\frac{m+h_{j+2}}{h_{j+1}},\frac{n+k_{j+2}}{k_{j+1}},\frac{(n-m)+(k_{j+2}-h_{j+2})}{k_{j+1}-h_{j+1}}\Big\}$$

$$=\begin{cases}\frac{m+h_{j+2}}{h_{j+1}} & \text{if } h_{j+1}n-k_{j+1}m\geq 1,\\ \frac{m+h_{j+2}}{h_{j+1}}=\frac{n+k_{j+2}}{k_{j+1}}=\frac{(n-m)+(k_{j+2}-h_{j+2})}{k_{j+1}-h_{j+1}} & \text{if } h_{j+1}n-k_{j+1}m=1,\\ \frac{(n-m)+(k_{j+2}-h_{j+2})}{k_{j+1}-h_{j+1}} & \text{if } h_{j+1}n-k_{j+1}m\leq 1\end{cases}$$

$$=\min\Big\{\frac{m+h_{j+2}}{h_{j+1}},\frac{(n-m)+(k_{j+2}-h_{j+2})}{k_{j+1}-h_{j+1}}\Big\}.$$

On the one hand, if $\min\{\frac{m+h_{j+2}}{h_{j+1}},\frac{n+k_{j+2}}{k_{j+1}}\}=\frac{m+h_{j+2}}{h_{j+1}}$, that is, $h_{j+1}n-k_{j+1}m\geq 1$, then we have

$$\Big\lfloor\frac{m+h_{j+2}}{h_{j+1}}\Big\rfloor=\Big\lfloor\frac{h_j+h_{j+2}+(m-h_j)}{h_{j+1}}\Big\rfloor=\frac{h_j+h_{j+2}}{h_{j+1}}+\Big\lfloor\frac{m-h_j}{h_{j+1}}\Big\rfloor.$$

But the summand $\lfloor\frac{m-h_j}{h_{j+1}}\rfloor$ vanishes, since $h_j+h_{j+1}>m$, in view of Remark 1.22 (i). As a consequence, we have

$$\frac{h_j+h_{j+2}}{h_{j+1}}=\Big\lfloor\frac{m+h_{j+2}}{h_{j+1}}\Big\rfloor.$$

On the other hand, if

$$\min\Big\{\frac{n+k_{j+2}}{k_{j+1}},\frac{(n-m)+(k_{j+2}-h_{j+2})}{k_{j+1}-h_{j+1}}\Big\}=\frac{(n-m)+(k_{j+2}-h_{j+2})}{k_{j+1}-h_{j+1}},$$

that is, $h_{j+1}n-k_{j+1}m\leq 1$, then

$$\Big\lfloor\frac{(n-m)+(k_{j+2}-h_{j+2})}{k_{j+1}-h_{j+1}}\Big\rfloor=\Big\lfloor\frac{(k_j-h_j)+(k_{j+2}-h_{j+2})+((n-m)-(k_j-h_j))}{k_{j+1}-h_{j+1}}\Big\rfloor$$

$$=\frac{(k_j-h_j)+(k_{j+2}-h_{j+2})}{k_{j+1}-h_{j+1}}+\Big\lfloor\frac{(n-m)-(k_j-h_j)}{k_{j+1}-h_{j+1}}\Big\rfloor.$$

But $\lfloor\frac{(n-m)-(k_j-h_j)}{k_{j+1}-h_{j+1}}\rfloor=0$, since $(k_j-h_j)+(k_{j+1}-h_{j+1})>n-m$, in view of Remark 1.22 (ii). As a consequence, we have

$$\frac{(k_j-h_j)+(k_{j+2}-h_{j+2})}{k_{j+1}-h_{j+1}}=\Big\lfloor\frac{(n-m)+(k_{j+2}-h_{j+2})}{k_{j+1}-h_{j+1}}\Big\rfloor.$$

Thus, we have

$$\begin{aligned}\frac{h_j+h_{j+2}}{h_{j+1}}=\frac{k_j+k_{j+2}}{k_{j+1}}&=\frac{(k_j-h_j)+(k_{j+2}-h_{j+2})}{k_{j+1}-h_{j+1}}\\&=:\gcd(h_j+h_{j+2},k_j+k_{j+2})\\&=\Big\lfloor\min\Big\{\frac{m+h_{j+2}}{h_{j+1}},\frac{(n-m)+(k_{j+2}-h_{j+2})}{k_{j+1}-h_{j+1}}\Big\}\Big\rfloor.\end{aligned}\tag{1.115}$$

Hence, (1.111) and (1.112) follow.

Similarly, we have

$$\begin{aligned}
&\gcd(h_j + h_{j+2}, k_j + k_{j+2}) \cdot h_{j+1} = h_j + h_{j+2} \le h_j + m,\\
&\gcd(h_j + h_{j+2}, k_j + k_{j+2}) \cdot k_{j+1} = k_j + k_{j+2} \le k_j + n,\\
&\gcd(h_j + h_{j+2}, k_j + k_{j+2}) \cdot (k_{j+1} - h_{j+1}) = (k_j - h_j) + (k_{j+2} - h_{j+2}) \le (k_j - h_j) + (n - m),
\end{aligned}$$

but note that

$$\begin{aligned}
&\min\left\{\frac{h_j + m}{h_{j+1}}, \frac{k_j + n}{k_{j+1}}, \frac{(k_j - h_j) + (n - m)}{k_{j+1} - h_{j+1}}\right\}\\
&= \begin{cases}
\frac{h_j+m}{h_{j+1}} & \text{if } k_{j+1}m - h_{j+1}n \le 1,\\
\frac{h_j+m}{h_{j+1}} = \frac{k_j+n}{k_{j+1}} = \frac{(k_j-h_j)+(n-m)}{k_{j+1}-h_{j+1}} & \text{if } k_{j+1}m - h_{j+1}n = 1,\\
\frac{(k_j-h_j)+(n-m)}{k_{j+1}-h_{j+1}} & \text{if } k_{j+1}m - h_{j+1}n \ge 1
\end{cases}\\
&= \min\left\{\frac{h_j + m}{h_{j+1}}, \frac{(k_j - h_j) + (n - m)}{k_{j+1} - h_{j+1}}\right\}.
\end{aligned}$$

On the one hand, if

$$\min\left\{\frac{h_j + m}{h_{j+1}}, \frac{(k_j - h_j) + (n - m)}{k_{j+1} - h_{j+1}}\right\} = \frac{h_j + m}{h_{j+1}},$$

that is, $k_{j+1}m - h_{j+1}n \le 1$, then we have

$$\frac{h_j + h_{j+2}}{h_{j+1}} = \left\lfloor \frac{h_j + m}{h_{j+1}} \right\rfloor,$$

since $\lfloor \frac{h_j+m}{h_{j+1}} \rfloor = \lfloor \frac{h_j+h_{j+2}+(m-h_{j+2})}{h_{j+1}} \rfloor$ and $h_j + h_{j+1} > m$, in view of Remark 1.22 (i).

On the other hand, if

$$\min\left\{\frac{h_j + m}{h_{j+1}}, \frac{(k_j - h_j) + (n - m)}{k_{j+1} - h_{j+1}}\right\} = \frac{(k_j - h_j) + (n - m)}{k_{j+1} - h_{j+1}},$$

that is, $k_{j+1}m - h_{j+1}n \ge 1$, then

$$\frac{(k_j - h_j) + (k_{j+2} - h_{j+2})}{k_{j+1} - h_{j+1}} = \left\lfloor \frac{(k_j - h_j) + (n - m)}{k_{j+1} - h_{j+1}} \right\rfloor,$$

since

$$\left\lfloor \frac{(k_j - h_j) + (n - m)}{k_{j+1} - h_{j+1}} \right\rfloor = \left\lfloor \frac{(k_j - h_j) + (k_{j+2} - h_{j+2}) + ((n - m) - (k_{j+2} - h_{j+2}))}{k_{j+1} - h_{j+1}} \right\rfloor$$

and $(k_{j+1} - h_{j+1}) + (k_{j+2} - h_{j+2}) > n - m$, in view of Remark 1.22 (ii).

Thus, we have

$$\gcd(h_j + h_{j+2}, k_j + k_{j+2}) = \left\lfloor \min\left\{\frac{h_j + m}{h_{j+1}}, \frac{(k_j - h_j) + (n - m)}{k_{j+1} - h_{j+1}}\right\} \right\rfloor. \tag{1.116}$$

Hence, (1.113) and (1.114) follow. □

Table 1.7. The number of fractions in Farey (sub)sequences.

Quantity	Formula, where $0 < m < n$	See
$\lvert\mathcal{F}_n\rvert$	$\frac{3}{2} + \frac{1}{2}\sum_{d=1}^{n}\mu(d)\lfloor\frac{n}{d}\rfloor^2$	Corollary 1.30
$\lvert\mathcal{F}_n^m\rvert = \lvert\mathcal{G}_n^{n-m}\rvert$	$\frac{3}{2} + \sum_{d\ge1}\mu(d)\lfloor\frac{m}{d}\rfloor(\lfloor\frac{n}{d}\rfloor - \frac{1}{2}\lfloor\frac{m}{d}\rfloor)$	Proposition 1.29
$\lvert\mathcal{G}_n^m\rvert = \lvert\mathcal{F}_n^{n-m}\rvert$	$\frac{3}{2} + \sum_{d\ge1}\mu(d)\lfloor\frac{n-m}{d}\rfloor(\lfloor\frac{n}{d}\rfloor - \frac{1}{2}\lfloor\frac{n-m}{d}\rfloor)$	Corollary 1.31
$\lvert\mathcal{F}_{2m}^m\rvert = \lvert\mathcal{G}_{2m}^m\rvert$	$\frac{3}{2} + \sum_{d=1}^{m}\mu(d)\lfloor\frac{m}{d}\rfloor(\lfloor\frac{2m}{d}\rfloor - \frac{1}{2}\lfloor\frac{m}{d}\rfloor)$	
$\lvert\mathcal{F}(\mathbb{B}(n), m)\rvert$	$2 + \sum_{d\ge1}\mu(d)\lfloor\frac{m}{d}\rfloor\lfloor\frac{n-m}{d}\rfloor$	Proposition 2.47
$\lvert\mathcal{F}(\mathbb{B}(2m), m)\rvert$	$2 + \sum_{d=1}^{m}\mu(d)\lfloor\frac{m}{d}\rfloor^2$	
$\lvert\mathcal{F}_n^m\rvert - \lvert\mathcal{F}(\mathbb{B}(n), m)\rvert$	$-\frac{1}{2} + \sum_{d=1}^{m}\mu(d)\lfloor\frac{m}{d}\rfloor(\lfloor\frac{n}{d}\rfloor - \frac{1}{2}\lfloor\frac{m}{d}\rfloor - \lfloor\frac{n-m}{d}\rfloor)$	Section 2.4.2
$\lvert\mathcal{G}_n^m\rvert - \lvert\mathcal{F}(\mathbb{B}(n), m)\rvert$	$-\frac{1}{2} + \sum_{d=1}^{m}\mu(d)\lfloor\frac{n-m}{d}\rfloor(\lfloor\frac{n}{d}\rfloor - \frac{1}{2}\lfloor\frac{n-m}{d}\rfloor - \lfloor\frac{m}{d}\rfloor)$	
$\lvert\mathcal{F}_{2m}^m\rvert - \lvert\mathcal{F}(\mathbb{B}(2m), m)\rvert$ $= \lvert\mathcal{G}_{2m}^m\rvert - \lvert\mathcal{F}(\mathbb{B}(2m), m)\rvert$	$-\frac{1}{2} + \sum_{d=1}^{m}\mu(d)\lfloor\frac{m}{d}\rfloor(\lfloor\frac{2m}{d}\rfloor - \frac{3}{2}\lfloor\frac{m}{d}\rfloor)$	
$\lvert\mathcal{F}_n\rvert - \lvert\mathcal{F}(\mathbb{B}(n), m)\rvert$	$-4 + \sum_{d\ge1}\mu(d)\lfloor\frac{n}{d}\rfloor(\lfloor\frac{m}{d}\rfloor + \lfloor\frac{n-m}{d}\rfloor)$	Section 2.4.2
$\lvert\mathcal{F}_{2m}\rvert - \lvert\mathcal{F}(\mathbb{B}(2m), m)\rvert$	$-4 + 2\sum_{d=1}^{m}\mu(d)\lfloor\frac{2m}{d}\rfloor\lfloor\frac{m}{d}\rfloor$	

1.6 The number of fractions in Farey (sub)sequences. I

How many fractions are there in Farey (sub)sequences? In this section and in Section 2.4 we find the number of fractions. The formulas for various quantities are collected in Table 1.7.

The number-theoretic *Möbius function* $\mu\colon \mathbb{P} \to \{-1, 0, 1\}$ is defined as follows:

- $\mu(1) := 1$.
- If $\mathfrak{p}^2 | d$ for some prime $\mathfrak{p}$, then $\mu(d) := 0$.
- If $d = \mathfrak{p}_1\mathfrak{p}_2\cdots\mathfrak{p}_s$ is the product of s distinct primes $\mathfrak{p}_1, \mathfrak{p}_2, \ldots, \mathfrak{p}_s$, then $\mu(d) := (-1)^s$.

The following identity is very useful for enumerative problems on Farey (sub)sequences.

For all positive integers m, we have (cf. (2.98))

$$\sum_{d=1}^{m}\mu(d)\Big\lfloor\frac{m}{d}\Big\rfloor = 1. \tag{1.117}$$

Given an interval of positive integers $[s', s''] \subset \mathbb{P}$, define

$$\varphi(n; [s', s'']) := \big|\{s \in [s', s''] : \gcd(n, s) = 1\}\big|.$$

Note that $\varphi(n; [1, n])$ is *Euler's totient function* $\varphi(n)$, which is defined in the literature in two ways:

$$\varphi(n) := \big|\{s \in \mathbb{Z} : 1 \le s \le n,\ \gcd(n, s) = 1\}\big| \tag{1.118}$$

$$:= \big|\{s \in \mathbb{Z} : 0 \le s \le n-1,\ \gcd(n, s) = 1\}\big|. \tag{1.119}$$

If $[i' + 1, i''] \subseteq [1, n]$, then

$$\varphi(n; [i' + 1, i'']) = \sum_{\substack{d \in [1, i']:\\ d | n}} \mu(d)\Big(\Big\lfloor \frac{i''}{d} \Big\rfloor - \Big\lfloor \frac{i'}{d} \Big\rfloor\Big). \tag{1.120}$$

1.6.1 The number of fractions in $\mathcal{F}_n^m$, $\mathcal{F}_n$ and $\mathcal{G}_n^m$

Let us count the number of fractions in the Farey (sub)sequences $\mathcal{F}_n^m$, $\mathcal{F}_n$ and $\mathcal{G}_n^m$. Recall that in the case $n > 1$, we have $\mathcal{F}_n^n = \mathcal{F}_n^{n-1} = \mathcal{F}_n$.

Suppose that $0 < m \le n$, and note that

$$|\mathcal{F}_n^m| = 1 + \sum_{j=1}^{n} \varphi(j; [1, \min\{m, j\}]) = 1 + \sum_{j=1}^{m} \varphi(j; [1, j]) + \sum_{j=m+1}^{n} \varphi(j; [1, m]),$$

To resolve a hidden recurrence in the sum

$$1 + \sum_{j=1}^{n} \varphi(j; [1, \min\{m, j\}]),$$

let us define an integer $\Phi(w, x)$, where $w, x \in \mathbb{R}$ and $w \le x$, by

$$\Phi(w, x) = \sum_{1 \le \ell \le x} \varphi(\ell; [1, \lfloor w \rfloor]).$$

Thus, $|\mathcal{F}_n^m| = 1 + \Phi(m, n)$.

On the one hand, we have

$$\big|\{\tfrac{h}{k} : 0 \le h \le w,\ 1 \le k \le x\}\big| = (\lfloor x \rfloor - \tfrac{1}{2}\lfloor w \rfloor)(\lfloor w \rfloor + 1).$$

On the other hand,

$$\big|\{\tfrac{h}{k} : 0 \le h \le w,\ 1 \le k \le x\}\big| = \sum_{d \ge 1} \Phi\Big(\Big\lfloor \frac{w}{d} \Big\rfloor, \Big\lfloor \frac{x}{d} \Big\rfloor\Big).$$

By means of Möbius inversion, we obtain

$$\Phi(w,x)=\sum_{d\geq 1}\mu(d)\Big(\Big\lfloor\frac{x}{d}\Big\rfloor-\frac{1}{2}\Big\lfloor\frac{w}{d}\Big\rfloor\Big)\Big(\Big\lfloor\frac{w}{d}\Big\rfloor+1\Big).$$

In view of (1.117), we come to the following conclusions.

Proposition 1.29. *The number of fractions in the Farey subsequence $\mathcal{F}_n^m$, $0<m\leq n$, is*

$$|\mathcal{F}_n^m|=\frac{3}{2}+\sum_{d\geq 1}\mu(d)\Big\lfloor\frac{m}{d}\Big\rfloor\Big(\Big\lfloor\frac{n}{d}\Big\rfloor-\frac{1}{2}\Big\lfloor\frac{m}{d}\Big\rfloor\Big).$$

Corollary 1.30. *The number of fractions in the Farey sequence $\mathcal{F}_n$ is*

$$|\mathcal{F}_n|=\frac{3}{2}+\frac{1}{2}\sum_{d=1}^{n}\mu(d)\Big\lfloor\frac{n}{d}\Big\rfloor^2 \tag{1.121}$$

and, by Möbius inversion, we have

$$|\mathcal{F}_n|=\frac{1}{2}(n+3)n-\sum_{d=2}^{n}|\mathcal{F}_{\lfloor n/d\rfloor}|. \tag{1.122}$$

Proposition 1.29 and the order-reversing bijections $\mathcal{G}_n^m \leftrightarrow \mathcal{F}_n^{n-m}$ defined in (1.8) allow us to give the following analogue of Proposition 1.29.

Corollary 1.31. *The number of fractions in the Farey subsequence $\mathcal{G}_n^m$, $0\leq m<n$, is*

$$|\mathcal{G}_n^m|=\frac{3}{2}+\sum_{d\geq 1}\mu(d)\Big\lfloor\frac{n-m}{d}\Big\rfloor\Big(\Big\lfloor\frac{n}{d}\Big\rfloor-\frac{1}{2}\Big\lfloor\frac{n-m}{d}\Big\rfloor\Big).$$

1.7 The position of a fraction in a Farey (sub)sequence. I

Given a fraction $f_t\in\mathcal{C}$ in a Farey (sub)sequence

$$\mathcal{C}=(f_0:=\tfrac{0}{1}<\cdots<f_t<\cdots<f_{|\mathcal{C}|-1}:=\tfrac{1}{1}),$$

in this section we turn to the question of how to find the index t; recall that the fractions are always indexed in this book starting with zero. For additional information, see Section 2.5.

1.7.1 The indices of fractions in $\mathcal{F}_n^m$ and $\mathcal{F}_n$

Let us find the positions of fractions in the Farey (sub)sequences $\mathcal{F}_n^m$ and $\mathcal{F}_n$.

If $f_t\in\mathcal{F}_n^m-\{\frac{1}{1}\}$, where $0<m\leq n$, then

$$1+t=\big|\{\tfrac{h}{k}\in\mathcal{F}_n : h\leq\min\{m,\lfloor kf_t\rfloor\}\}\big|=1+\big|\{\tfrac{h}{k}\in\mathcal{F}_n : 1\leq h\leq\min\{m,\lfloor kf_t\rfloor\}\}\big|,$$

that is,

$$t = \sum_{j=1}^{n} \varphi(j; [1, \min\{m, \lfloor jf_t \rfloor\}]). \tag{1.123}$$

Let us define an integer $\Phi(w, x; r)$, where $r \in \mathbb{R}$, $0 \le r < 1$, and $w, x \in \mathbb{R}$, $0 \le w \le x$, by

$$\Phi(w, x; r) = \sum_{1 \le \ell \le x} \varphi(\ell; [1, \lfloor \min\{w, \ell r\} \rfloor]).$$

Thus,

$$1 + t = \left|\left(\tfrac{h}{k} \in \mathcal{F}_n - \{\tfrac{1}{1}\} : h \le \min\{m, kf_t\}\right)\right| = \Phi(m, n; f_t).$$

On the one hand, we have

$$\left|\left\{\tfrac{h}{k} : 0 \le h \le \min\{w, kr\},\ 1 \le k \le x\right\}\right| = \lfloor x \rfloor + \sum_{1 \le j \le x} \lfloor \min\{w, jr\} \rfloor.$$

On the other hand,

$$\left|\left\{\tfrac{h}{k} : 0 \le h \le \min\{w, kr\},\ 1 \le k \le x\right\}\right| = \sum_{d \ge 1} \Phi\left(\left\lfloor \frac{w}{d} \right\rfloor, \left\lfloor \frac{x}{d} \right\rfloor; r\right).$$

By means of Möbius inversion, we obtain

$$\Phi(w, x; r) = \sum_{d \ge 1} \mu(d)\left(\left\lfloor \frac{x}{d} \right\rfloor + \sum_{j=1}^{\lfloor x/d \rfloor} \left\lfloor \min\left\{\frac{w}{d}, jr\right\} \right\rfloor\right).$$

Since $t = -1 + \Phi(m, n; f_t)$, using (1.117), we arrive at the following conclusion.

Proposition 1.32. *If $f_t \in \mathcal{F}_n^m - \{\tfrac{1}{1}\}$, where $0 < m \le n$, then*

$$t = \sum_{d \ge 1} \mu(d) \sum_{j=1}^{\lfloor n/d \rfloor} \left\lfloor \min\left\{\frac{m}{d}, jf_t\right\} \right\rfloor.$$

If $n > 1$, and $f_t \in \mathcal{F}_n - \{\tfrac{1}{1}\} = \mathcal{F}_n^{n-1} - \{\tfrac{1}{1}\}$, then

$$t = \sum_{j=1}^{n} \varphi(j; [1, \min\{n-1, \lfloor jf_t \rfloor\}]) = \sum_{j=1}^{n} \varphi(j; [1, \min\{n, \lfloor jf_t \rfloor\}]) = \sum_{j=1}^{n} \varphi(j; [1, \lfloor jf_t \rfloor]);$$

cf. (1.123). By this observation, we have the following result.

Corollary 1.33. *If $f_t \in \mathcal{F}_n - \{\tfrac{1}{1}\}$, where $n > 1$, then*

$$t = \sum_{d=1}^{n} \mu(d) \sum_{j=1}^{\lfloor n/d \rfloor} \lfloor jf_t \rfloor = \sum_{j=2}^{n} M\left(\frac{n}{j}\right) \lfloor jf_t \rfloor,$$

where $M(\cdot)$ is the Mertens function.

For $x \in \mathbb{R}$, the *Mertens function* $M(x)$ is defined to be the summatory function of the Möbius function as follows:

$$M(x) := \sum_{d\in\mathbb{P}:d\le x} \mu(d). \tag{1.124}$$

1.7.2 The indices of fractions in $\mathcal{G}_n^m$

Consider a fraction $\frac{h}{k} =: f_t \in \mathcal{G}_n^m - \{\frac{0}{1}\}$, where $0 \le m < n$. To find the index t, it suffices to compute the index s of its image $(1-f_t) = \frac{k-h}{k} =: f_s \in \mathcal{F}_n^{n-m}$ under the order-reversing bijection $\mathcal{G}_n^m \to \mathcal{F}_n^{n-m}$ defined in (1.8), with the help of Proposition 1.32. The number of fractions in $\mathcal{G}_n^m$ was earlier found in Corollary 1.31, and since $t = |\mathcal{G}_n^m| - s - 1$, we obtain the following result.

Corollary 1.34. *If $f_t \in \mathcal{G}_n^m - \{\frac{0}{1}\}$, where $0 \le m < n$, then*

$$t = \frac{1}{2} + \sum_{d\ge 1} \mu(d)\Big(\Big\lfloor\frac{n-m}{d}\Big\rfloor\Big(\Big\lfloor\frac{n}{d}\Big\rfloor - \frac{1}{2}\Big\lfloor\frac{n-m}{d}\Big\rfloor\Big) - \sum_{j=1}^{\lfloor n/d\rfloor} \Big\lfloor \min\Big\{\frac{n-m}{d}, j(1-f_t)\Big\}\Big\rfloor\Big).$$

1.8 The rank problem. I

Let $\mathcal{C}$ be a Farey (sub)sequence. Given a number $r \in \mathbb{R}$, $0 \le r \le 1$, we are concerned in this section, and in Section 2.6, with the *rank problem* of finding the index t of the fraction $f_t \in \mathcal{C}$, so that $f_t = \max\{\frac{h}{k} \in \mathcal{C} : \frac{h}{k} \le r\}$ or $f_t = \min\{\frac{h}{k} \in \mathcal{C} : r \le \frac{h}{k}\}$.

1.8.1 The rank problem for $\mathcal{F}_n^m$ and $\mathcal{F}_n$

In the context of the rank problem for the Farey (sub)sequences $\mathcal{F}_n^m$ and $\mathcal{F}_n$, Proposition 1.32 and Corollary 1.33 are restated as follows.

Remark 1.35. Let $r \in \mathbb{R}$, $0 \le r < 1$.

(i) If $f_t = \max\{\frac{h}{k} \in \mathcal{F}_n^m : \frac{h}{k} \le r\}$, where $0 < m \le n$, then

$$t = \sum_{d\ge 1} \mu(d) \sum_{j=1}^{\lfloor n/d\rfloor} \Big\lfloor \min\Big\{\frac{m}{d}, jr\Big\}\Big\rfloor.$$

(ii) If $f_t = \max\{\frac{h}{k} \in \mathcal{F}_n : \frac{h}{k} \le r\}$, where $n > 1$, then

$$t = \sum_{d=1}^{n} \mu(d) \sum_{j=1}^{\lfloor n/d\rfloor} \lfloor jr\rfloor = \sum_{j=2}^{n} M\Big(\frac{n}{j}\Big)\lfloor jr\rfloor.$$

1.8.2 The rank problem for $\mathcal{G}_n^m$

In the context of the rank problem for the Farey subsequence $\mathcal{G}_n^m$, Corollary 1.34 suggests the following observation.

Remark 1.36. Let $r \in \mathbb{R}$ be a number such that $0 < r \leq 1$. If $f_t = \min\{\frac{h}{k} \in \mathcal{G}_n^m : r \leq \frac{h}{k}\}$, where $0 \leq m < n$, then

$$t = \frac{1}{2} + \sum_{d\geq 1} \mu(d)\Big(\Big\lfloor \frac{n-m}{d} \Big\rfloor\Big(\Big\lfloor \frac{n}{d} \Big\rfloor - \frac{1}{2}\Big\lfloor \frac{n-m}{d} \Big\rfloor\Big) - \sum_{j=1}^{\lfloor n/d \rfloor} \Big\lfloor \min\Big\{\frac{n-m}{d}, j(1-r)\Big\}\Big\rfloor\Big).$$

1.9 A generating function of the Farey sequence $\mathcal{F}_n$

In this section we deal with the Farey sequence $\mathcal{F}_n$, and we show a way to give its *generating function*.

We begin with a generating function of the integers j, in an interval, that are relatively prime to a fixed integer i; see also (1.120).

If x is a formal variable, then for a positive integer i and for a nonempty interval $[i'+1, i''] \subset \mathbb{P}$, we have

$$\sum_{\substack{j\in[i'+1,i'']:\\ \gcd(i,j)=1}} \mathrm{x}^j = \sum_{\substack{d\in[1,i]:\\ d|i}} \mu(d) \frac{\mathrm{x}^{d\lceil \frac{i'+1}{d} \rceil} - \mathrm{x}^{d(\lfloor \frac{i''}{d} \rfloor + 1)}}{1-\mathrm{x}^d}. \tag{1.125}$$

That is,

$$\sum_{\substack{j\in[i'+1,i'']:\\ \gcd(i,j)=1}} \mathrm{x}^j = \sum_{\substack{d\in[1,i]:\\ d|i}} \mu(d) \frac{\mathrm{x}^{d(\lfloor \frac{i'}{d} \rfloor + 1)} - \mathrm{x}^{d(\lfloor \frac{i''}{d} \rfloor + 1)}}{1-\mathrm{x}^d}.$$

Proposition 1.37. *If* x *and* y *are formal variables, then*

$$\sum_{\frac{h}{k}\in\mathcal{F}_n} \mathrm{x}^h\mathrm{y}^k = \mathrm{y} + \sum_{d=1}^{n} \mu(d) \frac{\mathrm{x}^d}{1-\mathrm{x}^d}\Big(\frac{\mathrm{y}^d(1-\mathrm{y}^{d\lfloor \frac{n}{d} \rfloor})}{1-\mathrm{y}^d} - \frac{(\mathrm{xy})^d(1-(\mathrm{xy})^{d\lfloor \frac{n}{d} \rfloor})}{1-(\mathrm{xy})^d}\Big), \tag{1.126}$$

that is,

$$\begin{aligned}\sum_{\frac{h}{k}\in\mathcal{F}_n} \mathrm{x}^h\mathrm{y}^k &= \mathrm{y} + \sum_{d=1}^{n} \mu(d) \frac{((\mathrm{x}^{d\lfloor \frac{n}{d} \rfloor + 2d} - \mathrm{x}^{2d})\mathrm{y}^{2d} - (\mathrm{x}^{d\lfloor \frac{n}{d} \rfloor + 2d} - \mathrm{x}^d)\mathrm{y}^d)\mathrm{y}^{d\lfloor \frac{n}{d} \rfloor} + (\mathrm{x}^{2d} - \mathrm{x}^d)\mathrm{y}^d}{(\mathrm{x}^{2d} - \mathrm{x}^d)\mathrm{y}^{2d} - (\mathrm{x}^{2d} - 1)\mathrm{y}^d + \mathrm{x}^d - 1}.\end{aligned} \tag{1.127}$$

Proof. In view of (1.125), we have

$$\sum_{\frac{h}{k}\in\mathcal{F}_n-\{\frac{0}{1}\}} \mathrm{x}^h\mathrm{y}^k = \sum_{j=1}^{n} \mathrm{y}^j \sum_{\substack{i\in[1,j]:\\ \gcd(i,j)=1}} \mathrm{x}^i = \sum_{j=1}^{n} \mathrm{y}^j \sum_{\substack{d\in[1,j]:\\ d|j}} \mu(d)\frac{\mathrm{x}^d - \mathrm{x}^{d(\lfloor\frac{j}{d}\rfloor+1)}}{1-\mathrm{x}^d},$$

that is,

$$\sum_{\frac{h}{k}\in\mathcal{F}_n-\{\frac{0}{1}\}} \mathrm{x}^h\mathrm{y}^k = \sum_{d=1}^{n} \mu(d) \sum_{s=1}^{\lfloor n/d\rfloor} \frac{\mathrm{x}^d - \mathrm{x}^{d(s+1)}}{1-\mathrm{x}^d}\mathrm{y}^{ds}.$$

Further,

$$\begin{aligned}
\sum_{\frac{h}{k}\in\mathcal{F}_n-\{\frac{0}{1}\}} \mathrm{x}^h\mathrm{y}^k &= \sum_{d=1}^{n} \mu(d) \sum_{s=1}^{\lfloor n/d\rfloor} \Big(\frac{\mathrm{x}^d}{1-\mathrm{x}^d} - \frac{\mathrm{x}^{d(s+1)}}{1-\mathrm{x}^d}\Big)\mathrm{y}^{ds} \\
&= \sum_{d=1}^{n} \mu(d)\Big(\frac{\mathrm{x}^d}{1-\mathrm{x}^d}\sum_{s=1}^{\lfloor n/d\rfloor}\mathrm{y}^{ds} - \sum_{s=1}^{\lfloor n/d\rfloor}\frac{\mathrm{x}^{d(s+1)}\mathrm{y}^{ds}}{1-\mathrm{x}^d}\Big) \\
&= \sum_{d=1}^{n} \mu(d)\Big(\frac{\mathrm{x}^d}{1-\mathrm{x}^d}\cdot\frac{\mathrm{y}^d - \mathrm{y}^{d(\lfloor\frac{n}{d}\rfloor+1)}}{1-\mathrm{y}^d} - \frac{\mathrm{x}^d}{1-\mathrm{x}^d}\sum_{s=1}^{\lfloor n/d\rfloor}(\mathrm{xy})^{ds}\Big) \\
&= \sum_{d=1}^{n} \mu(d)\Big(\frac{\mathrm{x}^d}{1-\mathrm{x}^d}\cdot\frac{\mathrm{y}^d - \mathrm{y}^{d(\lfloor\frac{n}{d}\rfloor+1)}}{1-\mathrm{y}^d} - \frac{\mathrm{x}^d}{1-\mathrm{x}^d}\cdot\frac{(\mathrm{xy})^d - (\mathrm{xy})^{d(\lfloor\frac{n}{d}\rfloor+1)}}{1-(\mathrm{xy})^d}\Big) \\
&= \sum_{d=1}^{n} \mu(d)\frac{\mathrm{x}^d}{1-\mathrm{x}^d}\Big(\frac{\mathrm{y}^d - \mathrm{y}^{d(\lfloor\frac{n}{d}\rfloor+1)}}{1-\mathrm{y}^d} - \frac{(\mathrm{xy})^d - (\mathrm{xy})^{d(\lfloor\frac{n}{d}\rfloor+1)}}{1-(\mathrm{xy})^d}\Big),
\end{aligned}$$

that is,

$$\sum_{\frac{h}{k}\in\mathcal{F}_n} \mathrm{x}^h\mathrm{y}^k = \mathrm{y} + \sum_{d=1}^{n} \mu(d)\frac{\mathrm{x}^d}{1-\mathrm{x}^d}\Big(\frac{\mathrm{y}^d - \mathrm{y}^{d(\lfloor\frac{n}{d}\rfloor+1)}}{1-\mathrm{y}^d} - \frac{(\mathrm{xy})^d - (\mathrm{xy})^{d(\lfloor\frac{n}{d}\rfloor+1)}}{1-(\mathrm{xy})^d}\Big).$$

Hence, (1.126) follows. □

Notes

The *Farey sequence (series)* $\mathcal{F}_n$ is a fundamental construct in number theory. Its discovery and characterization by Charles Haros [73], Henry Goodwyn [66], John Farey, Sr. [55, 56], and Augustin-Louis Cauchy [44] is well described by the sentence [72, p. 44]:

The history of 'Farey series' is very curious.

We refer the reader to the book [67] for a historical and documentary panorama. See also, e.g., [18, pp. 168–169], [40], [45, pp. 2–3], [50, pp. 156–158], [64, Section 7] and [100, Section 6].

There are numerous research papers and popular scientific notes, video lectures, slides of talks, and web pages (the best way to begin surfing the Web is to visit the *On-Line Encyclopedia of Integer Sequences* [167]) that are concerned with Farey sequences and their applications.

The recent books [7, 74, 75] provide much information on the *Farey table*, Farey sequences, and the *Farey diagram*.

Farey sequences and Farey rows are similar collections of fractions, but they are differently structured. According to the definition of the Farey sequence $\mathcal{F}_n$ of order n, given in Table 1.1, the sequence $\mathcal{F}_{1+1}$ is the increasing sequence $(\frac{0}{1} < \frac{1}{2} < \frac{1}{1})$. By definition, the first *Farey row* of fractions in the *Farey table* also consists of the fractions $\frac{0}{1}$, $\frac{1}{2}$ and $\frac{1}{1}$, arranged in increasing order. For $n > 1$, the nth *Farey row* of fractions, arranged in increasing order, is composed of the fractions of the $(n-1)$th row, together with the mediants $\frac{a+a'}{b+b'}$ (it is a characteristic property of the Farey table that such mediants are always irreducible) inserted in all pairs of consecutive fractions $\frac{a}{b} < \frac{a'}{b'}$ (these are the so-called *Farey neighbors*) of the $(n-1)$th Farey row; see, e.g., [7, Section 3.2].

For instance, the Farey sequence of order 4 is

$$\mathcal{F}_{3+1} = (\tfrac{0}{1} < \tfrac{1}{4} < \tfrac{1}{3} < \tfrac{1}{2} < \tfrac{2}{3} < \tfrac{3}{4} < \tfrac{1}{1}),$$

while the third Farey row in the Farey table is

$$(\tfrac{0}{1} < \tfrac{1}{4} < \tfrac{1}{3} < \tfrac{2}{5} < \tfrac{1}{2} < \tfrac{3}{5} < \tfrac{2}{3} < \tfrac{3}{4} < \tfrac{1}{1}).$$

Standard frameworks of the Farey table are the so-called *Farey tree* and (the left-hand part of) *Stern–Brocot tree*, see, e.g., [7, pp. 53–54] and [65, Section 4.5].

Farey rows, as well as computational tree-like structures behind them, and related partitions of the unit interval $[\![0, 1]\!] := \{x \in \mathbb{R} : 0 \leq x \leq 1\}$ are out of the scope of this book.

The books [12, 41, 65, 72, 81, 101, 124, 137] are well-known standard references on Farey sequences. These sequences, as well as their applications, can be found in numerous sources, e.g., [10, 18, 27–30, 34, 42, 63, 79, 80, 86, 87, 91, 133, 144, 145, 154, 159].

Farey sequences, approximation problems and related impressive graph-theoretic concepts are surveyed in [130].

The Farey subsequence $\mathcal{F}_n^m$ was introduced by James Whitbread Lee Glaisher in [64, Section 6], see also [43, p. 288]; its properties and applications are also discussed, e.g., in [3, 4, 106], [108, Remark 7.10] and [153, Section 4].

The Farey subsequence $\mathcal{G}_n^m$ was briefly described in [109, Remark 2].

The Farey subsequence $\mathcal{F}(\mathbb{B}(n), m)$ was described in [108, Section 7]; its properties are also investigated in [107, 109–111].

Let $\boldsymbol{w} := (w_1, \ldots, w_n)$ be an ordered basis of a vector space $\mathbb{V}(n)$ with a distinguished m-dimensional subspace $\mathbf{A}$, $0 < m < n$. The linear algebraic technique that helps to see the equivalence of definitions (1.4) and (1.1) of the sequence $\mathcal{F}(\mathbb{B}(n), m)$

is that of *adapting* the basis $\boldsymbol{w}$ to the subspace **A**. Following the solution to [149, p. 188, Exercise 2.188], we refer the reader to [46, p. 153] for a description of this technique. The result of adapting is a new ordered basis $\boldsymbol{w}' := (w_{i_1}, \ldots, w_{i_{n-m}}, u_1, \ldots, u_m)$ of the space $\mathbb{V}(n)$ such that $(w_{i_1}, \ldots, w_{i_{n-m}})$ is a subsequence of $\boldsymbol{w}$, and the sequence $(u_1, \ldots, u_m)$ is an ordered basis of the subspace **A**. Now define the unordered sets $S := \{w : w \in \boldsymbol{w}'\}$ and $A := \{u_1, \ldots, u_m\}$. In order to obtain the sequence $\mathcal{F}(\mathbb{B}(n), m)$ as defined in (1.1), it remains to sort, in increasing order and without repetition, the fractions $\frac{|B \cap A|}{\gcd(|B \cap A|, |B|)} / \frac{|B|}{\gcd(|B \cap A|, |B|)}$ for all nonempty subsets B of the unordered basis S.

Every time we compute the fraction $\frac{\dim(\mathbf{B} \cap \mathbf{A})}{\gcd(\dim(\mathbf{B} \cap \mathbf{A}), \dim \mathbf{B})} / \frac{\dim \mathbf{B}}{\gcd(\dim(\mathbf{B} \cap \mathbf{A}), \dim \mathbf{B})}$ in definition (1.4) for two subspaces **A** and **B**, we deal with an elementary subspace arrangement in $\mathbb{V}(n)$.

Subspace arrangements (see [20]) and *hyperplane arrangements* (see [48, 51, 125, 126, 148]) are the area of an active cross-disciplinary study.

The order-reversing and bijective mapping $\frac{h}{k} \mapsto \frac{k-h}{k}$ in (1.8) is induced by the strictly decreasing mapping $x \mapsto 1 - x$ of the unit interval $[\![0, 1]\!]$ onto itself. Pairs of *symmetric* (or *complementary*) fractions $\frac{h}{k}$ and $\frac{k-h}{k}$ are specific to the Farey (sub)sequences $\mathcal{F}_n$, $\mathcal{F}(\mathbb{B}(2m), m)$, and to the rows of the *Farey table* [7, p. 51]. In the context of Farey sequences $\mathcal{F}_n$, the early reference on pairs of symmetric (and, as a consequence, equidistant from the central fraction $\frac{1}{2}$) fractions is [152]. In view of the bijection (1.8), the sum of fractions in $\mathcal{F}_n$ is easily computed, see (2.100).

Proposition 1.7, which describes neighboring fractions in $\mathcal{F}_n$, is inspired by [41, Theorem 253]; see also [72, Section 3.4].

Neighboring fractions should not be confused with Farey neighbors and general Farey pairs. Recall that two consecutive fractions $\frac{a}{b} < \frac{a'}{b'}$ in a row of the *Farey table* are called *Farey neighbors* (note that $ba' - ab' = 1$), see [7, p. 51].

An ordered pair of fractions $(\frac{a}{b} < \frac{a'}{b'})$ in the Farey sequence $\mathcal{F}_n$ is said to be a *normal pair* (see [91, p. 30]) or a *Farey pair* (see [154, p. 178]) if $ba' - ab' = 1$. Neighboring (that is, consecutive) fractions in $\mathcal{F}_n$ are examples of Farey pairs.

If $(\frac{a}{b} < \frac{a'}{b'})$ is a Farey pair of fractions, that is, $ba' - ab' = 1$, then $(\frac{a}{b} < \frac{a'}{b'})$ is a pair of consecutive fractions in all Farey sequences $\mathcal{F}_n$ of order n such that

$$\max\{b, b'\} \le n \le b + b' - 1,$$

see, e.g., [12, Theorem 5.3].

The coprimality criterion mentioned in a gray box on page 14 is a basic observation in elementary number theory; it is explained, e.g., in [65, p. 103] and [154, p. 67, Theorem 2.7].

Equations (1.13) and (1.18) are *linear Diophantine equations*. Such equations are investigated, e.g., in [8, Section 8.1], [9, Section 2.1], [17, Sections 1.3 and 1.5], [37, Section 2.4], [42, Section 2.5], [54, Chapter 3], [94, Section 3.5], [95, Section 1.9], [117, Chapter 2], [118, Example 1.15], [123, Section 1.6], [139, Theorem 4.1.4], [141,

Appendix 8], [142, Section 3.7], [150, Section 2.5], [151, Section 2.6], [161, Chapters 4 and 11], [164, Chapter 2], [163, Section 1.3.2] and [166, Section 1.2].

The very useful identities (1.15) and (1.16) for the floor and ceiling functions are verified, e.g., in the solution to [65, p. 507, Exercise 3.12]. On these functions, see also, e.g., [8, Section I.3], [74, Section 5.4], [118, Section 2.5] and [160, Section 9-1].

The observation made in Remark 1.8 is verified, e.g., in [101, Theorem 156], where it was proved that for any pair $\frac{h_j}{k_j} < \frac{h_{j+1}}{k_{j+1}}$ of consecutive fractions in the Farey sequence $\mathcal{F}_n$, we have

$$k_j + k_{j+1} > n, \quad \gcd(k_j, k_{j+1}) = 1.$$

On these relations, see also, e.g., the proof of [10, p. 75, Theorem 2.4.1], [41, Theorem 254], [42, p. 335, Proof of Theorem 15.10], [63, Exercise 3.4], [68, Lemma 1], [72, Theorem 30], [79, Theorem 10.2], [124, p. 300], [133, Exercise 6.30] and [137, Theorem 3].

Proposition 1.10 is a restatement of [3, Theorem 4]. In analogy to the proof of [3, p. 165, Theorem 4], in Corollary 1.11, we analyze the choice of two values of the multipliers t^* appearing in Proposition 1.10.

Remark 1.12 is a comment on the observation made in the proof of [3, p. 165, Theorem 4].

In the context of the Farey sequences $\mathcal{F}_n$, Proposition 1.23 goes back to [73] and [44], and it describes one of the two fundamental properties of these sequences, see [10, Theorem 2.4.2], [41, Theorem 254], [42, Theorem 15.10], [63, Exercise 3.3], [72, Theorem 28], [79, Theorem 10.2], [81, Lemma 1.2.1], [91, Theorem 1.29], [93, p. 157, Exercise 19], [101, Theorem 156], [124, Theorem 6.1], [133, Exercise 6.28], [137, Theorem 1] and [144, Theorem 2A].

In the context of the Farey subsequences $\mathcal{F}_n^m$, Proposition 1.23 goes back to [64], and it restates [3, Theorem 5].

The mediant property, the subject of Proposition 1.24, is discussed in [44, 55, 56, 73] and also verified for the sequences $\mathcal{F}_n$, e.g., in [10, Theorem 2.4.4], [41, Theorem 255], [63, Example 3.5], [72, Theorem 29], [79, Theorem 10.3], [91, Corollary 1.32], [133, Example 6.29] and [144, Corollary 2C].

In the context of the sequences $\mathcal{F}_n^m$, Proposition 1.24 again goes back to [64].

See [7, p. 51] on the analogous properties of rows of the *Farey table*.

Proposition 1.25 provides recurrent relations for any triple of consecutive fractions in the Farey sequences $\mathcal{F}_n$; cf., e.g., [10, Theorem 2.4.3] and the solution to [65, pp. 523–524, Exercise 4.61]. Such a triple should not be confused with a Farey triple. A *Farey triple* in a row of the *Farey table* is composed of a *Farey pair* taken from the preceding row, together with the inserted mediant of this pair, see [7, p. 51].

If

$$\frac{h_j}{k_j} < \frac{h_{j+1}}{k_{j+1}} < \frac{h_{j+2}}{k_{j+2}} \tag{1.128}$$

are three consecutive fractions in the Farey sequence $\mathcal{F}_n$, then the quantity

$$\nu_{j+1} := \nu\left(\frac{h_{j+1}}{k_{j+1}}\right) := \gcd(h_j + h_{j+2}, k_j + k_{j+2}) = \frac{h_j + h_{j+2}}{h_{j+1}} = \frac{k_j + k_{j+2}}{k_{j+1}}, \tag{1.129}$$

determined in (1.97) and (1.98), is called the *Farey index* of the fraction $\frac{h_{j+1}}{k_{j+1}}$ in the sequence $\mathcal{F}_n$. This key component in the recurrent descriptions of the triples of consecutive fractions, discussed in [52, p. 94], also appears in [18, p. 169], [50, p. 156] and [93, p. 157, Exercise 18]. On the Farey index, see, e.g., [26, 45, 69–71, 76].

For any $n \geq 2$, we have (see [70, Theorem 1])

$$2(n+2) + \sum_{\frac{h}{k} \in \mathcal{F}_n - \{\frac{0}{1}, \frac{1}{1}\}} \nu\left(\frac{h}{k}\right) = 3|\mathcal{F}_n|.$$

Thus, in view of Corollary 1.30, we have

$$2n + \sum_{\frac{h}{k} \in \mathcal{F}_n - \{\frac{0}{1}, \frac{1}{1}\}} \nu\left(\frac{h}{k}\right) = \frac{1}{2} + \frac{3}{2} \sum_{d=1}^{n} \mu(d) \left\lfloor \frac{n}{d} \right\rfloor^2.$$

More precisely, the following can be said:

For any $n \geq 2$, we have (see [69, Theorem 1])

$$2 + \sum_{\substack{\frac{h}{k} \in \mathcal{F}_n - \{\frac{0}{1}, \frac{1}{1}\}: \\ k \text{ even}}} \nu\left(\frac{h}{k}\right) = |\mathcal{F}_n| = \frac{3}{2} + \frac{1}{2} \sum_{d=1}^{n} \mu(d) \left\lfloor \frac{n}{d} \right\rfloor^2,$$

$$2(n+1) + \sum_{\substack{\frac{h}{k} \in \mathcal{F}_n - \{\frac{0}{1}, \frac{1}{1}\}: \\ k \text{ odd}}} \nu\left(\frac{h}{k}\right) = 2|\mathcal{F}_n| = 3 + \sum_{d=1}^{n} \mu(d) \left\lfloor \frac{n}{d} \right\rfloor^2.$$

Recall that the quantities ν_{j+1} defined by (1.129) are determined in (1.103) and (1.104) for fractions $\frac{h_{j+1}}{k_{j+1}} \in \mathcal{F}_n^m$, in (1.109) and (1.110) for fractions $\frac{h_{j+1}}{k_{j+1}} \in \mathcal{G}_n^m$, and in (1.115) and (1.116) for fractions $\frac{h_{j+1}}{k_{j+1}} \in \mathcal{F}(\mathbb{B}(n), m)$.

Following [76, p. 97], note that for the triple of consecutive fractions (1.128), in any Farey (sub)sequence of the form $\mathcal{F}_n$, $\mathcal{F}_n^m$, $\mathcal{G}_n^m$ or $\mathcal{F}(\mathbb{B}(n), m)$, we have

$$\nu_{j+1} = k_j h_{j+2} - h_j k_{j+2}.$$

Let $\mathcal{T} \subset \mathbb{R}^2$ be the *Farey triangle* defined by

$$\mathcal{T} := \left\{ \left[\begin{smallmatrix} x \\ y \end{smallmatrix}\right] \in \mathbb{R}^2 : 0 < x \leq y,\ 0 < y \leq 1,\ x + y > 1 \right\}.$$

The *Boca–Cobeli–Zaharescu (BCZ) map* $\Upsilon\colon \mathcal{T} \to \mathcal{T}$ is defined by

$$\Upsilon(x, y) := \Big(y, \Big\lfloor \frac{1+x}{y} \Big\rfloor y - x\Big). \tag{1.130}$$

This map was presented in [24, Section 4]; cf. (1.96). The BCZ map is a bijective area-preserving transformation of the Farey triangle $\mathcal{T}$; its inverse map is (cf. (1.94))

$$\Upsilon^{-1}(x, y) := \Big(\Big\lfloor \frac{1+y}{x} \Big\rfloor x - y, x\Big).$$

If $\frac{h_j}{k_j} < \frac{h_{j+1}}{k_{j+1}} < \frac{h_{j+2}}{k_{j+2}}$ is a triple of consecutive fractions in the Farey sequence $\mathcal{F}_n$, then, according to the formula (1.96) of Proposition 1.25, we have $k_{j+2} = \nu_{j+1}k_{j+1} - k_j$, where the Farey index ν_{j+1} of the fraction $\frac{h_{j+1}}{k_{j+1}}$ is $\lfloor \frac{n+k_j}{k_{j+1}} \rfloor = \lfloor \frac{1+(k_j/n)}{k_{j+1}/n} \rfloor$. We obtain $\frac{k_{j+2}}{n} = \nu_{j+1}\frac{k_{j+1}}{n} - \frac{k_j}{n}$. In other words (see [24, 25, 162] and [15, 78]),

$$\Upsilon\Big(\frac{k_j}{n}, \frac{k_{j+1}}{n}\Big) = \Big(\frac{k_{j+1}}{n}, \frac{k_{j+2}}{n}\Big).$$

Thus, the BCZ map can be used for generating a Farey sequence $\mathcal{F}_n$ of any order n (see [25, Remark 1.1(2)]):

$$n\Upsilon^j\Big(\frac{1}{n}, 1\Big) = (k_j, k_{j+1}), \quad 1 \le j \le |\mathcal{F}_n| - 2.$$

The numerators of the fractions $\frac{h_j}{k_j}, \frac{h_{j+1}}{k_{j+1}} \in \mathcal{F}_n$ can be retrieved from any denominator pair (k_j, k_{j+1}) as the solution $(\alpha_0, \beta_0) =: (h_j, h_{j+1})$ to the linear Diophantine equation $k_j\beta - k_{j+1}\alpha = 1$, in the unknowns α and β, with $1 \le \alpha \le k_j$ and $1 \le \beta \le k_{j+1}$.

The characteristic properties of the Farey sequences $\mathcal{F}_n$ are derived in [143], in the opposite direction, from their definition by the recurrent relations for the triples of consecutive fractions.

Returning to the number-theoretic Möbius function $\mu(d)$, note that $\mu(d) = \boldsymbol{\mu}_P(\hat{0}, d)$, where P denotes the set of positive integers $\mathbb{P}$, partially ordered by divisibility, $\hat{0} := 1$ and, for any locally finite poset Q, the poset-theoretic *Möbius function* $\boldsymbol{\mu}_Q\colon Q \times Q \to \mathbb{Z}$ is defined as follows: $\boldsymbol{\mu}_Q(x, x) := 1$ for any $x \in Q$; further, if $z \in Q$ and $x < z$ in Q, then

$$\boldsymbol{\mu}_Q(x, z) := -\sum_{y \in Q: x \le y < z} \boldsymbol{\mu}_Q(x, y);$$

if $x \nleq z$ in Q, then $\boldsymbol{\mu}_Q(x,z) := 0$. Recall that the poset P is isomorphic to the family of all finite *multisets* of the set $\mathbb{P}$, partially ordered by inclusion; see, e.g., [149, Examples 1.1.14 and 3.8.4].

Möbius inversion is a fundamental subject in number theory and combinatorics.

On number-theoretic *Möbius inversion*, see, e.g., [8, Section 6.1], [11, Section 2.7], [16, Section 2.4], [37, Section 4.4], [42, Section 6.2], [61, Section 3.6], [95, Section 12.2], [101, Chapter IV], [103, Chapter 6], [105, Section 4.7], [118, Section 2.3], [119, Chapters 2 and 3], [123, Section 6.3], [137, Chapter 12], [121, Sections I.1.1 and II.1.1], [139, Section 2.2], [142, Section 7.4], [145, Chapter 21], [154, Section 3.3], [155, Section 4.3] and [164, Section 2.3]. One statement is as follows (see [119, p. 29]):

If $\mathfrak{f}, \mathfrak{g} \colon \mathbb{P} \to \mathbb{C}$ are *arithmetic functions*, then

$$\mathfrak{g}(n) = \sum_{d=1}^{n} \mathfrak{f}\left(\left\lfloor \frac{n}{d} \right\rfloor\right) \quad \text{for all } n \in \mathbb{P} \tag{1.131}$$

if and only if

$$\mathfrak{f}(n) = \sum_{d=1}^{n} \mu(d)\, \mathfrak{g}\left(\left\lfloor \frac{n}{d} \right\rfloor\right) \quad \text{for all } n \in \mathbb{P}. \tag{1.132}$$

The poset-theoretic *Möbius inversion* is discussed, e.g., in [6, Section 5.2], [5, Chapter IV], [32, Section 16.2], [33, Section 1.5.5], [39, Section 6.6], [96, Section 3.1], [105, Section 4.9], [113, Section 8.5], [149, Section 3.7] and [158, Chapter 10].

The identity (1.117) can be found, e.g., in [101, Theorem 36], [118, Corollary 2.5], [119, p. 10 and p. 28], the proof of [123, p. 220, Theorem 6.16] and [147, p. 193]. The exercise to prove this identity $\sum_{d=1}^{n} \mu(d)\lfloor \frac{n}{d} \rfloor = 1$ for all $n \geq 1$, given in [122, p. 45], mentions the early reference [116], as well as [38, 104]. Cf. the identity (2.98); see also Table 1.7.

The family of the sums $\sum_{d=1}^{n} \mu(d)\lfloor \frac{n}{d} \rfloor^{\ell}$ is investigated, e.g., in [131, 146, 157].

The definition of Euler's totient function $\varphi(n)$, given by (1.118), can be found, e.g., in [16, p. 9] and [147, p. 245]; the definition of $\varphi(n)$ (1.119) is given, e.g., in [65, p. 133] and [123, p. 54], to name just a few references.

The basic problem of enumerating the *k-subsets* $A \subseteq X$ of a finite nonempty set of integers X, *relatively prime* to an integer n, that is, $\gcd(A \cup \{n\}) = 1$, is reviewed in [131]. Formula (1.120), the most elementary observation of this kind, can be found, e.g., in [107, p. 4] and [108, p. 395].

The approach to the problem of enumerating the fractions in Section 1.6.1, as well as to the problem of finding the index of a fraction in a Farey (sub)sequence in Section 1.7.1, draws heavily on the argument given in the context of the Farey sequences $\mathcal{F}_n$ in [65, pp. 138–139].

The index of a fraction should not be confused with the Farey index: the *index* of a fraction $\frac{h_j}{k_j}$ in a Farey subsequence is simply its position number, that is, the subscript j. Recall that $\frac{0}{1} =: \frac{h_0}{k_0}$, by convention.

If $\frac{1}{k} =: f_t \in \mathcal{F}_n$ for some $k > 1$, then Corollary 1.33 implies that $t = \sum_{j=k}^{n} M(\frac{n}{j})\lfloor \frac{j}{k} \rfloor$. The combined behaviour of the indices t of the fractions $f_t := \frac{1}{k}$ and the values of Euler's totient function $\varphi(k)$ is investigated in [156].

The recursive relation given in (1.122) for the number of fractions in the Farey sequence $\mathcal{F}_n$ can be found, e.g., in [63, Exercise 3.13] and [167, A005728]. To see the inverse equivalence of the expressions (1.121) and (1.122), it suffices to take, in (1.131) and (1.132), $\mathfrak{f}(\lfloor \frac{n}{d} \rfloor) := |\mathcal{F}_{\lfloor n/d \rfloor}| - \frac{3}{2}$ and $\mathfrak{g}(\lfloor \frac{n}{d} \rfloor) := \frac{1}{2}\lfloor \frac{n}{d} \rfloor^2$.

On the Mertens function, defined by (1.124), see, e.g., [14, p. 198], [37, pp. 150–152], [118, p. 98] and [140, p. 172].

Algorithms to solve the rank problem and the *order statistic problem* (given an integer j, find the fraction $\frac{h_j}{k_j} \in \mathcal{F}_n$) for the Farey sequence $\mathcal{F}_n$ are designed in the works [127–129].

The generating function (1.125) of the integers contained in an interval, which are relatively prime to a fixed integer, can be found in [110].

We applied the `simplify_rational()` command provided by the *Sage Math Cloud* platform [168] to the expression to the right of $\mu(d)$ in (1.126); the result is formula (1.127).

In addition to the generating function of the fractions in the Farey sequence $\mathcal{F}_n$, given in (1.126), generating functions for the half sequences of $\mathcal{F}_n$ can be found in [110]: if x and y are formal variables, then we have

$$\sum_{\substack{\frac{h}{k} \in \mathcal{F}_n:\\ \frac{0}{1} < \frac{h}{k} < \frac{1}{2}}} \mathrm{x}^h \mathrm{y}^k = \sum_{d=1}^{\lceil n/2 \rceil - 1} \mu(d) \frac{\mathrm{y}^d}{1-\mathrm{y}^d}\left(\frac{\mathrm{x}^d\mathrm{y}^{2d} - \mathrm{x}^{d\lceil \frac{n}{2d} \rceil}\mathrm{y}^{2d\lceil \frac{n}{2d} \rceil}}{1-\mathrm{x}^d\mathrm{y}^{2d}} - \frac{\mathrm{x}^d - \mathrm{x}^{d\lceil \frac{n}{2d} \rceil}}{1-\mathrm{x}^d}\mathrm{y}^{d\lfloor \frac{n}{d} \rfloor}\right).$$

For example,

$$\sum_{\substack{\frac{h}{k} \in \mathcal{F}_5:\\ \frac{0}{1} < \frac{h}{k} < \frac{1}{2}}} \mathrm{x}^h \mathrm{y}^k = \mathrm{x}\mathrm{y}^5 + \mathrm{x}\mathrm{y}^4 + \mathrm{x}\mathrm{y}^3 + \mathrm{x}^2\mathrm{y}^5 = \frac{\mathrm{y}}{1-\mathrm{y}}\left(\frac{\mathrm{x}\mathrm{y}^2 - \mathrm{x}^3\mathrm{y}^6}{1-\mathrm{x}\mathrm{y}^2} - \frac{\mathrm{x} - \mathrm{x}^3}{1-\mathrm{x}}\mathrm{y}^5\right).$$

Similarly,

$$\sum_{\substack{\frac{h}{k} \in \mathcal{F}_n:\\ \frac{1}{2} < \frac{h}{k} < \frac{1}{1}}} \mathrm{x}^h \mathrm{y}^k = \sum_{d=1}^{\lceil n/2 \rceil - 1} \mu(d) \frac{\mathrm{x}^d\mathrm{y}^d}{1-\mathrm{x}^d\mathrm{y}^d}\left(\frac{\mathrm{x}^d\mathrm{y}^{2d} - \mathrm{x}^{d\lceil \frac{n}{2d} \rceil}\mathrm{y}^{2d\lceil \frac{n}{2d} \rceil}}{1-\mathrm{x}^d\mathrm{y}^{2d}} - \frac{\mathrm{x}^{d(\lfloor \frac{n}{d} \rfloor - \lceil \frac{n}{2d} \rceil + 1)} - \mathrm{x}^{d\lfloor \frac{n}{d} \rfloor}}{1-\mathrm{x}^d}\mathrm{y}^{d\lfloor \frac{n}{d} \rfloor}\right).$$

2 Farey duality

The Farey subsequences $\mathcal{F}(\mathbb{B}(n), m)$, $\mathcal{F}^{\leq\frac{1}{2}}(\mathbb{B}(n), m)$ and $\mathcal{F}^{\geq\frac{1}{2}}(\mathbb{B}(n), m)$, which are closely related to finite sets, Boolean lattices and finite-dimensional vector spaces, were defined in Chapter 1 by

$$\mathcal{F}(\mathbb{B}(n), m) := (\tfrac{h}{k} \in \mathcal{F}_n : m + k - n \leq h \leq m),$$
$$\mathcal{F}^{\leq\frac{1}{2}}(\mathbb{B}(n), m) := (\tfrac{h}{k} \in \mathcal{F}(\mathbb{B}(n), m) : \tfrac{h}{k} \leq \tfrac{1}{2})$$

and

$$\mathcal{F}^{\geq\frac{1}{2}}(\mathbb{B}(n), m) := (\tfrac{h}{k} \in \mathcal{F}(\mathbb{B}(n), m) : \tfrac{h}{k} \geq \tfrac{1}{2}),$$

where $n > 1$, and $0 < m < n$.

By *Farey duality*, we mean the existence of the following *monotone* and *bijective* maps:

$$\mathcal{F}^{\leq\frac{1}{2}}(\mathbb{B}(n), m) \leftrightarrow \mathcal{F}^{m}_{n-m} := (\tfrac{h}{k} \in \mathcal{F}_{n-m} : h \leq m) \quad \text{(order-preserving)},$$
$$\mathcal{F}^{\geq\frac{1}{2}}(\mathbb{B}(n), m) \leftrightarrow \mathcal{F}^{n-m}_{m} := (\tfrac{h}{k} \in \mathcal{F}_{m} : h \leq n - m) \quad \text{(order-reversing)}$$

and, in particular,

$$\mathcal{F}^{\leq\frac{1}{2}}(\mathbb{B}(2m), m) \leftrightarrow \mathcal{F}_m \leftrightarrow \mathcal{F}^{\geq\frac{1}{2}}(\mathbb{B}(2m), m).$$

In this chapter we first discuss *duality properties* of Farey (sub)sequences, describe their connection with the *Farey map*, and then we show that almost all results of Chapter 1 have useful dual formulations.

2.1 Duality properties of Farey (sub)sequences

In this section we establish explicitly monotone bijections between the sequence $\mathcal{F}^{\leq\frac{1}{2}}(\mathbb{B}(n), m)$ (respectively, $\mathcal{F}^{\geq\frac{1}{2}}(\mathbb{B}(n), m)$) and its dual sequence $\mathcal{F}^{m}_{n-m}$ (respectively, $\mathcal{F}^{n-m}_{m}$). Of course, the order-reversing bijections $\mathcal{F}^{m}_{n-m} \leftrightarrow \mathcal{G}^{n-2m}_{n-m}$ and $\mathcal{F}^{n-m}_{m} \leftrightarrow \mathcal{G}^{2m-n}_{m}$ provided by the mapping $\frac{h}{k} \mapsto \frac{k-h}{k}$ in (1.8) double the list of observations.

2.1.1 The sequences $\mathcal{F}^{\leq\frac{1}{2}}(\mathbb{B}(n), m)$, $\mathcal{F}^{\geq\frac{1}{2}}(\mathbb{B}(n), m)$, and their duals

In this subsection we state the duality characterization of Farey (sub)sequences in its general form.

We will see in Section 2.1.3 that the corresponding relations have perhaps the most elegant formulation for the half sequences of the sequence $\mathcal{F}(\mathbb{B}(2m), m)$ and for their dual sequence $\mathcal{F}_m$.

In the context of the sequence $\mathcal{F}(\mathbb{B}(n), m)$, where $n \neq 2m$, the Farey duality results of this subsection will be restated with more details in Section 2.1.4.

DOI 10.1515/9783110547665-003

Theorem 2.1. *Consider a Farey subsequence $\mathcal{F}(\mathbb{B}(n), m)$. The maps*

$$\mathcal{F}^{\le\frac{1}{2}}(\mathbb{B}(n), m) \to \mathcal{F}^{m}_{n-m}, \quad \tfrac{h}{k} \mapsto \tfrac{h}{k-h}, \quad \left[\begin{smallmatrix} h \\ k \end{smallmatrix}\right] \mapsto \left[\begin{smallmatrix} 1 & 0 \\ -1 & 1 \end{smallmatrix}\right] \cdot \left[\begin{smallmatrix} h \\ k \end{smallmatrix}\right], \tag{2.1}$$

$$\mathcal{F}^{m}_{n-m} \to \mathcal{F}^{\le\frac{1}{2}}(\mathbb{B}(n), m), \quad \tfrac{h}{k} \mapsto \tfrac{h}{h+k}, \quad \left[\begin{smallmatrix} h \\ k \end{smallmatrix}\right] \mapsto \left[\begin{smallmatrix} 1 & 0 \\ 1 & 1 \end{smallmatrix}\right] \cdot \left[\begin{smallmatrix} h \\ k \end{smallmatrix}\right], \tag{2.2}$$

$$\mathcal{F}^{\ge\frac{1}{2}}(\mathbb{B}(n), m) \to \mathcal{G}^{2m-n}_{m}, \quad \tfrac{h}{k} \mapsto \tfrac{2h-k}{h}, \quad \left[\begin{smallmatrix} h \\ k \end{smallmatrix}\right] \mapsto \left[\begin{smallmatrix} 2 & -1 \\ 1 & 0 \end{smallmatrix}\right] \cdot \left[\begin{smallmatrix} h \\ k \end{smallmatrix}\right], \tag{2.3}$$

$$\mathcal{G}^{2m-n}_{m} \to \mathcal{F}^{\ge\frac{1}{2}}(\mathbb{B}(n), m), \quad \tfrac{h}{k} \mapsto \tfrac{k}{2k-h}, \quad \left[\begin{smallmatrix} h \\ k \end{smallmatrix}\right] \mapsto \left[\begin{smallmatrix} 0 & 1 \\ -1 & 2 \end{smallmatrix}\right] \cdot \left[\begin{smallmatrix} h \\ k \end{smallmatrix}\right], \tag{2.4}$$

are order-preserving and bijective, and the maps

$$\mathcal{F}^{\le\frac{1}{2}}(\mathbb{B}(n), m) \to \mathcal{G}^{n-2m}_{n-m}, \quad \tfrac{h}{k} \mapsto \tfrac{k-2h}{k-h}, \quad \left[\begin{smallmatrix} h \\ k \end{smallmatrix}\right] \mapsto \left[\begin{smallmatrix} -2 & 1 \\ -1 & 1 \end{smallmatrix}\right] \cdot \left[\begin{smallmatrix} h \\ k \end{smallmatrix}\right], \tag{2.5}$$

$$\mathcal{G}^{n-2m}_{n-m} \to \mathcal{F}^{\le\frac{1}{2}}(\mathbb{B}(n), m), \quad \tfrac{h}{k} \mapsto \tfrac{k-h}{2k-h}, \quad \left[\begin{smallmatrix} h \\ k \end{smallmatrix}\right] \mapsto \left[\begin{smallmatrix} -1 & 1 \\ -1 & 2 \end{smallmatrix}\right] \cdot \left[\begin{smallmatrix} h \\ k \end{smallmatrix}\right], \tag{2.6}$$

$$\mathcal{F}^{\ge\frac{1}{2}}(\mathbb{B}(n), m) \to \mathcal{F}^{n-m}_{m}, \quad \tfrac{h}{k} \mapsto \tfrac{k-h}{h}, \quad \left[\begin{smallmatrix} h \\ k \end{smallmatrix}\right] \mapsto \left[\begin{smallmatrix} -1 & 1 \\ 1 & 0 \end{smallmatrix}\right] \cdot \left[\begin{smallmatrix} h \\ k \end{smallmatrix}\right], \tag{2.7}$$

$$\mathcal{F}^{n-m}_{m} \to \mathcal{F}^{\ge\frac{1}{2}}(\mathbb{B}(n), m), \quad \tfrac{h}{k} \mapsto \tfrac{k}{h+k}, \quad \left[\begin{smallmatrix} h \\ k \end{smallmatrix}\right] \mapsto \left[\begin{smallmatrix} 0 & 1 \\ 1 & 1 \end{smallmatrix}\right] \cdot \left[\begin{smallmatrix} h \\ k \end{smallmatrix}\right], \tag{2.8}$$

are order-reversing and bijective.

Proof. For any integer h, $1 \le h \le m$, according to (1.120), we have

$$\begin{aligned}
\left|\left\{\tfrac{h}{k} \in \mathcal{F}(\mathbb{B}(n), m) : \tfrac{h}{k} < \tfrac{1}{2}\right\}\right| &= \varphi(h; [2h+1, h+n-m]) \\
&= \sum_{d \in [1,h]:\, d|h} \mu(d)\left(\left\lfloor \frac{h+n-m}{d} \right\rfloor - \frac{2h}{d}\right) \\
&= \sum_{d \in [1,h]:\, d|h} \mu(d)\left(\left\lfloor \frac{n-m}{d} \right\rfloor - \frac{h}{d}\right) \\
&= \varphi(h; [h+1, n-m]) \\
&= \left|\left\{\tfrac{h}{k} \in \mathcal{F}^{m}_{n-m} : \tfrac{h}{k} < \tfrac{1}{1}\right\}\right|,
\end{aligned}$$

and we see that the sequences $\mathcal{F}^{\le\frac{1}{2}}(\mathbb{B}(n), m)$ and $\mathcal{F}^{m}_{n-m}$ have the same cardinality; but this conclusion also implies

$$|\mathcal{F}^{\ge\frac{1}{2}}(\mathbb{B}(n), m)| = |\mathcal{F}^{n-m}_{m}|,$$

because of the bijection $\mathcal{F}^{\ge\frac{1}{2}}(\mathbb{B}(n), m) \leftarrow \mathcal{F}^{\le\frac{1}{2}}(\mathbb{B}(n), n-m)$ defined in (1.8). The proof of the assertions on the maps (2.1), (2.2), (2.7) and (2.8) is completed by checking that, on the one hand, the fraction $\frac{h_j}{k_j}$ precedes the fraction $\frac{h_{j+1}}{k_{j+1}}$ in $\mathcal{F}^{m}_{n-m}$ if and only if the fraction $\frac{h_j}{k_j+h_j}$ precedes the fraction $\frac{h_{j+1}}{k_{j+1}+h_{j+1}}$ in $\mathcal{F}^{\le\frac{1}{2}}(\mathbb{B}(n), m)$; on the other hand, the fraction $\frac{h_j}{k_j}$ precedes the fraction $\frac{h_{j+1}}{k_{j+1}}$ in $\mathcal{F}^{n-m}_{m}$ if and only if $\frac{k_{j+1}}{k_{j+1}+h_{j+1}}$ precedes $\frac{k_j}{k_j+h_j}$ in $\mathcal{F}^{\ge\frac{1}{2}}(\mathbb{B}(n), m)$. Hence, thanks to Remark 1.3, the remaining assertions of the theorem follow. □

Example 2.2. Suppose $n := 6$ and $m := 4$. Then we have

$$\begin{aligned}
\mathcal{F}^{m}_{n-m} &= \underbrace{(\tfrac{0}{1} < \tfrac{1}{2} < \tfrac{1}{1})}, \\
&\qquad (2.1) \uparrow\downarrow (2.2) \\
\mathcal{F}(\mathbb{B}(n), m) &= (\overbrace{\tfrac{0}{1} < \tfrac{1}{3} < \tfrac{1}{2}} < \tfrac{3}{5} < \tfrac{2}{3} < \tfrac{3}{4} < \tfrac{4}{5} < \tfrac{1}{1}), \\
&\qquad\qquad\qquad (2.3) \downarrow\uparrow (2.4) \\
\mathcal{G}^{2m-n}_{m} &= \qquad\quad \overbrace{(\tfrac{0}{1} < \tfrac{1}{3} < \tfrac{1}{2} < \tfrac{2}{3} < \tfrac{3}{4} < \tfrac{1}{1})}, \\
\mathcal{G}^{n-2m}_{n-m} &= \underbrace{(\tfrac{0}{1} < \tfrac{1}{2} < \tfrac{1}{1})}, \\
&\qquad (2.5) \uparrow\downarrow (2.6) \\
\mathcal{F}(\mathbb{B}(n), m) &= (\overbrace{\tfrac{0}{1} < \tfrac{1}{3} < \tfrac{1}{2}} < \tfrac{3}{5} < \tfrac{2}{3} < \tfrac{3}{4} < \tfrac{4}{5} < \tfrac{1}{1}), \\
&\qquad\qquad\qquad (2.7) \downarrow\uparrow (2.8) \\
\mathcal{F}^{n-m}_{m} &= \qquad\quad \overbrace{(\tfrac{0}{1} < \tfrac{1}{4} < \tfrac{1}{3} < \tfrac{1}{2} < \tfrac{2}{3} < \tfrac{1}{1})}.
\end{aligned}$$

(In the original, the underbraces span $\tfrac{1}{2} < \dots < \tfrac{1}{1}$ in the lines for $\mathcal{F}(\mathbb{B}(n), m)$.)

Remark 2.3. An interesting manifestation of the Farey duality is that the sequence of the numerators of fractions in $\mathcal{F}^{m}_{n-m}$ is the sequence of the numerators of fractions in the left half sequence $\mathcal{F}^{\leq\frac{1}{2}}(\mathbb{B}(n), m)$, in the same order. The sequence of the denominators of fractions in $\mathcal{F}^{n-m}_{m}$ is the sequence of the numerators of fractions in the right half sequence $\mathcal{F}^{\geq\frac{1}{2}}(\mathbb{B}(n), m)$, in the reverse order.

An obvious restatement of these observations in the context of the Farey subsequence $\mathcal{F}(\mathbb{B}(2m), m)$ is given in Remark 2.7.

2.1.2 The Farey duality and the Farey map

No matter which Farey (sub)sequence we consider in this book, it is a certain finite and ordered collection of rational numbers contained in the unit interval $\llbracket 0, 1 \rrbracket$.

The *Farey map* $T\colon \llbracket 0, 1 \rrbracket \to \llbracket 0, 1 \rrbracket$ is defined by

$$T(x) := \begin{cases} T_0(x) & \text{if } x \in \llbracket 0, \tfrac{1}{2} \rrbracket, \\ T_1(x) & \text{if } x \in \llbracket \tfrac{1}{2}, 1 \rrbracket, \end{cases} \tag{2.9}$$

where

$$T_0(x) := \frac{x}{1-x}, \quad T_1(x) := \frac{1-x}{x}.$$

For a fraction $\frac{h}{k}$ in a Farey subsequence $\mathcal{F}(\mathbb{B}(n), m) \subset \llbracket 0, 1 \rrbracket$, we have

$$T(\tfrac{h}{k}) = \begin{cases} T_0(\tfrac{h}{k}) = \tfrac{h}{k-h} & \text{if } \tfrac{h}{k} \in \llbracket 0, \tfrac{1}{2} \rrbracket, \\ T_1(\tfrac{h}{k}) = \tfrac{k-h}{h} & \text{if } \tfrac{h}{k} \in \llbracket \tfrac{1}{2}, 1 \rrbracket, \end{cases}$$

that is, $T_0(\frac{h}{k}) \in \mathcal{F}^m_{n-m} \subset [\![0,1]\!]$ and $T_1(\frac{h}{k}) \in \mathcal{F}^{n-m}_m \subset [\![0,1]\!]$ are the images of the fraction $\frac{h}{k}$ under the order-preserving map (2.1) and the order-reversing map (2.7) mentioned in Theorem 2.1.

Consider the inverse branches

$$T_0^{-1}(x) = \frac{x}{1+x}, \quad T_1^{-1}(x) = \frac{1}{1+x}. \tag{2.10}$$

Note that we have

$$T_0^{-1}\left(\frac{h}{k}\right) = \frac{h}{h+k}$$

for a fraction $\frac{h}{k} \in \mathcal{F}^m_{n-m} \subset [\![0,1]\!]$, and

$$T_1^{-1}\left(\frac{h}{k}\right) = \frac{k}{h+k}$$

for a fraction $\frac{h}{k} \in \mathcal{F}^{n-m}_m \subset [\![0,1]\!]$, but these are precisely the images of $\frac{h}{k}$ under the order-preserving map (2.2) and the order-reversing map (2.8).

The *modified Farey map* $F\colon [\![0,1]\!] \to [\![0,1]\!]$ is defined by

$$F(x) := \begin{cases} F_0(x) & \text{if } x \in [\![0, \frac{1}{2}]\!], \\ F_1(x) & \text{if } x \in [\![\frac{1}{2}, 1]\!], \end{cases} \tag{2.11}$$

where

$$F_0(x) := \frac{x}{1-x}, \quad F_1(x) := \frac{2x-1}{x}.$$

If $\frac{h}{k} \in \mathcal{F}(\mathbb{B}(n), m) \subset [\![0,1]\!]$, then we have

$$F\left(\frac{h}{k}\right) = \begin{cases} F_0(\frac{h}{k}) = \frac{h}{k-h} & \text{if } \frac{h}{k} \in [\![0, \frac{1}{2}]\!], \\ F_1(\frac{h}{k}) = \frac{2h-k}{h} & \text{if } \frac{h}{k} \in [\![\frac{1}{2}, 1]\!], \end{cases}$$

that is, $F_0(\frac{h}{k}) \in \mathcal{F}^m_{n-m} \subset [\![0,1]\!]$ and $F_1(\frac{h}{k}) \in \mathcal{G}^{2m-n}_m \subset [\![0,1]\!]$ are the images of the fraction $\frac{h}{k}$ under the order-preserving maps (2.1) and (2.3).

Since

$$F_0^{-1}(x) = \frac{x}{1+x}, \quad F_1^{-1}(x) = \frac{1}{2-x},$$

we have

$$F_0^{-1}\left(\frac{h}{k}\right) = \frac{h}{h+k}$$

for a fraction $\frac{h}{k} \in \mathcal{F}^m_{n-m} \subset [\![0,1]\!]$, and

$$F_1^{-1}\left(\frac{h}{k}\right) = \frac{k}{2k-h}$$

for a fraction $\frac{h}{k} \in \mathcal{G}^{2m-n}_m \subset [\![0,1]\!]$, but these are precisely the images of $\frac{h}{k}$ under the order-preserving maps (2.2) and (2.4).

Let us restate Theorem 2.1 in terms of strictly monotone maps between finite subsets of the intervals $[\![0,1]\!]$, $[\![0,\frac{1}{2}]\!]$ and $[\![\frac{1}{2},1]\!]$; note that in Theorem 2.1 all maps

$$\mathcal{C}' \to \mathcal{C}'', \quad \left[\begin{smallmatrix} h \\ k \end{smallmatrix}\right] \mapsto \left[\begin{smallmatrix} a & b \\ c & d \end{smallmatrix}\right] \cdot \left[\begin{smallmatrix} h \\ k \end{smallmatrix}\right],$$

between the Farey (sub)sequences $\mathcal{C}'$ and $\mathcal{C}''$, determined by 2×2 matrices $\left[\begin{smallmatrix} a & b \\ c & d \end{smallmatrix}\right]$ with integer entries and with determinant 1 or −1, are induced by the restrictions of the real *linear fractional transformations*

$$\mathbb{R} \to \mathbb{R}, \quad x \mapsto \frac{ax+b}{cx+d},$$

to intervals.

Proposition 2.4. *Consider a Farey subsequence $\mathcal{F}(\mathbb{B}(n),m) \subset [\![0,1]\!]$. The maps*

$$\begin{aligned}
&[\![0,\tfrac{1}{2}]\!] \supset \mathcal{F}^{\le\frac{1}{2}}(\mathbb{B}(n),m) \to \mathcal{F}^{m}_{n-m} \subset [\![0,1]\!], && x \mapsto \tfrac{x}{1-x},\\
&[\![0,1]\!] \supset \mathcal{F}^{m}_{n-m} \to \mathcal{F}^{\le\frac{1}{2}}(\mathbb{B}(n),m) \subset [\![0,\tfrac{1}{2}]\!], && x \mapsto \tfrac{x}{1+x},\\
&[\![\tfrac{1}{2},1]\!] \supset \mathcal{F}^{\ge\frac{1}{2}}(\mathbb{B}(n),m) \to \mathcal{G}^{2m-n}_{m} \subset [\![0,1]\!], && x \mapsto \tfrac{2x-1}{x},\\
&[\![0,1]\!] \supset \mathcal{G}^{2m-n}_{m} \to \mathcal{F}^{\ge\frac{1}{2}}(\mathbb{B}(n),m) \subset [\![\tfrac{1}{2},1]\!], && x \mapsto \tfrac{1}{2-x},
\end{aligned}$$

are strictly increasing, and the maps

$$\begin{aligned}
&[\![0,\tfrac{1}{2}]\!] \supset \mathcal{F}^{\le\frac{1}{2}}(\mathbb{B}(n),m) \to \mathcal{G}^{n-2m}_{n-m} \subset [\![0,1]\!], && x \mapsto \tfrac{1-2x}{1-x},\\
&[\![0,1]\!] \supset \mathcal{G}^{n-2m}_{n-m} \to \mathcal{F}^{\le\frac{1}{2}}(\mathbb{B}(n),m) \subset [\![0,\tfrac{1}{2}]\!], && x \mapsto \tfrac{1-x}{2-x},\\
&[\![\tfrac{1}{2},1]\!] \supset \mathcal{F}^{\ge\frac{1}{2}}(\mathbb{B}(n),m) \to \mathcal{F}^{n-m}_{m} \subset [\![0,1]\!], && x \mapsto \tfrac{1-x}{x},\\
&[\![0,1]\!] \supset \mathcal{F}^{n-m}_{m} \to \mathcal{F}^{\ge\frac{1}{2}}(\mathbb{B}(n),m) \subset [\![\tfrac{1}{2},1]\!], && x \mapsto \tfrac{1}{1+x},
\end{aligned}$$

are strictly decreasing.

2.1.3 The sequences $\mathcal{F}^{\le\frac{1}{2}}(\mathbb{B}(2m),m)$, $\mathcal{F}^{\ge\frac{1}{2}}(\mathbb{B}(2m),m)$, and their dual $\mathcal{F}_m$

In the context of the sequences $\mathcal{F}(\mathbb{B}(2m),m)$, the characterization of Farey duality given by Theorem 2.1 is restated as follows.

Corollary 2.5. *Consider a Farey subsequence $\mathcal{F}(\mathbb{B}(2m),m)$. The maps*

$$\mathcal{F}^{\le\frac{1}{2}}(\mathbb{B}(2m),m) \to \mathcal{F}_m, \quad \tfrac{h}{k} \mapsto \tfrac{h}{k-h}, \quad \left[\begin{smallmatrix} h \\ k \end{smallmatrix}\right] \mapsto \left[\begin{smallmatrix} 1 & 0 \\ -1 & 1 \end{smallmatrix}\right] \cdot \left[\begin{smallmatrix} h \\ k \end{smallmatrix}\right], \tag{2.12}$$

$$\mathcal{F}_m \to \mathcal{F}^{\le\frac{1}{2}}(\mathbb{B}(2m),m), \quad \tfrac{h}{k} \mapsto \tfrac{h}{h+k}, \quad \left[\begin{smallmatrix} h \\ k \end{smallmatrix}\right] \mapsto \left[\begin{smallmatrix} 1 & 0 \\ 1 & 1 \end{smallmatrix}\right] \cdot \left[\begin{smallmatrix} h \\ k \end{smallmatrix}\right], \tag{2.13}$$

$$\mathcal{F}^{\ge\frac{1}{2}}(\mathbb{B}(2m),m) \to \mathcal{F}_m, \quad \tfrac{h}{k} \mapsto \tfrac{2h-k}{h}, \quad \left[\begin{smallmatrix} h \\ k \end{smallmatrix}\right] \mapsto \left[\begin{smallmatrix} 2 & -1 \\ 1 & 0 \end{smallmatrix}\right] \cdot \left[\begin{smallmatrix} h \\ k \end{smallmatrix}\right], \tag{2.14}$$

$$\mathcal{F}_m \to \mathcal{F}^{\ge\frac{1}{2}}(\mathbb{B}(2m),m), \quad \tfrac{h}{k} \mapsto \tfrac{k}{2k-h}, \quad \left[\begin{smallmatrix} h \\ k \end{smallmatrix}\right] \mapsto \left[\begin{smallmatrix} 0 & 1 \\ -1 & 2 \end{smallmatrix}\right] \cdot \left[\begin{smallmatrix} h \\ k \end{smallmatrix}\right], \tag{2.15}$$

are order-preserving and bijective, and the maps

$$\mathcal{F}^{\leq\frac{1}{2}}(\mathbb{B}(2m), m) \to \mathcal{F}_m, \quad \tfrac{h}{k} \mapsto \tfrac{k-2h}{k-h}, \quad \left[\begin{smallmatrix} h \\ k \end{smallmatrix}\right] \mapsto \left[\begin{smallmatrix} -2 & 1 \\ -1 & 1 \end{smallmatrix}\right] \cdot \left[\begin{smallmatrix} h \\ k \end{smallmatrix}\right], \tag{2.16}$$

$$\mathcal{F}_m \to \mathcal{F}^{\leq\frac{1}{2}}(\mathbb{B}(2m), m), \quad \tfrac{h}{k} \mapsto \tfrac{k-h}{2k-h}, \quad \left[\begin{smallmatrix} h \\ k \end{smallmatrix}\right] \mapsto \left[\begin{smallmatrix} -1 & 1 \\ -1 & 2 \end{smallmatrix}\right] \cdot \left[\begin{smallmatrix} h \\ k \end{smallmatrix}\right], \tag{2.17}$$

$$\mathcal{F}^{\geq\frac{1}{2}}(\mathbb{B}(2m), m) \to \mathcal{F}_m, \quad \tfrac{h}{k} \mapsto \tfrac{k-h}{h}, \quad \left[\begin{smallmatrix} h \\ k \end{smallmatrix}\right] \mapsto \left[\begin{smallmatrix} -1 & 1 \\ 1 & 0 \end{smallmatrix}\right] \cdot \left[\begin{smallmatrix} h \\ k \end{smallmatrix}\right], \tag{2.18}$$

$$\mathcal{F}_m \to \mathcal{F}^{\geq\frac{1}{2}}(\mathbb{B}(2m), m), \quad \tfrac{h}{k} \mapsto \tfrac{k}{h+k}, \quad \left[\begin{smallmatrix} h \\ k \end{smallmatrix}\right] \mapsto \left[\begin{smallmatrix} 0 & 1 \\ 1 & 1 \end{smallmatrix}\right] \cdot \left[\begin{smallmatrix} h \\ k \end{smallmatrix}\right], \tag{2.19}$$

are order-reversing and bijective.

Example 2.6. Suppose $m := 3$. Then we have

$$\mathcal{F}_m = \left(\tfrac{0}{1} < \tfrac{1}{3} < \tfrac{1}{2} < \tfrac{2}{3} < \tfrac{1}{1}\right),$$

$$(2.12),(2.16) \uparrow \quad \downarrow (2.13),(2.17)$$

$$\mathcal{F}(\mathbb{B}(2m), m) = \left(\tfrac{0}{1} < \tfrac{1}{4} < \tfrac{1}{3} < \tfrac{2}{5} < \tfrac{1}{2} < \tfrac{3}{5} < \tfrac{2}{3} < \tfrac{3}{4} < \tfrac{1}{1}\right),$$

$$(2.14),(2.18) \downarrow \quad \uparrow (2.15),(2.19)$$

$$\mathcal{F}_m = \left(\tfrac{0}{1} < \tfrac{1}{3} < \tfrac{1}{2} < \tfrac{2}{3} < \tfrac{1}{1}\right).$$

Remark 2.7 (see also Remark 2.3). The sequence of the numerators of fractions in $\mathcal{F}_m$ is precisely the sequence of the numerators of fractions in the left half sequence $\mathcal{F}^{\leq\frac{1}{2}}(\mathbb{B}(2m), m)$. The sequence of the denominators of fractions in $\mathcal{F}_m$ is precisely the sequence of the numerators of fractions in the right half sequence $\mathcal{F}^{\geq\frac{1}{2}}(\mathbb{B}(2m), m)$.

Thus, the numerators of fractions in the sequence $\mathcal{F}(\mathbb{B}(2m), m)$ are nothing but a complete and explicit description of the Farey sequence $\mathcal{F}_m$.

Note that Corollary 2.5 and the order-reversing bijection $\mathcal{F}_m \to \mathcal{F}_m$ defined in (1.8) guarantee that in the case $m > 1$, we have

$$|\mathcal{F}(\mathbb{B}(2m), m)| \equiv 1 \pmod 4.$$

2.1.4 More on the sequences $\mathcal{F}^{\leq\frac{1}{2}}(\mathbb{B}(n), m)$, $\mathcal{F}^{\geq\frac{1}{2}}(\mathbb{B}(n), m)$ and their duals; $n \neq 2m$

The Farey duality results in Theorem 2.1 applied to the half sequences of the sequence $\mathcal{F}(\mathbb{B}(n), m)$, with $n \neq 2m$, can be refined as follows.

Corollary 2.8. *Consider a Farey subsequence $\mathcal{F}(\mathbb{B}(n), m)$, where $n \neq 2m$.*

(i) *Suppose that $n < 2m$. Then the maps*

$$\mathcal{F}^{\leq\frac{1}{2}}(\mathbb{B}(n), m) \to \mathcal{F}_{n-m}, \quad \tfrac{h}{k} \mapsto \tfrac{h}{k-h}, \quad \left[\begin{smallmatrix} h \\ k \end{smallmatrix}\right] \mapsto \left[\begin{smallmatrix} 1 & 0 \\ -1 & 1 \end{smallmatrix}\right] \cdot \left[\begin{smallmatrix} h \\ k \end{smallmatrix}\right], \tag{2.20}$$

$$\mathcal{F}_{n-m} \to \mathcal{F}^{\leq\frac{1}{2}}(\mathbb{B}(n), m), \quad \tfrac{h}{k} \mapsto \tfrac{h}{h+k}, \quad \left[\begin{smallmatrix} h \\ k \end{smallmatrix}\right] \mapsto \left[\begin{smallmatrix} 1 & 0 \\ 1 & 1 \end{smallmatrix}\right] \cdot \left[\begin{smallmatrix} h \\ k \end{smallmatrix}\right], \tag{2.21}$$

$$\mathcal{F}^{\geq\frac{1}{2}}(\mathbb{B}(n), m) \to \mathcal{G}_m^{2m-n}, \quad \tfrac{h}{k} \mapsto \tfrac{2h-k}{h}, \quad \left[\begin{smallmatrix} h \\ k \end{smallmatrix}\right] \mapsto \left[\begin{smallmatrix} 2 & -1 \\ 1 & 0 \end{smallmatrix}\right] \cdot \left[\begin{smallmatrix} h \\ k \end{smallmatrix}\right],$$

$$\mathcal{G}_m^{2m-n} \to \mathcal{F}^{\geq\frac{1}{2}}(\mathbb{B}(n), m), \quad \tfrac{h}{k} \mapsto \tfrac{k}{2k-h}, \quad \left[\begin{smallmatrix} h \\ k \end{smallmatrix}\right] \mapsto \left[\begin{smallmatrix} 0 & 1 \\ -1 & 2 \end{smallmatrix}\right] \cdot \left[\begin{smallmatrix} h \\ k \end{smallmatrix}\right],$$

are order-preserving and bijective, and the maps

$$\mathcal{F}^{\leq\frac{1}{2}}(\mathbb{B}(n),m) \to \mathcal{F}_{n-m}, \quad \tfrac{h}{k} \mapsto \tfrac{k-2h}{k-h}, \quad \left[\begin{smallmatrix} h \\ k \end{smallmatrix}\right] \mapsto \left[\begin{smallmatrix} -2 & 1 \\ -1 & 1 \end{smallmatrix}\right] \cdot \left[\begin{smallmatrix} h \\ k \end{smallmatrix}\right],$$

$$\mathcal{F}_{n-m} \to \mathcal{F}^{\leq\frac{1}{2}}(\mathbb{B}(n),m), \quad \tfrac{h}{k} \mapsto \tfrac{k-h}{2k-h}, \quad \left[\begin{smallmatrix} h \\ k \end{smallmatrix}\right] \mapsto \left[\begin{smallmatrix} -1 & 1 \\ -1 & 2 \end{smallmatrix}\right] \cdot \left[\begin{smallmatrix} h \\ k \end{smallmatrix}\right], \tag{2.22}$$

$$\mathcal{F}^{\geq\frac{1}{2}}(\mathbb{B}(n),m) \to \mathcal{F}_m^{n-m}, \quad \tfrac{h}{k} \mapsto \tfrac{k-h}{h}, \quad \left[\begin{smallmatrix} h \\ k \end{smallmatrix}\right] \mapsto \left[\begin{smallmatrix} -1 & 1 \\ 1 & 0 \end{smallmatrix}\right] \cdot \left[\begin{smallmatrix} h \\ k \end{smallmatrix}\right], \tag{2.23}$$

$$\mathcal{F}_m^{n-m} \to \mathcal{F}^{\geq\frac{1}{2}}(\mathbb{B}(n),m), \quad \tfrac{h}{k} \mapsto \tfrac{k}{h+k}, \quad \left[\begin{smallmatrix} h \\ k \end{smallmatrix}\right] \mapsto \left[\begin{smallmatrix} 0 & 1 \\ 1 & 1 \end{smallmatrix}\right] \cdot \left[\begin{smallmatrix} h \\ k \end{smallmatrix}\right], \tag{2.24}$$

are order-reversing and bijective.

(ii) *Suppose that $n > 2m$. Then the maps*

$$\mathcal{F}^{\leq\frac{1}{2}}(\mathbb{B}(n),m) \to \mathcal{F}_{n-m}^{m}, \quad \tfrac{h}{k} \mapsto \tfrac{h}{k-h}, \quad \left[\begin{smallmatrix} h \\ k \end{smallmatrix}\right] \mapsto \left[\begin{smallmatrix} 1 & 0 \\ -1 & 1 \end{smallmatrix}\right] \cdot \left[\begin{smallmatrix} h \\ k \end{smallmatrix}\right], \tag{2.25}$$

$$\mathcal{F}_{n-m}^{m} \to \mathcal{F}^{\leq\frac{1}{2}}(\mathbb{B}(n),m), \quad \tfrac{h}{k} \mapsto \tfrac{h}{h+k}, \quad \left[\begin{smallmatrix} h \\ k \end{smallmatrix}\right] \mapsto \left[\begin{smallmatrix} 1 & 0 \\ 1 & 1 \end{smallmatrix}\right] \cdot \left[\begin{smallmatrix} h \\ k \end{smallmatrix}\right], \tag{2.26}$$

$$\mathcal{F}^{\geq\frac{1}{2}}(\mathbb{B}(n),m) \to \mathcal{F}_m, \quad \tfrac{h}{k} \mapsto \tfrac{2h-k}{h}, \quad \left[\begin{smallmatrix} h \\ k \end{smallmatrix}\right] \mapsto \left[\begin{smallmatrix} 2 & -1 \\ 1 & 0 \end{smallmatrix}\right] \cdot \left[\begin{smallmatrix} h \\ k \end{smallmatrix}\right], \tag{2.27}$$

$$\mathcal{F}_m \to \mathcal{F}^{\geq\frac{1}{2}}(\mathbb{B}(n),m), \quad \tfrac{h}{k} \mapsto \tfrac{k}{2k-h}, \quad \left[\begin{smallmatrix} h \\ k \end{smallmatrix}\right] \mapsto \left[\begin{smallmatrix} 0 & 1 \\ -1 & 2 \end{smallmatrix}\right] \cdot \left[\begin{smallmatrix} h \\ k \end{smallmatrix}\right], \tag{2.28}$$

are order-preserving and bijective, and the maps

$$\mathcal{F}^{\leq\frac{1}{2}}(\mathbb{B}(n),m) \to \mathcal{G}_{n-m}^{n-2m}, \quad \tfrac{h}{k} \mapsto \tfrac{k-2h}{k-h}, \quad \left[\begin{smallmatrix} h \\ k \end{smallmatrix}\right] \mapsto \left[\begin{smallmatrix} -2 & 1 \\ -1 & 1 \end{smallmatrix}\right] \cdot \left[\begin{smallmatrix} h \\ k \end{smallmatrix}\right],$$

$$\mathcal{G}_{n-m}^{n-2m} \to \mathcal{F}^{\leq\frac{1}{2}}(\mathbb{B}(n),m), \quad \tfrac{h}{k} \mapsto \tfrac{k-h}{2k-h}, \quad \left[\begin{smallmatrix} h \\ k \end{smallmatrix}\right] \mapsto \left[\begin{smallmatrix} -1 & 1 \\ -1 & 2 \end{smallmatrix}\right] \cdot \left[\begin{smallmatrix} h \\ k \end{smallmatrix}\right],$$

$$\mathcal{F}^{\geq\frac{1}{2}}(\mathbb{B}(n),m) \to \mathcal{F}_m, \quad \tfrac{h}{k} \mapsto \tfrac{k-h}{h}, \quad \left[\begin{smallmatrix} h \\ k \end{smallmatrix}\right] \mapsto \left[\begin{smallmatrix} -1 & 1 \\ 1 & 0 \end{smallmatrix}\right] \cdot \left[\begin{smallmatrix} h \\ k \end{smallmatrix}\right], \tag{2.29}$$

$$\mathcal{F}_m \to \mathcal{F}^{\geq\frac{1}{2}}(\mathbb{B}(n),m), \quad \tfrac{h}{k} \mapsto \tfrac{k}{h+k}, \quad \left[\begin{smallmatrix} h \\ k \end{smallmatrix}\right] \mapsto \left[\begin{smallmatrix} 0 & 1 \\ 1 & 1 \end{smallmatrix}\right] \cdot \left[\begin{smallmatrix} h \\ k \end{smallmatrix}\right], \tag{2.30}$$

are order-reversing and bijective.

2.2 Pairs of neighboring fractions. II

Applications require a variety of easily describable pairs of neighboring fractions within Farey (sub)sequences which per se are interesting and can be involved in the efficient recurrent computation of other fractions.

In this section we first revisit the results of Section 1.4, which were surveyed in Tables 1.3, 1.4 and 1.5. By imposing additional restrictions on auxiliary integer quantities that appeared in those computations, we arrive in Sections 2.2.1, 2.2.3 and 2.2.4 at further results on pairs of neighboring fractions in the Farey (sub)sequences $\mathcal{F}_m$, $\mathcal{F}_n^m$ and $\mathcal{G}_n^m$, in perhaps their simplest form.

The Farey duality, which is the central topic of Chapter 2, will be applied to the above-mentioned preparatory results in Sections 2.2.2 and 2.2.5; see Tables 2.1 and 2.2 for formulas.

The remainder of the section will be concerned with numerous selected pairs of neighboring fractions.

2.2.1 Neighboring fractions in $\mathcal{F}_m$

In this subsection we pursue the investigation of pairs of neighboring fractions in the Farey sequences initiated in Section 1.4.1.

Lemma 2.9. (i) *Let $\frac{h}{k} \in \mathcal{F}_m - \{\frac{0}{1}\}$. Let a and b be two integers such that*

$$ka \equiv -1 \pmod{h}, \quad \left\lceil \frac{hm}{k} \right\rceil - h \le a \le \left\lceil \frac{hm}{k} \right\rceil - 1, \tag{2.31}$$

$$hb \equiv 1 \pmod{k}, \quad m - k + 1 \le b \le m. \tag{2.32}$$

Then the fraction $a/\frac{ka+1}{h} = \frac{hb-1}{k}/b$ precedes $\frac{h}{k}$ in $\mathcal{F}_m$.

(ii) *Let $\frac{h}{k} \in \mathcal{F}_m - \{\frac{0}{1}, \frac{1}{1}\}$. Let a and b be two integers such that*

$$ka \equiv 1 \pmod{h}, \quad \left\lceil \frac{hm+2}{k} \right\rceil - h \le a \le \left\lceil \frac{hm+2}{k} \right\rceil - 1, \tag{2.33}$$

$$hb \equiv -1 \pmod{k}, \quad m - k + 1 \le b \le m.$$

Then the fraction $a/\frac{ka-1}{h} = \frac{hb+1}{k}/b$ succeeds $\frac{h}{k}$ in $\mathcal{F}_m$.

Proof. The assertions of the lemma follow directly from the respective assertions of Proposition 1.7. For instance, it suffices to note that the multiplier $\lfloor \frac{hn-kx_0-1}{hk} \rfloor$ in (1.11) vanishes whenever $\lceil \frac{hn}{k} \rceil - h \le x_0 \le \lceil \frac{hn}{k} \rceil - 1$, and so on. □

Remark 2.10. Lemma 2.9 implies that

$$\frac{\lfloor (m-1)/2 \rfloor}{2\lfloor (m-1)/2 \rfloor + 1} = \frac{\lceil m/2 \rceil - 1}{2\lceil m/2 \rceil - 1}$$

precedes $\frac{1}{2}$ in $\mathcal{F}_m$, and

$$\frac{\lceil m/2 \rceil}{2\lceil m/2 \rceil - 1} = \frac{\lfloor (m-1)/2 \rfloor + 1}{2\lfloor (m-1)/2 \rfloor + 1}$$

succeeds $\frac{1}{2}$ in $\mathcal{F}_m$.

Thus, if m is even, then the fractions

$$\frac{(m-2)/2}{m-1} < \frac{1}{2} < \frac{m/2}{m-1}$$

are consecutive in $\mathcal{F}_m$.

If m is odd, then the fractions

$$\frac{(m-1)/2}{m} < \frac{1}{2} < \frac{(m+1)/2}{m}$$

are consecutive in $\mathcal{F}_m$. See also Table 2.5.

2.2.2 Neighboring fractions in $\mathcal{F}(\mathbb{B}(2m), m)$

In this subsection we adapt, via duality, the results obtained in the previous subsection to the Farey subsequence $\mathcal{F}(\mathbb{B}(2m), m)$.

Table 2.1. The fractions of the form $a/\frac{ka+1}{h}$ that precede fractions $\frac{h}{k}$ in Farey (sub)sequences, derived from the expressions listed in Table 1.3.

$a/\frac{ka+1}{h}$ **precedes** $\frac{h}{k}$ **Sequence and restrictions on** a**, with** $ka \equiv -1 \pmod h$		**See**
$\mathcal{F}_m - \{\frac{0}{1}\}$		
$\lceil \frac{hm}{k} \rceil - h \le a \le \lceil \frac{hm}{k} \rceil - 1$		Lemma 2.9 (i)
$\mathcal{F}_m^{\ell} - \{\frac{0}{1}\}$, $0 < \ell < m$		
$\ell - h + 1 \le a \le \ell$	if $hm - k\ell \ge 1$	Lemma 2.13 (i) (a)
$\lceil \frac{hm}{k} \rceil - h \le a \le \lceil \frac{hm}{k} \rceil - 1$	if $hm - k\ell \le 1$	Lemma 2.13 (i) (b)
$\mathcal{G}_m^{\ell} - \{\frac{0}{1}\}$, $0 < \ell < m$		
$\lceil \frac{hm}{k} \rceil - h \le a \le \lceil \frac{hm}{k} \rceil - 1$	if $hm - k\ell \ge 1$	Lemma 2.15 (i) (a)
$\lceil \frac{h(m-\ell)}{k-h} \rceil - h \le a \le \lceil \frac{h(m-\ell)}{k-h} \rceil - 1$	if $hm - k\ell \le 1$	Lemma 2.15 (i) (b)
$\mathcal{F}^{\le\frac{1}{2}}(\mathbb{B}(2m), m) - \{\frac{0}{1}, \frac{1}{2}\}$		
$\lceil \frac{hm}{k-h} \rceil - h \le a \le \lceil \frac{hm}{k-h} \rceil - 1$		Proposition 2.12 (i) (a)
$\mathcal{F}^{\ge\frac{1}{2}}(\mathbb{B}(2m), m) - \{\frac{1}{2}, \frac{1}{1}\}$		
$m - h + 1 \le a \le m$		Proposition 2.12 (ii) (a)
$\mathcal{F}^{\le\frac{1}{2}}(\mathbb{B}(n), m) - \{\frac{0}{1}, \frac{1}{2}\}$, where $n < 2m$		
$\lceil \frac{h(n-m)}{k-h} \rceil - h \le a \le \lceil \frac{h(n-m)}{k-h} \rceil - 1$		Proposition 2.18 (i) (a)
$\mathcal{F}^{\ge\frac{1}{2}}(\mathbb{B}(n), m) - \{\frac{1}{2}, \frac{1}{1}\}$, where $n < 2m$		
$m - h + 1 \le a \le m$	if $hn - km \ge 1$	Proposition 2.18 (ii) (a)
$\lceil \frac{h(n-m)}{k-h} \rceil - h \le a \le \lceil \frac{h(n-m)}{k-h} \rceil - 1$	if $hn - km \le 1$	Proposition 2.18 (ii) (a)
$\mathcal{F}^{\le\frac{1}{2}}(\mathbb{B}(n), m) - \{\frac{0}{1}, \frac{1}{2}\}$, where $n > 2m$		
$m - h + 1 \le a \le m$	if $hn - km \ge 1$	Proposition 2.19 (i) (a)
$\lceil \frac{h(n-m)}{k-h} \rceil - h \le a \le \lceil \frac{h(n-m)}{k-h} \rceil - 1$	if $hn - km \le 1$	Proposition 2.19 (i) (a)
$\mathcal{F}^{\ge\frac{1}{2}}(\mathbb{B}(n), m) - \{\frac{1}{2}, \frac{1}{1}\}$, where $n > 2m$		
$m - h + 1 \le a \le m$		Proposition 2.19 (ii) (a)

Recall that $\frac{m}{m+1}$ precedes $\frac{1}{1}$, and $\frac{1}{m+1}$ succeeds $\frac{0}{1}$ in $\mathcal{F}(\mathbb{B}(2m), m)$, see Remark 1.17 and Table 1.5.

Remark 2.11. Since $\frac{m-1}{m}$ precedes $\frac{1}{1}$, and $\frac{1}{m}$ succeeds $\frac{0}{1}$ in $\mathcal{F}_m$, the order-preserving bijections given in (2.13) and (2.15) guarantee that $\frac{m-1}{2m-1}$ precedes $\frac{1}{2}$, and $\frac{m}{2m-1}$ succeeds $\frac{1}{2}$ in $\mathcal{F}(\mathbb{B}(2m), m)$. In other words, the three fractions

$$\frac{m-1}{2m-1} < \frac{1}{2} < \frac{m}{2m-1}$$

are consecutive in $\mathcal{F}(\mathbb{B}(2m), m)$. See also Table 2.5.

Proposition 2.12. (i) *Consider a fraction* $\frac{h}{k} \in \mathcal{F}^{\leq\frac{1}{2}}(\mathbb{B}(2m), m) - \{\frac{0}{1}, \frac{1}{2}\}$.

(a) *Let* a *and* b *be two integers such that*

$$ka \equiv -1 \pmod{h}, \qquad \left\lceil \frac{hm}{k-h} \right\rceil - h \leq a \leq \left\lceil \frac{hm}{k-h} \right\rceil - 1,$$
$$hb \equiv 1 \pmod{(k-h)}, \quad m - k + h + 1 \leq b \leq m.$$

Then the fraction $a / \frac{ka+1}{h} = \frac{hb-1}{k-h} / \frac{kb-1}{k-h}$ *precedes* $\frac{h}{k}$ *in* $\mathcal{F}(\mathbb{B}(2m), m)$.

(b) *Let* a *and* b *be two integers such that*

$$ka \equiv 1 \pmod{h}, \qquad \left\lceil \frac{hm+2}{k-h} \right\rceil - h \leq a \leq \left\lceil \frac{hm+2}{k-h} \right\rceil - 1,$$
$$hb \equiv -1 \pmod{(k-h)}, \quad m - k + h + 1 \leq b \leq m.$$

Then the fraction $a / \frac{ka-1}{h} = \frac{hb+1}{k-h} / \frac{kb+1}{k-h}$ *succeeds* $\frac{h}{k}$ *in* $\mathcal{F}(\mathbb{B}(2m), m)$.

(ii) *Consider a fraction* $\frac{h}{k} \in \mathcal{F}^{\geq\frac{1}{2}}(\mathbb{B}(2m), m) - \{\frac{1}{2}, \frac{1}{1}\}$.

(a) *Let* a *and* b *be two integers such that*

$$ka \equiv -1 \pmod{h}, \qquad m - h + 1 \leq a \leq m,$$
$$hb \equiv 1 \pmod{(k-h)}, \quad \left\lceil \frac{km+2}{h} \right\rceil - m - k + h \leq b \leq \left\lceil \frac{km+2}{h} \right\rceil - m - 1.$$

Then the fraction $a / \frac{ka+1}{h} = \frac{hb-1}{k-h} / \frac{kb-1}{k-h}$ *precedes* $\frac{h}{k}$ *in* $\mathcal{F}(\mathbb{B}(2m), m)$.

(b) *Let* a *and* b *be two integers such that*

$$ka \equiv 1 \pmod{h}, \qquad m - h + 1 \leq a \leq m,$$
$$hb \equiv -1 \pmod{(k-h)}, \quad \left\lceil \frac{km}{h} \right\rceil - m - k + h \leq b \leq \left\lceil \frac{km}{h} \right\rceil - m - 1.$$

Then the fraction $a / \frac{ka-1}{h} = \frac{hb+1}{k-h} / \frac{kb+1}{k-h}$ *succeeds* $\frac{h}{k}$ *in* $\mathcal{F}(\mathbb{B}(2m), m)$.

Proof. To prove assertion (i) (a), first note that $\frac{h}{k-h} \in \mathcal{F}_m$ is the image of $\frac{h}{k}$ under the order-preserving bijection (2.12); then use Lemma 2.9 (i) to find the predecessor

$$a \Big/ \frac{(k-h)a+1}{h} = \frac{hb-1}{k-h} \Big/ b \tag{2.34}$$

of $\frac{h}{k-h}$ in the sequence $\mathcal{F}_m$, where a and b are integers such that $(k-h)a \equiv -1 \pmod{h}$ or, equivalently, $ka \equiv -1 \pmod{h}$, under the restriction

$$\left\lceil \frac{hm}{k-h} \right\rceil - h \leq a \leq \left\lceil \frac{hm}{k-h} \right\rceil - 1,$$

and

$$hb \equiv 1 \pmod{(k-h)}, \quad m - (k-h) + 1 \leq b \leq m.$$

Finally, send this fraction (2.34) back to the sequence $\mathcal{F}(\mathbb{B}(2m), m)$ by means of the order-preserving bijection (2.13).

Assertion (i) (b) is proved in a similar way, by the application of Lemma 2.9 (ii), with the help of the order-preserving bijections (2.12) and (2.13).

Table 2.2. The fractions of the form $\frac{hb-1}{k}/b$ and $\frac{hb-1}{k-h}/\frac{kb-1}{k-h}$ that precede fractions $\frac{h}{k}$ in Farey (sub)sequences, derived from the expressions listed in Table 1.3.

$\frac{hb-1}{k}/b$ **precedes** $\frac{h}{k}$ **Sequence and restrictions on** b**, with** $hb \equiv 1 \pmod{k}$		**See**
$\mathcal{F}_m - \{\frac{0}{1}\}$		
$m-k+1 \le b \le m$		Lemma 2.9 (i)
$\mathcal{F}_m^{\ell} - \{\frac{0}{1}\}, 0 < \ell < m$		
$\lceil\frac{k\ell+2}{h}\rceil - k \le b \le \lceil\frac{k\ell+2}{h}\rceil - 1$	if $hm - k\ell \ge 1$	Lemma 2.13 (i) (a)
$m-k+1 \le b \le m$	if $hm - k\ell \le 1$	Lemma 2.13 (i) (b)
$\mathcal{G}_m^{\ell} - \{\frac{0}{1}\}, 0 < \ell < m$		
$m-k+1 \le b \le m$	if $hm - k\ell \ge 1$	Lemma 2.15 (i) (a)
$\lceil\frac{k(m-\ell)}{k-h}\rceil - k \le b \le \lceil\frac{k(m-\ell)}{k-h}\rceil - 1$	if $hm - k\ell \le 1$	Lemma 2.15 (i) (b)

$\frac{hb-1}{k-h}/\frac{kb-1}{k-h}$ **precedes** $\frac{h}{k}$ **Sequence and restrictions on** b**, with** $hb \equiv 1 \pmod{(k-h)}$		**See**
$\mathcal{F}^{\le\frac{1}{2}}(\mathbb{B}(2m), m) - \{\frac{0}{1}, \frac{1}{2}\}$		
$m-k+h+1 \le b \le m$		Proposition 2.12 (i) (a)
$\mathcal{F}^{\ge\frac{1}{2}}(\mathbb{B}(2m), m) - \{\frac{1}{2}, \frac{1}{1}\}$		
$\lceil\frac{km+2}{h}\rceil - m - k + h \le b \le \lceil\frac{km+2}{h}\rceil - m - 1$		Proposition 2.12 (ii) (a)
$\mathcal{F}^{\le\frac{1}{2}}(\mathbb{B}(n), m) - \{\frac{0}{1}, \frac{1}{2}\}$, where $n < 2m$		
$n-m-k+h+1 \le b \le n-m$		Proposition 2.18 (i) (a)
$\mathcal{F}^{\ge\frac{1}{2}}(\mathbb{B}(n), m) - \{\frac{1}{2}, \frac{1}{1}\}$, where $n < 2m$		
$\lceil\frac{km+2}{h}\rceil - m - k + h \le b \le \lceil\frac{km+2}{h}\rceil - m - 1$	if $hn - km \ge 1$	Proposition 2.18 (ii) (a)
$n-m-k+h+1 \le b \le n-m$	if $hn - km \le 1$	Proposition 2.18 (ii) (a)
$\mathcal{F}^{\le\frac{1}{2}}(\mathbb{B}(n), m) - \{\frac{0}{1}, \frac{1}{2}\}$, where $n > 2m$		
$\lceil\frac{km+2}{h}\rceil - m - k + h \le b \le \lceil\frac{km+2}{h}\rceil - m - 1$	if $hn - km \ge 1$	Proposition 2.19 (i) (a)
$n-m-k+h+1 \le b \le n-m$	if $hn - km \le 1$	Proposition 2.19 (i) (a)
$\mathcal{F}^{\ge\frac{1}{2}}(\mathbb{B}(n), m) - \{\frac{1}{2}, \frac{1}{1}\}$, where $n > 2m$		
$\lceil\frac{km+2}{h}\rceil - m - k + h \le b \le \lceil\frac{km+2}{h}\rceil - m - 1$		Proposition 2.19 (ii) (a)

To prove assertion (ii) (a), first note that $\frac{k-h}{h} \in \mathcal{F}_m$ is the image of $\frac{h}{k}$ under the order-reversing bijection (2.18). Then use Lemma 2.9 (ii) to find the successor of $\frac{k-h}{h}$ in $\mathcal{F}_m$, and send it to $\mathcal{F}(\mathbb{B}(2m), m)$ by means of the order-reversing bijection (2.19).

Assertion (ii) (b) is proved in a similar way, by the application of Lemma 2.9 (i), with the help of the order-reversing bijections (2.18) and (2.19). □

2.2.3 Neighboring fractions in $\mathcal{F}_m^\ell$

In order to derive the simplest formulas for finding the neighbors of fractions in the Farey subsequence $\mathcal{F}_m^\ell$, we will now simplify Corollary 1.11. Recall that the same approach was used in Lemma 2.9, derived from Proposition 1.7. See also Lemma 2.15, where similar results are obtained for the Farey subsequence $\mathcal{G}_m^\ell$.

Lemma 2.13. (i) *Let* $\frac{h}{k} \in \mathcal{F}_m^\ell - \{\frac{0}{1}\}$, *where* $0 < \ell < m$.

(a) *Suppose that* $hm - k\ell \geq 1$. *Let* a *and* b *be two integers such that*

$$ka \equiv -1 \pmod{h}, \quad \ell - h + 1 \leq a \leq \ell,$$

$$hb \equiv 1 \pmod{k}, \quad \left\lceil \frac{k\ell + 2}{h} \right\rceil - k \leq b \leq \left\lceil \frac{k\ell + 2}{h} \right\rceil - 1.$$

Then the fraction $a / \frac{ka+1}{h} = \frac{hb-1}{k} / b$ *precedes the fraction* $\frac{h}{k}$ *in* $\mathcal{F}_m^\ell$.

(b) *Suppose that* $hm - k\ell \leq 1$. *Let* a *and* b *be two integers such that*

$$ka \equiv -1 \pmod{h}, \quad \left\lceil \frac{hm}{k} \right\rceil - h \leq a \leq \left\lceil \frac{hm}{k} \right\rceil - 1,$$

$$hb \equiv 1 \pmod{k}, \quad m - k + 1 \leq b \leq m.$$

Then the fraction $a / \frac{ka+1}{h} = \frac{hb-1}{k} / b$ *precedes the fraction* $\frac{h}{k}$ *in* $\mathcal{F}_m^\ell$.

(ii) *Let* $\frac{h}{k} \in \mathcal{F}_m^\ell - \{\frac{0}{1}, \frac{1}{1}\}$, *where* $0 < \ell < m$.

(a) *Suppose that* $k\ell - hm \geq 1$. *Let* a *and* b *be two integers such that*

$$ka \equiv 1 \pmod{h}, \quad \left\lceil \frac{hm + 2}{k} \right\rceil - h \leq a \leq \left\lceil \frac{hm + 2}{k} \right\rceil - 1,$$

$$hb \equiv -1 \pmod{k}, \quad m - k + 1 \leq b \leq m.$$

Then the fraction $a / \frac{ka-1}{h} = \frac{hb+1}{k} / b$ *succeeds the fraction* $\frac{h}{k}$ *in* $\mathcal{F}_m^\ell$.

(b) *Suppose that* $k\ell - hm \leq 1$. *Let* a *and* b *be two integers such that*

$$ka \equiv 1 \pmod{h}, \quad \ell - h + 1 \leq a \leq \ell,$$

$$hb \equiv -1 \pmod{k}, \quad \left\lceil \frac{k\ell}{h} \right\rceil - k \leq b \leq \left\lceil \frac{k\ell}{h} \right\rceil - 1.$$

Then the fraction $a / \frac{ka-1}{h} = \frac{hb+1}{k} / b$ *succeeds the fraction* $\frac{h}{k}$ *in* $\mathcal{F}_m^\ell$.

Proof. The assertions of the lemma follow from the respective assertions of Corollary 1.11.

Our goal is to turn the multipliers of the form $\lfloor \cdot \rfloor$ that appear in (1.30)–(1.33) of Corollary 1.11 into zero by pointing out at suitable search intervals for the integers x_0 and y_0.

For instance, in order to verify assertion (i) (a), note that the multipliers $\lfloor \frac{m-x_0}{h} \rfloor$ and $\lfloor \frac{km-hy_0+1}{hk} \rfloor$ in (1.30) vanish whenever

$$m - h + 1 \leq x_0 \leq m \quad \text{and} \quad \left\lceil \frac{km + 2}{h} \right\rceil - k \leq y_0 \leq \left\lceil \frac{km + 2}{h} \right\rceil - 1. \qquad \square$$

Table 2.3. The fractions of the form $a/\frac{ka-1}{h}$ that succeed fractions $\frac{h}{k}$ in Farey (sub)sequences, derived from the expressions listed in Table 1.4.

$a/\frac{ka-1}{h}$ succeeds $\frac{h}{k}$ Sequence and restrictions on a, with $ka \equiv 1 \pmod{h}$		See
$\mathcal{F}_m - \{\frac{0}{1}, \frac{1}{1}\}$		
$\lceil\frac{hm+2}{k}\rceil - h \le a \le \lceil\frac{hm+2}{k}\rceil - 1$		Lemma 2.9 (ii)
$\mathcal{F}_m^\ell - \{\frac{0}{1}, \frac{1}{1}\}, 0 < \ell < m$		
$\lceil\frac{hm+2}{k}\rceil - h \le a \le \lceil\frac{hm+2}{k}\rceil - 1$	if $k\ell - hm \ge 1$	Lemma 2.13 (ii) (a)
$\ell - h + 1 \le a \le \ell$	if $k\ell - hm \le 1$	Lemma 2.13 (ii) (b)
$\mathcal{G}_m^\ell - \{\frac{0}{1}, \frac{1}{1}\}, 0 < \ell < m$		
$\lceil\frac{h(m-\ell)+2}{k-h}\rceil - h \le a \le \lceil\frac{h(m-\ell)+2}{k-h}\rceil - 1$	if $k\ell - hm \ge 1$	Lemma 2.15 (ii) (a)
$\lceil\frac{hm+2}{k}\rceil - h \le a \le \lceil\frac{hm+2}{k}\rceil - 1$	if $k\ell - hm \le 1$	Lemma 2.15 (ii) (b)
$\mathcal{F}^{\le\frac{1}{2}}(\mathbb{B}(2m), m) - \{\frac{0}{1}, \frac{1}{2}\}$		
$\lceil\frac{hm+2}{k-h}\rceil - h \le a \le \lceil\frac{hm+2}{k-h}\rceil - 1$		Proposition 2.12 (i) (b)
$\mathcal{F}^{\ge\frac{1}{2}}(\mathbb{B}(2m), m) - \{\frac{1}{2}, \frac{1}{1}\}$		
$m - h + 1 \le a \le m$		Proposition 2.12 (ii) (b)
$\mathcal{F}^{\le\frac{1}{2}}(\mathbb{B}(n), m) - \{\frac{0}{1}, \frac{1}{2}\}$, where $n < 2m$		
$\lceil\frac{h(n-m)+2}{k-h}\rceil - h \le a \le \lceil\frac{h(n-m)+2}{k-h}\rceil - 1$		Proposition 2.18 (i) (b)
$\mathcal{F}^{\ge\frac{1}{2}}(\mathbb{B}(n), m) - \{\frac{1}{2}, \frac{1}{1}\}$, where $n < 2m$		
$\lceil\frac{h(n-m)+2}{k-h}\rceil - h \le a \le \lceil\frac{h(n-m)+2}{k-h}\rceil - 1$	if $km - hn \ge 1$	Proposition 2.18 (ii) (b)
$m - h + 1 \le a \le m$	if $km - hn \le 1$	Proposition 2.18 (ii) (b)
$\mathcal{F}^{\le\frac{1}{2}}(\mathbb{B}(n), m) - \{\frac{0}{1}, \frac{1}{2}\}$, where $n > 2m$		
$\lceil\frac{h(n-m)+2}{k-h}\rceil - h \le a \le \lceil\frac{h(n-m)+2}{k-h}\rceil - 1$	if $km - hn \ge 1$	Proposition 2.19 (i) (b)
$m - h + 1 \le a \le m$	if $km - hn \le 1$	Proposition 2.19 (i) (b)
$\mathcal{F}^{\ge\frac{1}{2}}(\mathbb{B}(n), m) - \{\frac{1}{2}, \frac{1}{1}\}$, where $n > 2m$		
$m - h + 1 \le a \le m$		Proposition 2.19 (ii) (b)

Remark 2.14 (see also Table 2.5). If the parameters m and ℓ of a Farey subsequence $\mathcal{F}_m^\ell$ satisfy the relation $m - 2\ell \ge 1$, then Lemma 2.13 (i) (a) guarantees that $\frac{\ell}{2\ell+1}$ precedes $\frac{1}{2}$ in $\mathcal{F}_m^\ell$. If $m - 2\ell \le 1$, then, according to Lemma 2.13 (i) (b),

$$\frac{\lfloor(m-1)/2\rfloor}{2\lfloor(m-1)/2\rfloor+1} = \frac{\lceil m/2\rceil - 1}{2\lceil m/2\rceil - 1}$$

precedes $\frac{1}{2}$ in $\mathcal{F}_m^\ell$.

Thus, if $m - 2\ell \le 1$ and m is even, then $\frac{(m-2)/2}{m-1}$ precedes $\frac{1}{2}$ in $\mathcal{F}_m^\ell$. If $m - 2\ell \le 1$ and m is odd, then $\frac{(m-1)/2}{m}$ precedes $\frac{1}{2}$ in $\mathcal{F}_m^\ell$.

If $2\ell - m \geq 1$, then Lemma 2.13 (ii) (a) guarantees that

$$\frac{\lceil m/2 \rceil}{2\lceil m/2 \rceil - 1} = \frac{\lfloor (m-1)/2 \rfloor + 1}{2\lfloor (m-1)/2 \rfloor + 1}$$

succeeds $\frac{1}{2}$ in $\mathcal{F}_m^\ell$.

Thus, if $2\ell - m \geq 1$ and m is even, then $\frac{m/2}{m-1}$ succeeds $\frac{1}{2}$ in $\mathcal{F}_m^\ell$. If $2\ell - m \geq 1$ and m is odd, then $\frac{(m+1)/2}{m}$ succeeds $\frac{1}{2}$ in $\mathcal{F}_m^\ell$.

If $2\ell - m \leq 1$, then, according to Lemma 2.13 (ii) (b), $\frac{\ell}{2\ell-1}$ succeeds $\frac{1}{2}$ in $\mathcal{F}_m^\ell$.

2.2.4 Neighboring fractions in $\mathcal{G}_m^\ell$

In the same way that we obtained in Lemma 2.13 the simplest formulas for finding the neighbors of fractions in the Farey subsequence $\mathcal{F}_m^\ell$, we will now simplify Corollary 1.15 in order to obtain the similar results for the sequence $\mathcal{G}_m^\ell$. The same approach was used in Lemma 2.9, derived from Proposition 1.7.

Lemma 2.15. (i) *Let* $\frac{h}{k} \in \mathcal{G}_m^\ell - \{\frac{0}{1}, \frac{1}{1}\}$, *where* $0 < \ell < m$.

(a) *Suppose that* $hm - k\ell \geq 1$. *Let* a *and* b *be two integers such that*

$$ka \equiv -1 \pmod{h}, \quad \left\lceil \frac{hm}{k} \right\rceil - h \leq a \leq \left\lceil \frac{hm}{k} \right\rceil - 1,$$
$$hb \equiv 1 \pmod{k}, \quad m - k + 1 \leq b \leq m.$$

Then the fraction $a/\frac{ka+1}{h} = \frac{hb-1}{k}/b$ *precedes the fraction* $\frac{h}{k}$ *in* $\mathcal{G}_m^\ell$.

(b) *Suppose that* $hm - k\ell \leq 1$. *Let* a *and* b *be two integers such that*

$$ka \equiv -1 \pmod{h}, \quad \left\lceil \frac{h(m-\ell)}{k-h} \right\rceil - h \leq a \leq \left\lceil \frac{h(m-\ell)}{k-h} \right\rceil - 1,$$
$$hb \equiv 1 \pmod{k}, \quad \left\lceil \frac{k(m-\ell)}{k-h} \right\rceil - k \leq b \leq \left\lceil \frac{k(m-\ell)}{k-h} \right\rceil - 1.$$

Then the fraction $a/\frac{ka+1}{h} = \frac{hb-1}{k}/b$ *precedes the fraction* $\frac{h}{k}$ *in* $\mathcal{G}_m^\ell$.

(ii) *Let* $\frac{h}{k} \in \mathcal{G}_m^\ell - \{\frac{0}{1}, \frac{1}{1}\}$, *where* $0 < \ell < m$.

(a) *Suppose that* $k\ell - hm \geq 1$. *Let* a *and* b *be two integers such that*

$$ka \equiv 1 \pmod{h}, \quad \left\lceil \frac{h(m-\ell)+2}{k-h} \right\rceil - h \leq a \leq \left\lceil \frac{h(m-\ell)+2}{k-h} \right\rceil - 1,$$
$$hb \equiv -1 \pmod{k}, \quad \left\lceil \frac{k(m-\ell)+2}{k-h} \right\rceil - k \leq b \leq \left\lceil \frac{k(m-\ell)+2}{k-h} \right\rceil - 1.$$

Then the fraction $a/\frac{ka-1}{h} = \frac{hb+1}{k}/b$ *succeeds the fraction* $\frac{h}{k}$ *in* $\mathcal{G}_m^\ell$.

(b) *Suppose that* $k\ell - hm \leq 1$. *Let* a *and* b *be two integers such that*

$$ka \equiv 1 \pmod{h}, \quad \left\lceil \frac{hm+2}{k} \right\rceil - h \leq a \leq \left\lceil \frac{hm+2}{k} \right\rceil - 1,$$
$$hb \equiv -1 \pmod{k}, \quad m - k + 1 \leq b \leq m.$$

Then the fraction $a/\frac{ka-1}{h} = \frac{hb+1}{k}/b$ *succeeds the fraction* $\frac{h}{k}$ *in* $\mathcal{G}_m^\ell$.

Table 2.4. The fractions of the form $\frac{hb+1}{k}/b$ and $\frac{hb+1}{k-h}/\frac{kb+1}{k-h}$ that succeed fractions $\frac{h}{k}$ in Farey (sub)sequences, derived from the expressions listed in Table 1.4.

$\frac{hb+1}{k}/b$ **succeeds** $\frac{h}{k}$ **Sequence and restrictions on** b**, with** $hb \equiv -1 \pmod{k}$		**See**
$\mathcal{F}_m - \{\frac{0}{1}, \frac{1}{1}\}$		
$m - k + 1 \le b \le m$		Lemma 2.9 (ii)
$\mathcal{F}_m^{\ell} - \{\frac{0}{1}, \frac{1}{1}\}, 0 < \ell < m$		
$m - k + 1 \le b \le m$	if $k\ell - hm \ge 1$	Lemma 2.13 (ii) (a)
$\lceil \frac{k\ell}{h} \rceil - k \le b \le \lceil \frac{k\ell}{h} \rceil - 1$	if $k\ell - hm \le 1$	Lemma 2.13 (ii) (b)
$\mathcal{G}_m^{\ell} - \{\frac{0}{1}, \frac{1}{1}\}, 0 < \ell < m$		
$\lceil \frac{k(m-\ell)+2}{k-h} \rceil - k \le b \le \lceil \frac{k(m-\ell)+2}{k-h} \rceil - 1$	if $k\ell - hm \ge 1$	Lemma 2.15 (ii) (a)
$m - k + 1 \le b \le m$	if $k\ell - hm \le 1$	Lemma 2.15 (ii) (b)

$\frac{hb+1}{k-h}/\frac{kb+1}{k-h}$ **succeeds** $\frac{h}{k}$ **Sequence and restrictions on** b**, with** $hb \equiv -1 \pmod{(k-h)}$		**See**
$\mathcal{F}^{\le\frac{1}{2}}(\mathbb{B}(2m), m) - \{\frac{0}{1}, \frac{1}{2}\}$		
$m - k + h + 1 \le b \le m$		Proposition 2.12 (i) (b)
$\mathcal{F}^{\ge\frac{1}{2}}(\mathbb{B}(2m), m) - \{\frac{1}{2}, \frac{1}{1}\}$		
$\lceil \frac{km}{h} \rceil - m - k + h \le b \le \lceil \frac{km}{h} \rceil - m - 1$		Proposition 2.12 (ii) (b)
$\mathcal{F}^{\le\frac{1}{2}}(\mathbb{B}(n), m) - \{\frac{0}{1}, \frac{1}{2}\}$, where $n < 2m$		
$n - m - k + h + 1 \le b \le n - m$		Proposition 2.18 (i) (b)
$\mathcal{F}^{\ge\frac{1}{2}}(\mathbb{B}(n), m) - \{\frac{1}{2}, \frac{1}{1}\}$, where $n < 2m$		
$n - m - k + h + 1 \le b \le n - m$	if $km - hn \ge 1$	Proposition 2.18 (ii) (b)
$\lceil \frac{km}{h} \rceil - m - k + h \le b \le \lceil \frac{km}{h} \rceil - m - 1$	if $km - hn \le 1$	Proposition 2.18 (ii) (b)
$\mathcal{F}^{\le\frac{1}{2}}(\mathbb{B}(n), m) - \{\frac{0}{1}, \frac{1}{2}\}$, where $n > 2m$		
$n - m - k + h + 1 \le b \le n - m$	if $km - hn \ge 1$	Proposition 2.19 (i) (b)
$\lceil \frac{km}{h} \rceil - m - k + h \le b \le \lceil \frac{km}{h} \rceil - m - 1$	if $km - hn \le 1$	Proposition 2.19 (i) (b)
$\mathcal{F}^{\ge\frac{1}{2}}(\mathbb{B}(n), m) - \{\frac{1}{2}, \frac{1}{1}\}$, where $n > 2m$		
$\lceil \frac{km}{h} \rceil - m - k + h \le b \le \lceil \frac{km}{h} \rceil - m - 1$		Proposition 2.19 (ii) (b)

Proof. The assertions of the lemma follow from the respective assertions of Corollary 1.15. □

Our goal is to turn the multipliers of the form $\lfloor \cdot \rfloor$ that appear in (1.49)–(1.52) of Corollary 1.15 into zero by pointing out at suitable search intervals for the integers x_0 and y_0.

Table 2.5. The neighbors of the fraction $\frac{1}{2}$ in Farey (sub)sequences.

Sequence	The predecessor of $\frac{1}{2}$		The successor of $\frac{1}{2}$		See
$\mathcal{F}_n$	$\frac{\lfloor (n-1)/2\rfloor}{2\lfloor (n-1)/2\rfloor+1}$		$\frac{\lceil n/2\rceil}{2\lceil n/2\rceil-1}$		Remark 2.10
$\mathcal{F}_n^m,\ 0<m<n$	$\frac{m}{2m+1}$	if $n-2m\geq 1$	$\frac{\lceil n/2\rceil}{2\lceil n/2\rceil-1}$	if $2m-n\geq 1$	Remark 2.14
	$\frac{\lfloor (n-1)/2\rfloor}{2\lfloor (n-1)/2\rfloor+1}$	if $n-2m\leq 1$	$\frac{m}{2m-1}$	if $2m-n\leq 1$	Remark 2.14
$\mathcal{G}_n^m,\ 0<m<n$	$\frac{\lfloor (n-1)/2\rfloor}{2\lfloor (n-1)/2\rfloor+1}$	if $n-2m\geq 1$	$\frac{n-m+1}{2(n-m)+1}$	if $2m-n\geq 1$	Remark 2.16
	$\frac{n-m-1}{2(n-m)-1}$	if $n-2m\leq 1$	$\frac{\lceil n/2\rceil}{2\lceil n/2\rceil-1}$	if $2m-n\leq 1$	Remark 2.16
$\mathcal{F}(\mathbb{B}(2m),m)$	$\frac{m-1}{2m-1}$		$\frac{m}{2m-1}$		Remark 2.11
$\mathcal{F}(\mathbb{B}(n),m),\ n<2m$	$\frac{n-m-1}{2(n-m)-1}$		$\frac{n-m+1}{2(n-m)+1}$		Remark 2.17
$\mathcal{F}(\mathbb{B}(n),m),\ n>2m$	$\frac{m}{2m+1}$		$\frac{m}{2m-1}$		Remark 2.17

For instance, in order to check assertion (i) (a), note that the multipliers $\lfloor\frac{hn-kx_0-1}{hk}\rfloor$ and $\lfloor\frac{n-y_0}{k}\rfloor$ in (1.49) vanish whenever

$$\left\lceil\frac{hn}{k}\right\rceil-h\leq x_0\leq\left\lceil\frac{hn}{k}\right\rceil-1 \quad\text{and}\quad n-k+1\leq y_0\leq n. \qquad\square$$

Remark 2.16 (see also Table 2.5). If the parameters m and ℓ of a Farey subsequence $\mathcal{G}_m^\ell$ satisfy the relation $m-2\ell\geq 1$, then Lemma 2.15 (i) (a) guarantees that

$$\frac{\lfloor (m-1)/2\rfloor}{2\lfloor (m-1)/2\rfloor+1}=\frac{\lceil m/2\rceil-1}{2\lceil m/2\rceil-1}$$

precedes $\frac{1}{2}$ in $\mathcal{G}_m^\ell$.

Thus, if $m-2\ell\geq 1$ and m is even, then $\frac{(m-2)/2}{m-1}$ precedes $\frac{1}{2}$ in $\mathcal{G}_m^\ell$. If $m-2\ell\geq 1$ and m is odd, then $\frac{(m-1)/2}{m}$ precedes $\frac{1}{2}$ in $\mathcal{G}_m^\ell$.

If $m-2\ell\leq 1$, then, according to Lemma 2.15 (i) (b), $\frac{m-\ell-1}{2(m-\ell)-1}$ precedes $\frac{1}{2}$ in $\mathcal{G}_m^\ell$.

If $2\ell-m\geq 1$, then Lemma 2.15 (ii) (a) guarantees that $\frac{m-\ell+1}{2(m-\ell)+1}$ succeeds $\frac{1}{2}$ in $\mathcal{G}_m^\ell$. If $2\ell-m\leq 1$, then, according to Lemma 2.15 (ii) (b),

$$\frac{\lceil m/2\rceil}{2\lceil m/2\rceil-1}=\frac{\lfloor (m-1)/2\rfloor+1}{2\lfloor (m-1)/2\rfloor+1}$$

succeeds $\frac{1}{2}$ in $\mathcal{G}_m^\ell$.

Thus, if $2\ell-m\leq 1$ and m is even, then $\frac{m/2}{m-1}$ succeeds $\frac{1}{2}$ in $\mathcal{G}_m^\ell$. If $2\ell-m\leq 1$ and m is odd, then $\frac{(m+1)/2}{m}$ succeeds $\frac{1}{2}$ in $\mathcal{G}_m^\ell$.

2.2.5 Neighboring fractions in $\mathcal{F}(\mathbb{B}(n),m)$, $n\neq 2m$

In this subsection we adapt, via duality, the results obtained in the two previous subsections to the Farey subsequence $\mathcal{F}(\mathbb{B}(n),m)$, with $n\neq 2m$.

Recall that $\frac{m}{m+1}$ precedes $\frac{1}{1}$, and $\frac{1}{n-m+1}$ succeeds $\frac{0}{1}$ in $\mathcal{F}(\mathbb{B}(n), m)$, see Remark 1.17 and Table 1.5.

Remark 2.17 (see also Table 2.5). Suppose that $n < 2m$. Recall that $\frac{n-m-1}{n-m}$ precedes $\frac{1}{1}$ in $\mathcal{F}_{n-m}$, and $\frac{n-m}{n-m+1}$ precedes $\frac{1}{1}$ in $\mathcal{F}_m^{n-m}$. As a consequence, the monotone bijections in (2.21) and (2.24) guarantee that $\frac{n-m-1}{2(n-m)-1}$ precedes $\frac{1}{2}$, and $\frac{n-m+1}{2(n-m)+1}$ succeeds $\frac{1}{2}$ in $\mathcal{F}(\mathbb{B}(n), m)$.

In other words, if $n < 2m$, then the three fractions

$$\frac{n-m-1}{2(n-m)-1} < \frac{1}{2} < \frac{n-m+1}{2(n-m)+1}$$

are consecutive in $\mathcal{F}(\mathbb{B}(n), m)$.

Now suppose that $n > 2m$. Recall that $\frac{m}{m+1}$ precedes $\frac{1}{1}$ in $\mathcal{F}_{n-m}^m$, and $\frac{m-1}{m}$ precedes $\frac{1}{1}$ in $\mathcal{F}_m$. As a consequence, the monotone bijections in (2.26) and (2.30) guarantee that $\frac{m}{2m+1}$ precedes $\frac{1}{2}$, and $\frac{m}{2m-1}$ succeeds $\frac{1}{2}$ in $\mathcal{F}(\mathbb{B}(n), m)$.

In other words, if $n > 2m$, then the three fractions

$$\frac{m}{2m+1} < \frac{1}{2} < \frac{m}{2m-1}$$

are consecutive in $\mathcal{F}(\mathbb{B}(n), m)$.

The following two statements refine Proposition 1.21.

Proposition 2.18. *Let $\mathcal{F}(\mathbb{B}(n), m)$ be a Farey subsequence such that $n < 2m$.*

(i) *Consider a fraction $\frac{h}{k} \in \mathcal{F}^{\le\frac{1}{2}}(\mathbb{B}(n), m) - \{\frac{0}{1}, \frac{1}{2}\}$.*

(a) *Let a and b be two integers such that*

$$ka \equiv -1 \pmod{h}, \qquad \left\lceil \frac{h(n-m)}{k-h} \right\rceil - h \le a \le \left\lceil \frac{h(n-m)}{k-h} \right\rceil - 1,$$
$$hb \equiv 1 \pmod{(k-h)}, \quad n-m-k+h+1 \le b \le n-m.$$

Then the fraction

$$a \Big/ \frac{ka+1}{h} = \frac{hb-1}{k-h} \Big/ \frac{kb-1}{k-h} \tag{2.35}$$

precedes $\frac{h}{k}$ in $\mathcal{F}(\mathbb{B}(n), m)$.

(b) *Let a and b be two integers such that*

$$ka \equiv 1 \pmod{h}, \qquad \left\lceil \frac{h(n-m)+2}{k-h} \right\rceil - h \le a \le \left\lceil \frac{h(n-m)+2}{k-h} \right\rceil - 1,$$
$$hb \equiv -1 \pmod{(k-h)}, \quad n-m-k+h+1 \le b \le n-m.$$

Then the fraction $a/\frac{ka-1}{h} = \frac{hb+1}{k-h}/\frac{kb+1}{k-h}$ succeeds $\frac{h}{k}$ in $\mathcal{F}(\mathbb{B}(n), m)$.

(ii) *Consider a fraction $\frac{h}{k} \in \mathcal{F}^{\ge\frac{1}{2}}(\mathbb{B}(n), m) - \{\frac{1}{2}, \frac{1}{1}\}$.*

(a) *If $hn - km \ge 1$, let a and b be two integers such that*

$$ka \equiv -1 \pmod{h}, \qquad m-h+1 \le a \le m,$$
$$hb \equiv 1 \pmod{(k-h)}, \quad \left\lceil \frac{km+2}{h} \right\rceil - m-k+h \le b \le \left\lceil \frac{km+2}{h} \right\rceil - m - 1.$$

Then the fraction $a/\frac{ka+1}{h} = \frac{hb-1}{k-h}/\frac{kb-1}{k-h}$ precedes $\frac{h}{k}$ in $\mathcal{F}(\mathbb{B}(n), m)$.

If $hn - km \leq 1$*, let* a *and* b *be two integers such that*

$$\begin{aligned} ka &\equiv -1 \pmod{h}, & \left\lceil \frac{h(n-m)}{k-h} \right\rceil - h \leq a \leq \left\lceil \frac{h(n-m)}{k-h} \right\rceil - 1, \\ hb &\equiv 1 \pmod{(k-h)}, & n - m - k + h + 1 \leq b \leq n - m. \end{aligned}$$

Then the fraction $a/\frac{ka+1}{h} = \frac{hb-1}{k-h}/\frac{kb-1}{k-h}$ *precedes* $\frac{h}{k}$ *in* $\mathcal{F}(\mathbb{B}(n), m)$.

(b) *If* $km - hn \geq 1$*, let* a *and* b *be two integers such that*

$$\begin{aligned} ka &\equiv 1 \pmod{h}, & \left\lceil \frac{h(n-m)+2}{k-h} \right\rceil - h \leq a \leq \left\lceil \frac{h(n-m)+2}{k-h} \right\rceil - 1, \\ hb &\equiv -1 \pmod{(k-h)}, & n - m - k + h + 1 \leq b \leq n - m. \end{aligned}$$

Then the fraction $a/\frac{ka-1}{h} = \frac{hb+1}{k-h}/\frac{kb+1}{k-h}$ *succeeds* $\frac{h}{k}$ *in* $\mathcal{F}(\mathbb{B}(n), m)$.
If $km - hn \leq 1$*, let* a *and* b *be two integers such that*

$$\begin{aligned} ka &\equiv 1 \pmod{h}, & m - h + 1 \leq a \leq m, \\ hb &\equiv -1 \pmod{(k-h)}, & \left\lceil \frac{km}{h} \right\rceil - m - k + h \leq b \leq \left\lceil \frac{km}{h} \right\rceil - m - 1. \end{aligned}$$

Then the fraction $a/\frac{ka-1}{h} = \frac{hb+1}{k-h}/\frac{kb+1}{k-h}$ *succeeds* $\frac{h}{k}$ *in* $\mathcal{F}(\mathbb{B}(n), m)$.

Proof. To prove assertion (i), first note that $\frac{h}{k-h} \in \mathcal{F}_{n-m}$ is the image of $\frac{h}{k}$ under the order-preserving bijection (2.20); then use Lemma 2.9 to find the predecessor and successor of $\frac{h}{k-h}$ in $\mathcal{F}_{n-m}$, and then send them back to $\mathcal{F}(\mathbb{B}(n), m)$ by means of the order-preserving bijection (2.21).

For instance, according to Lemma 2.9 (i), the fraction

$$a/\frac{(k-h)a+1}{h} = \frac{hb-1}{k-h}/b \tag{2.36}$$

is the predecessor of $\frac{h}{k-h}$ in the sequence $\mathcal{F}_{n-m}$, where a and b are integers such that $(k-h)a \equiv -1 \pmod{h}$ or, equivalently, $ka \equiv -1 \pmod{h}$ under the restriction

$$\left\lceil \frac{h(n-m)}{k-h} \right\rceil - h \leq a \leq \left\lceil \frac{h(n-m)}{k-h} \right\rceil - 1,$$

and

$$hb \equiv 1 \pmod{(k-h)}, \quad (n-m) - (k-h) + 1 \leq b \leq n - m.$$

As a consequence, the image (2.35) of the fraction (2.36) under the order-preserving bijection (2.21) precedes the fraction $\frac{h}{k}$ in the sequence $\mathcal{F}(\mathbb{B}(n), m)$.

(ii) (a) Note that $\frac{k-h}{h} \in \mathcal{F}_m^{n-m}$ is the image of $\frac{h}{k}$ under the order-reversing bijection (2.23). We will use Lemma 2.13 (ii) for finding the fraction that succeeds $\frac{k-h}{h}$ in $\mathcal{F}_m^{n-m}$.

Let us consider the quantity $h(n-m) - (k-h)m = hn - km$.

If $hn - km \geq 1$, then we turn to Lemma 2.13 (ii) (a). Let α and β be two integers such that

$$h\alpha \equiv 1 \pmod{(k-h)}, \qquad \left\lceil \frac{(k-h)m+2}{h} \right\rceil - (k-h) \leq \alpha \leq \left\lceil \frac{(k-h)m+2}{h} \right\rceil - 1,$$
$$(k-h)\beta \equiv -1 \pmod{h}, \quad m - h + 1 \leq \beta \leq m.$$

The fraction

$$\alpha \Big/ \frac{h\alpha - 1}{k-h} = \frac{(k-h)\beta + 1}{h} \Big/ \beta$$

succeeds the fraction $\frac{k-h}{k}$ in $\mathcal{F}_m^{n-m}$. As a consequence, its image

$$\frac{h\alpha - 1}{k-h} \Big/ \frac{k\alpha - 1}{k-h} = \beta \Big/ \frac{k\beta + 1}{h}$$

under the order-reversing bijection (2.24) precedes $\frac{h}{k}$ in $\mathcal{F}(\mathbb{B}(n), m)$. It remains to substitute β by a, and α by b.

If $hn - km \leq 1$, then we turn to Lemma 2.13 (ii) (b). Let α and β be two integers such that

$$h\alpha \equiv 1 \pmod{(k-h)}, \qquad (n-m) - (k-h) + 1 \leq \alpha \leq n - m,$$
$$(k-h)\beta \equiv -1 \pmod{h}, \quad \left\lceil \frac{h(n-m)}{k-h} \right\rceil - h \leq \beta \leq \left\lceil \frac{h(n-m)}{k-h} \right\rceil - 1.$$

The fraction

$$\alpha \Big/ \frac{h\alpha - 1}{k-h} = \frac{(k-h)\beta + 1}{h} \Big/ \beta$$

succeeds the fraction $\frac{k-h}{k}$ in $\mathcal{F}_m^{n-m}$. As a consequence, its image

$$\frac{h\alpha - 1}{k-h} \Big/ \frac{k\alpha - 1}{k-h} = \beta \Big/ \frac{k\beta + 1}{h}$$

under the order-reversing bijection (2.24) precedes $\frac{h}{k}$ in $\mathcal{F}(\mathbb{B}(n), m)$. It remains to substitute β by a, and α by b.

(ii) (b) Note that $\frac{k-h}{h} \in \mathcal{F}_m^{n-m}$ is the image of $\frac{h}{k}$ under the order-reversing bijection (2.23). We will use Lemma 2.13 (i) to find the fraction that precedes $\frac{k-h}{h}$ in $\mathcal{F}_m^{n-m}$.

Let us consider the quantity $(k-h)m - h(n-m) = km - hn$.

If $km - hn \geq 1$, then we turn to Lemma 2.13 (i) (a). Let α and β be two integers such that

$$h\alpha \equiv -1 \pmod{(k-h)}, \quad (n-m) - (k-h) + 1 \leq \alpha \leq n - m,$$
$$(k-h)\beta \equiv 1 \pmod{h}, \qquad \left\lceil \frac{h(n-m)+2}{k-h} \right\rceil - h \leq \beta \leq \left\lceil \frac{h(n-m)+2}{k-h} \right\rceil - 1.$$

The fraction

$$\alpha \Big/ \frac{h\alpha + 1}{k-h} = \frac{(k-h)\beta - 1}{h} \Big/ \beta$$

precedes the fraction $\frac{k-h}{h}$ in $\mathcal{F}_m^{n-m}$. As a consequence, its image

$$\frac{h\alpha+1}{k-h}\Big/\frac{k\alpha+1}{k-h}=\beta\Big/\frac{k\beta-1}{h}$$

under the order-reversing bijection (2.24) succeeds $\frac{h}{k}$ in $\mathcal{F}(\mathbb{B}(n), m)$. It remains to substitute β by a, and α by b.

If $km-hn \le 1$, then we turn to Lemma 2.13 (i) (b). Let α and β be two integers such that

$$\begin{aligned}
&h\alpha \equiv -1 \pmod{(k-h)}, && \left\lceil\frac{(k-h)m}{h}\right\rceil-(k-h)\le\alpha\le\left\lceil\frac{(k-h)m}{h}\right\rceil-1,\\
&(k-h)\beta \equiv 1 \pmod{h}, && m-h+1\le\beta\le m.
\end{aligned}$$

The fraction

$$\alpha\Big/\frac{h\alpha+1}{k-h}=\frac{(k-h)\beta-1}{h}\Big/\beta$$

precedes the fraction $\frac{k-h}{h}$ in $\mathcal{F}_m^{n-m}$. As a consequence, its image

$$\frac{h\alpha+1}{k-h}\Big/\frac{k\alpha+1}{k-h}=\beta\Big/\frac{k\beta-1}{h}$$

under the order-reversing bijection (2.24) succeeds $\frac{h}{k}$ in $\mathcal{F}(\mathbb{B}(n), m)$. It remains to substitute β by a, and α by b. □

Proposition 2.19. *Let $\mathcal{F}(\mathbb{B}(n), m)$ be a Farey subsequence such that $n > 2m$.*

(i) *Consider a fraction $\frac{h}{k} \in \mathcal{F}^{\le\frac{1}{2}}(\mathbb{B}(n), m) - \{\frac{0}{1}, \frac{1}{2}\}$.*

(a) *If $hn-km \ge 1$, let a and b be two integers such that*

$$\begin{aligned}
&ka \equiv -1 \pmod{h}, && m-h+1\le a\le m,\\
&hb \equiv 1 \pmod{(k-h)}, && \left\lceil\frac{km+2}{h}\right\rceil-m-k+h\le b\le\left\lceil\frac{km+2}{h}\right\rceil-m-1.
\end{aligned}$$

Then the fraction $a/\frac{ka+1}{h}=\frac{hb-1}{k-h}/\frac{kb-1}{k-h}$ precedes $\frac{h}{k}$ in $\mathcal{F}(\mathbb{B}(n), m)$.

If $hn-km \le 1$, let a and b be two integers such that

$$\begin{aligned}
&ka \equiv -1 \pmod{h}, && \left\lceil\frac{h(n-m)}{k-h}\right\rceil-h\le a\le\left\lceil\frac{h(n-m)}{k-h}\right\rceil-1,\\
&hb \equiv 1 \pmod{(k-h)}, && n-m-k+h+1\le b\le n-m.
\end{aligned}$$

Then the fraction $a/\frac{ka+1}{h}=\frac{hb-1}{k-h}/\frac{kb-1}{k-h}$ precedes $\frac{h}{k}$ in $\mathcal{F}(\mathbb{B}(n), m)$.

(b) *If $km-hn \ge 1$, let a and b be two integers such that*

$$\begin{aligned}
&ka \equiv 1 \pmod{h}, && \left\lceil\frac{h(n-m)+2}{k-h}\right\rceil-h\le a\le\left\lceil\frac{h(n-m)+2}{k-h}\right\rceil-1,\\
&hb \equiv -1 \pmod{(k-h)}, && n-m-k+h+1\le b\le n-m.
\end{aligned}$$

Then the fraction $a/\frac{ka-1}{h}=\frac{hb+1}{k-h}/\frac{kb+1}{k-h}$ succeeds $\frac{h}{k}$ in $\mathcal{F}(\mathbb{B}(n), m)$.

If $km - hn \le 1$, let a and b be two integers such that

$$ka \equiv 1 \pmod{h}, \qquad m - h + 1 \le a \le m,$$
$$hb \equiv -1 \pmod{(k-h)}, \quad \left\lceil \frac{km}{h} \right\rceil - m - k + h \le b \le \left\lceil \frac{km}{h} \right\rceil - m - 1.$$

Then the fraction $a / \frac{ka-1}{h} = \frac{hb+1}{k-h} / \frac{kb+1}{k-h}$ succeeds $\frac{h}{k}$ in $\mathcal{F}(\mathbb{B}(n), m)$.

(ii) *Consider a fraction $\frac{h}{k} \in \mathcal{F}^{\ge \frac{1}{2}}(\mathbb{B}(n), m) - \{\frac{1}{2}, \frac{1}{1}\}$.*

(a) *Let a and b be two integers such that*

$$ka \equiv -1 \pmod{h}, \qquad m - h + 1 \le a \le m,$$
$$hb \equiv 1 \pmod{(k-h)}, \quad \left\lceil \frac{km+2}{h} \right\rceil - m - k + h \le b \le \left\lceil \frac{km+2}{h} \right\rceil - m - 1.$$

Then the fraction $a / \frac{ka+1}{h} = \frac{hb-1}{k-h} / \frac{kb-1}{k-h}$ precedes $\frac{h}{k}$ in $\mathcal{F}(\mathbb{B}(n), m)$.

(b) *Let a and b be two integers such that*

$$ka \equiv 1 \pmod{h}, \qquad m - h + 1 \le a \le m,$$
$$hb \equiv -1 \pmod{(k-h)}, \quad \left\lceil \frac{km}{h} \right\rceil - m - k + h \le b \le \left\lceil \frac{km}{h} \right\rceil - m - 1.$$

Then the fraction $a / \frac{ka-1}{h} = \frac{hb+1}{k-h} / \frac{kb+1}{k-h}$ succeeds $\frac{h}{k}$ in $\mathcal{F}(\mathbb{B}(n), m)$.

Proof. (i) (a) Note that $\frac{h}{k-h} \in \mathcal{F}^m_{n-m}$ is the image of $\frac{h}{k}$ under the order-preserving bijection (2.25). To find the predecessor of $\frac{h}{k-h}$ in $\mathcal{F}^m_{n-m}$, we use Lemma 2.13 (i). Let us consider the quantity $h(n-m) - (k-h)m = hn - km$.

If $hn - km \ge 1$, then we use Lemma 2.13 (i) (a). Let a and b be two integers such that $(k-h)a \equiv -1 \pmod{h}$ or, equivalently, $ka \equiv -1 \pmod{h}$, under the restriction

$$m - h + 1 \le a \le m,$$

and

$$hb \equiv 1 \pmod{(k-h)}, \quad \left\lceil \frac{(k-h)m+2}{h} \right\rceil - (k-h) \le b \le \left\lceil \frac{(k-h)m+2}{h} \right\rceil - 1.$$

The fraction

$$a \Big/ \frac{(k-h)a+1}{h} = \frac{hb-1}{k-h} \Big/ b$$

precedes $\frac{h}{k-h}$ in $\mathcal{F}^m_{n-m}$, and its image

$$a \Big/ \frac{ka+1}{h} = \frac{hb-1}{k-h} \Big/ \frac{kb-1}{k-h}$$

under the order-preserving bijection (2.26) precedes $\frac{h}{k}$ in $\mathcal{F}(\mathbb{B}(n), m)$.

If $hn - km \le 1$, then we use Lemma 2.13 (i) (b). Let a and b be two integers such that

$$(k-h)a \equiv -1 \pmod{h}, \quad \left\lceil \frac{h(n-m)}{k-h} \right\rceil - h \le a \le \left\lceil \frac{h(n-m)}{k-h} \right\rceil - 1,$$
$$hb \equiv 1 \pmod{(k-h)}, \quad (n-m) - (k-h) + 1 \le b \le n - m.$$

The fraction

$$a\Big/\frac{(k-h)a+1}{h} = \frac{hb-1}{k-h}\Big/b$$

precedes $\frac{h}{k-h}$ in $\mathcal{F}^m_{n-m}$, and its image

$$a\Big/\frac{ka+1}{h} = \frac{hb-1}{k-h}\Big/\frac{kb-1}{k-h}$$

under the order-preserving bijection (2.26) precedes $\frac{h}{k}$ in $\mathcal{F}(\mathbb{B}(n), m)$.

(i) (b) To find the successor of $\frac{h}{k-h}$ in $\mathcal{F}^m_{n-m}$, we use Lemma 2.13 (ii). Let us consider the quantity $(k-h)m - h(n-m) = km - hn$.

If $km - hn \geq 1$, then we use Lemma 2.13 (ii) (a). Let a and b be two integers such that

$$(k-h)a \equiv 1 \pmod{h}, \qquad \left\lceil \frac{h(n-m)+2}{k-h} \right\rceil - h \leq a \leq \left\lceil \frac{h(n-m)+2}{k-h} \right\rceil - 1,$$
$$hb \equiv -1 \pmod{(k-h)}, \qquad (n-m) - (k-h) + 1 \leq b \leq n-m.$$

The fraction

$$a\Big/\frac{(k-h)a-1}{h} = \frac{hb+1}{k-h}\Big/b$$

succeeds $\frac{h}{k-h}$ in $\mathcal{F}^m_{n-m}$, and its image

$$a\Big/\frac{ka-1}{h} = \frac{hb+1}{k-h}\Big/\frac{kb+1}{k-h}$$

under the order-preserving bijection (2.26) succeeds $\frac{h}{k}$ in $\mathcal{F}(\mathbb{B}(n), m)$.

If $km - hn \leq 1$, then we use Lemma 2.13 (ii) (b). Let a and b be two integers such that

$$(k-h)a \equiv 1 \pmod{h}, \qquad m-h+1 \leq a \leq m,$$
$$hb \equiv -1 \pmod{(k-h)}, \qquad \left\lceil \frac{(k-h)m}{h} \right\rceil - (k-h) \leq b \leq \left\lceil \frac{(k-h)m}{h} \right\rceil - 1.$$

The fraction

$$a\Big/\frac{(k-h)a-1}{h} = \frac{hb+1}{k-h}\Big/b$$

succeeds $\frac{h}{k-h}$ in $\mathcal{F}^m_{n-m}$, and its image

$$a\Big/\frac{ka-1}{h} = \frac{hb+1}{k-h}\Big/\frac{kb+1}{k-h}$$

under the order-preserving bijection (2.26) succeeds $\frac{h}{k}$ in $\mathcal{F}(\mathbb{B}(n), m)$.

To prove assertion (ii), first note that $\frac{k-h}{h} \in \mathcal{F}_m$ is the image of $\frac{h}{k}$ under the order-reversing bijection (2.29); then use Lemma 2.9 to find the successor and predecessor of $\frac{k-h}{h}$ in $\mathcal{F}_m$, and then send them back to $\mathcal{F}(\mathbb{B}(n), m)$ by means of the order-reversing bijection (2.30).

For instance, according to Lemma 2.9 (ii), the fraction

$$\alpha\Big/\frac{h\alpha-1}{k-h} = \frac{(k-h)\beta+1}{h}\Big/\beta \tag{2.37}$$

is the successor of $\frac{k-h}{h}$ in the sequence $\mathcal{F}_m$, where α and β are integers such that

$$h\alpha \equiv 1 \ (\mathrm{mod}\,(k-h)), \qquad \left\lceil \frac{(k-h)m+2}{h} \right\rceil - (k-h) \le \alpha \le \left\lceil \frac{(k-h)m+2}{h} \right\rceil - 1,$$
$$(k-h)\beta \equiv -1 \ (\mathrm{mod}\, h), \quad m-h+1 \le \beta \le m.$$

As a consequence, the image of the fraction (2.37)

$$\frac{h\alpha - 1}{k-h} \Big/ \frac{k\alpha - 1}{k-h} = \beta \Big/ \frac{k\beta + 1}{h}$$

under the order-reversing bijection (2.30) precedes the fraction $\frac{h}{k}$ in the sequence $\mathcal{F}(\mathbb{B}(n), m)$. To verify assertion (ii) (a), it remains to substitute β by a, and α by b. □

2.2.6 More on neighboring fractions in $\mathcal{F}_m$

Now that in Lemma 2.9 we have obtained the simplest formulas for finding the neighbors of fractions in the Farey sequences, in this subsection we will describe explicitly the neighbors of fractions of the form $\frac{1}{j}$, $\frac{j-1}{j}$, $\frac{2}{j}$ and $\frac{j-2}{j}$. Similar results in the context of the Farey subsequences $\mathcal{F}_m^{\ell}$ and $\mathcal{G}_m^{\ell}$ will be presented in Section 2.2.8. See Table 2.6 on the selected pairs of neighboring fractions that are considered in this book.

2.2.6.1 The neighbors of $\frac{1}{j}$ and $\frac{j-1}{j}$

Let us find in the Farey sequence $\mathcal{F}_m$ the neighbors of fractions of the form $\frac{1}{j}$ and $\frac{j-1}{j}$.

Corollary 2.20. (i) *Consider a fraction $\frac{1}{j} \in \mathcal{F}_m$. The fraction*

$$\frac{\lceil \frac{m}{j} \rceil - 1}{j(\lceil \frac{m}{j} \rceil - 1) + 1} \tag{2.38}$$

precedes $\frac{1}{j}$ in $\mathcal{F}_m$. If $j > 1$, then the fraction

$$\frac{\lceil \frac{m+2}{j} \rceil - 1}{j(\lceil \frac{m+2}{j} \rceil - 1) - 1}$$

succeeds $\frac{1}{j}$ in $\mathcal{F}_m$.

(ii) *Consider a fraction $\frac{j-1}{j} \in \mathcal{F}_m$. If $j > 1$, then the fraction*

$$\frac{(j-1)(\lceil \frac{m+2}{j} \rceil - 1) - 1}{j(\lceil \frac{m+2}{j} \rceil - 1) - 1}$$

precedes $\frac{j-1}{j}$ in $\mathcal{F}_m$, and the fraction

$$\frac{(j-1)(\lceil \frac{m}{j} \rceil - 1) + 1}{j(\lceil \frac{m}{j} \rceil - 1) + 1}$$

succeeds $\frac{j-1}{j}$ in $\mathcal{F}_m$.

Table 2.6. The neighbors of selected fractions in Farey (sub)sequences; see also Tables 2.7 and 2.8.

Sequence	Neighbors of the fraction	See
$\mathcal{F}_m$	$\frac{1}{j}$	Corollary 2.20 (i)
	$\frac{j-1}{j}$	Corollary 2.20 (ii)
	$\frac{2}{j}$	Corollary 2.21 (i)
	$\frac{j-2}{j}$	Corollary 2.21 (ii)
$\mathcal{F}_m^\ell$, $0 < \ell < m$	$\frac{1}{j}$	Corollary 2.26
	$\frac{j-1}{j}$	Corollary 2.29
	$\frac{2}{j}$	Corollary 2.30
	$\frac{j-2}{j}$	Corollary 2.33
$\mathcal{G}_m^\ell$, $0 < \ell < m$	$\frac{1}{j}$	Corollary 2.27
	$\frac{j-1}{j}$	Corollary 2.28
	$\frac{2}{j}$	Corollary 2.31
	$\frac{j-2}{j}$	Corollary 2.32
$\mathcal{F}^{\leq\frac{1}{2}}(\mathbb{B}(2m), m)$	$\frac{1}{j+1}$	Corollary 2.22 (i)
	$\frac{j-1}{2j-1}$	Corollary 2.22 (ii)
	$\frac{2}{j+2}$	Corollary 2.23 (i)
	$\frac{j-2}{2(j-1)}$	Corollary 2.23 (ii)
$\mathcal{F}^{\geq\frac{1}{2}}(\mathbb{B}(2m), m)$	$\frac{j}{2j-1}$	Corollary 2.22 (iii)
	$\frac{j}{j-1}$	Corollary 2.22 (iv)
	$\frac{j}{2(j-1)}$	Corollary 2.23 (iii)
	$\frac{j}{j+2}$	Corollary 2.23 (iv)
$\mathcal{F}^{\leq\frac{1}{2}}(\mathbb{B}(n), m)$, $n \neq 2m$	$\frac{1}{j+1}$	Corollary 2.34
	$\frac{j-1}{2j-1}$	Corollary 2.36
	$\frac{2}{j+2}$	Corollary 2.38
	$\frac{j-2}{2(j-1)}$	Corollary 2.40
$\mathcal{F}^{\geq\frac{1}{2}}(\mathbb{B}(n), m)$, $n \neq 2m$	$\frac{j}{j+1}$	Corollary 2.35
	$\frac{j}{2j-1}$	Corollary 2.37
	$\frac{j}{j+2}$	Corollary 2.39
	$\frac{j}{2(j-1)}$	Corollary 2.41

Proof. (i) It suffices to note that if we consider in Lemma 2.9 fractions $\frac{h}{k}$ of the form $\frac{1}{j}$, then the search intervals for integers a in the constraints (2.31) and (2.33) turn into singleton sets.

To verify assertion (ii), note that $\frac{1}{j}$ is the image of $\frac{j-1}{j}$ under the order-reversing bijection defined in (1.8), and find the images of the neighbors of $\frac{1}{j}$, given in assertion (i), under the same bijection. □

2.2.6.2 The neighbors of $\frac{2}{j}$ and $\frac{j-2}{j}$

The following statement summarizes the results of routine computations similar to those suggested in the proof of Corollary 2.20, now for fractions of the form $\frac{2}{j}$ and $\frac{j-2}{j}$ in the Farey sequence $\mathcal{F}_m$.

Corollary 2.21. (i) *If $\frac{2}{j} \in \mathcal{F}_m$ for some odd j, then the fraction*

$$\begin{cases} (\lceil \frac{2m}{j} \rceil - 1) \Big/ \dfrac{j(\lceil \frac{2m}{j} \rceil - 1) + 1}{2} & \text{if } \lceil \frac{2m}{j} \rceil \text{ is even,} \\ (\lceil \frac{2m}{j} \rceil - 2) \Big/ \dfrac{j(\lceil \frac{2m}{j} \rceil - 2) + 1}{2} & \text{if } \lceil \frac{2m}{j} \rceil \text{ is odd} \end{cases} \tag{2.39}$$

precedes $\frac{2}{j}$ in $\mathcal{F}_m$, and the fraction

$$\begin{cases} (\lceil \frac{2(m+1)}{j} \rceil - 1) \Big/ \dfrac{j(\lceil \frac{2(m+1)}{j} \rceil - 1) - 1}{2} & \text{if } \lceil \frac{2(m+1)}{j} \rceil \text{ is even,} \\ (\lceil \frac{2(m+1)}{j} \rceil - 2) \Big/ \dfrac{j(\lceil \frac{2(m+1)}{j} \rceil - 2) - 1}{2} & \text{if } \lceil \frac{2(m+1)}{j} \rceil \text{ is odd} \end{cases} \tag{2.40}$$

succeeds $\frac{2}{j}$ in $\mathcal{F}_m$.

(ii) *If $\frac{j-2}{j} \in \mathcal{F}_m$ for some odd j, then the fraction*

$$\begin{cases} \dfrac{(j-2)(\lceil \frac{2(m+1)}{j} \rceil - 1) - 1}{2} \Big/ \dfrac{j(\lceil \frac{2(m+1)}{j} \rceil - 1) - 1}{2} & \text{if } \lceil \frac{2(m+1)}{j} \rceil \text{ is even,} \\ \dfrac{(j-2)(\lceil \frac{2(m+1)}{j} \rceil - 2) - 1}{2} \Big/ \dfrac{j(\lceil \frac{2(m+1)}{j} \rceil - 2) - 1}{2} & \text{if } \lceil \frac{2(m+1)}{j} \rceil \text{ is odd} \end{cases} \tag{2.41}$$

precedes $\frac{j-2}{j}$ in $\mathcal{F}_m$, and the fraction

$$\begin{cases} \dfrac{(j-2)(\lceil \frac{2m}{j} \rceil - 1) + 1}{2} \Big/ \dfrac{j(\lceil \frac{2m}{j} \rceil - 1) + 1}{2} & \text{if } \lceil \frac{2m}{j} \rceil \text{ is even,} \\ \dfrac{(j-2)(\lceil \frac{2m}{j} \rceil - 2) + 1}{2} \Big/ \dfrac{j(\lceil \frac{2m}{j} \rceil - 2) + 1}{2} & \text{if } \lceil \frac{2m}{j} \rceil \text{ is odd} \end{cases} \tag{2.42}$$

succeeds $\frac{j-2}{j}$ in $\mathcal{F}_m$.

Proof. (i) Note that, according to Lemma 2.9 (i), the fraction $\frac{2}{j}$ succeeds the fraction $a/\frac{ja+1}{2}$ in $\mathcal{F}_m$, where

$$ja \equiv -1 \pmod 2), \quad \left\lceil \frac{2m}{j} \right\rceil - 2 \leq a \leq \left\lceil \frac{2m}{j} \right\rceil - 1.$$

Since the fraction $\frac{j-2}{j} \in \mathcal{F}_m$ is the image of $\frac{2}{j}$ under the order-reversing bijection defined in (1.8), assertion (ii) follows directly from assertion (i). □

2.2.7 More on neighboring fractions in $\mathcal{F}(\mathbb{B}(2m), m)$

In this subsection we apply, via duality, the observations made in Corollaries 2.20 and 2.21 to some fractions in the Farey subsequence $\mathcal{F}(\mathbb{B}(2m), m)$; those are fractions of the form $\frac{1}{j+1}$, $\frac{j-1}{2j-1}$, $\frac{j}{2j-1}$, $\frac{j}{j+1}$, $\frac{2}{j+2}$, $\frac{j-2}{2(j-1)}$, $\frac{j}{2(j-1)}$ and $\frac{j}{j+2}$. See also Table 2.6.

2.2.7.1 The neighbors of $\frac{1}{j+1}$, $\frac{j-1}{2j-1}$, $\frac{j}{2j-1}$ and $\frac{j}{j+1}$

Let us find in the Farey subsequence $\mathcal{F}(\mathbb{B}(2m), m)$ the neighbors of fractions of the form $\frac{1}{j+1}$, $\frac{j-1}{2j-1}$, $\frac{j}{2j-1}$ and $\frac{j}{j+1}$.

Corollary 2.22. (i) *If $\frac{1}{j+1} \in \mathcal{F}^{\leq\frac{1}{2}}(\mathbb{B}(2m), m) - \{\frac{1}{2}\}$ for some j, then the fraction*

$$\frac{\lceil \frac{m}{j} \rceil - 1}{(j+1)(\lceil \frac{m}{j} \rceil - 1) + 1} \tag{2.43}$$

precedes $\frac{1}{j+1}$ in $\mathcal{F}(\mathbb{B}(2m), m)$, and the fraction

$$\frac{\lceil \frac{m+2}{j} \rceil - 1}{(j+1)(\lceil \frac{m+2}{j} \rceil - 1) - 1} \tag{2.44}$$

succeeds $\frac{1}{j+1}$ in $\mathcal{F}(\mathbb{B}(2m), m)$.

(ii) *If $\frac{j-1}{2j-1} \in \mathcal{F}^{\leq\frac{1}{2}}(\mathbb{B}(2m), m)$ for some j, then the fraction*

$$\frac{(j-1)(\lceil \frac{m+2}{j} \rceil - 1) - 1}{(2j-1)(\lceil \frac{m+2}{j} \rceil - 1) - 2} \tag{2.45}$$

precedes $\frac{j-1}{2j-1}$ in $\mathcal{F}(\mathbb{B}(2m), m)$, and the fraction

$$\frac{(j-1)(\lceil \frac{m}{j} \rceil - 1) + 1}{(2j-1)(\lceil \frac{m}{j} \rceil - 1) + 2} \tag{2.46}$$

succeeds $\frac{j-1}{2j-1}$ in $\mathcal{F}(\mathbb{B}(2m), m)$.

(iii) *If $\frac{j}{2j-1} \in \mathcal{F}^{\geq\frac{1}{2}}(\mathbb{B}(2m), m) - \{\frac{1}{1}\}$ for some j, then the fraction*

$$\frac{j(\lceil \frac{m}{j} \rceil - 1) + 1}{(2j-1)(\lceil \frac{m}{j} \rceil - 1) + 2}$$

precedes $\frac{j}{2j-1}$ in $\mathcal{F}(\mathbb{B}(2m), m)$, and the fraction

$$\frac{j(\lceil \frac{m+2}{j} \rceil - 1) - 1}{(2j-1)(\lceil \frac{m+2}{j} \rceil - 1) - 2}$$

succeeds $\frac{j}{2j-1}$ in $\mathcal{F}(\mathbb{B}(2m), m)$.

(iv) *If* $\frac{j}{j+1} \in \mathcal{F}^{\geq\frac{1}{2}}(\mathbb{B}(2m), m) - \{\frac{1}{2}\}$ *for some* j*, then the fraction*

$$\frac{j(\lceil\frac{m+2}{j}\rceil - 1) - 1}{(j+1)(\lceil\frac{m+2}{j}\rceil - 1) - 1}$$

precedes $\frac{j}{j+1}$ *in* $\mathcal{F}(\mathbb{B}(2m), m)$*, and the fraction*

$$\frac{j(\lceil\frac{m}{j}\rceil - 1) + 1}{(j+1)(\lceil\frac{m}{j}\rceil - 1) + 1}$$

succeeds $\frac{j}{j+1}$ *in* $\mathcal{F}(\mathbb{B}(2m), m)$.

Proof. (i) According to Corollary 2.5, the fraction $\frac{1}{j} \in \mathcal{F}_m$ is the image of the fraction $\frac{1}{j+1}$ under the order-preserving bijection (2.12).

Recall that, according to Corollary 2.20 (i), the fraction $\frac{\lceil m/j\rceil-1}{j(\lceil m/j\rceil-1)+1}$ precedes $\frac{1}{j}$ in $\mathcal{F}_m$. As a consequence, its image (2.43) under the order-preserving bijection (2.13) precedes $\frac{1}{j+1}$ in the sequence $\mathcal{F}(\mathbb{B}(2m), m)$.

Similarly, according to Corollary 2.20 (i), the fraction $\frac{\lceil (m+2)/j\rceil-1}{j(\lceil (m+2)/j\rceil-1)-1}$ succeeds $\frac{1}{j}$ in $\mathcal{F}_m$. As a consequence, its image (2.44) under the order-preserving bijection (2.13) succeeds $\frac{1}{j+1}$ in $\mathcal{F}(\mathbb{B}(2m), m)$.

Let us verify assertion (iv). Since $\frac{1}{j+1} \in \mathcal{F}^{\leq\frac{1}{2}}(\mathbb{B}(2m), m)$ is the image of the fraction $\frac{j}{j+1} \in \mathcal{F}^{\geq\frac{1}{2}}(\mathbb{B}(2m), m)$ under the order-reversing bijection $\mathcal{F}(\mathbb{B}(2m), m) \to \mathcal{F}(\mathbb{B}(2m), m)$ defined in (1.8), assertion (iv) follows directly from assertion (i) on the neighbors of the fraction $\frac{1}{j+1}$.

(ii) According to Corollary 2.5, the fraction $\frac{j-1}{j} \in \mathcal{F}_m$ is the image of the fraction $\frac{j-1}{2j-1}$ under the order-preserving bijection (2.12).

Recall that, according to Corollary 2.20 (ii), the fraction $\frac{(j-1)(\lceil (m+2)/j\rceil-1)-1}{j(\lceil (m+2)/j\rceil-1)-1}$ precedes $\frac{j-1}{j}$ in $\mathcal{F}_m$. As a consequence, its image (2.45) under the order-preserving bijection (2.13) precedes $\frac{j-1}{2j-1}$ in $\mathcal{F}(\mathbb{B}(2m), m)$.

Similarly, according to Corollary 2.20 (ii), the fraction $\frac{(j-1)(\lceil m/j\rceil-1)+1}{j(\lceil m/j\rceil-1)+1}$ succeeds $\frac{j-1}{j}$ in $\mathcal{F}_m$. As a consequence, its image (2.46) under the order-preserving bijection (2.13) succeeds $\frac{j-1}{2j-1}$ in $\mathcal{F}(\mathbb{B}(2m), m)$.

Let us verify assertion (iii). Since $\frac{j-1}{2j-1} \in \mathcal{F}^{\leq\frac{1}{2}}(\mathbb{B}(2m), m)$ is the image of the fraction $\frac{j}{2j-1} \in \mathcal{F}^{\geq\frac{1}{2}}(\mathbb{B}(2m), m)$ under the order-reversing bijection $\mathcal{F}(\mathbb{B}(2m), m) \to \mathcal{F}(\mathbb{B}(2m), m)$ defined in (1.8), assertion (iii) follows directly from assertion (ii) on the neighbors of the fraction $\frac{j-1}{2j-1}$. □

2.2.7.2 The neighbors of $\frac{2}{j+2}$, $\frac{j-2}{2(j-1)}$, $\frac{j}{2(j-1)}$ and $\frac{j}{j+2}$

We will now find in the Farey subsequence $\mathcal{F}(\mathbb{B}(2m), m)$ the neighbors of fractions of the form $\frac{2}{j+2}$, $\frac{j-2}{2(j-1)}$, $\frac{j}{2(j-1)}$ and $\frac{j}{j+2}$.

Corollary 2.23. (i) *If* $\frac{2}{j+2} \in \mathcal{F}^{\leq\frac{1}{2}}(\mathbb{B}(2m), m)$ *for some* j*, then the fraction*

$$\begin{cases} (\lceil \frac{2m}{j} \rceil - 1) \Big/ \frac{(j+2)(\lceil \frac{2m}{j} \rceil - 1) + 1}{2} & \text{if } \lceil \frac{2m}{j} \rceil \text{ is even,} \\ (\lceil \frac{2m}{j} \rceil - 2) \Big/ \frac{(j+2)(\lceil \frac{2m}{j} \rceil - 2) + 1}{2} & \text{if } \lceil \frac{2m}{j} \rceil \text{ is odd} \end{cases} \tag{2.47}$$

precedes $\frac{2}{j+2}$ *in* $\mathcal{F}(\mathbb{B}(2m), m)$*, and the fraction*

$$\begin{cases} (\lceil \frac{2(m+1)}{j} \rceil - 1) \Big/ \frac{(j+2)(\lceil \frac{2(m+1)}{j} \rceil - 1) - 1}{2} & \text{if } \lceil \frac{2(m+1)}{j} \rceil \text{ is even,} \\ (\lceil \frac{2(m+1)}{j} \rceil - 2) \Big/ \frac{(j+2)(\lceil \frac{2(m+1)}{j} \rceil - 2) - 1}{2} & \text{if } \lceil \frac{2(m+1)}{j} \rceil \text{ is odd} \end{cases} \tag{2.48}$$

succeeds $\frac{2}{j+2}$ *in* $\mathcal{F}(\mathbb{B}(2m), m)$.

(ii) *If* $\frac{j-2}{2(j-1)} \in \mathcal{F}^{\leq\frac{1}{2}}(\mathbb{B}(2m), m)$ *for some* j*, then the fraction*

$$\begin{cases} \frac{(j-2)(\lceil \frac{2(m+1)}{j} \rceil - 1) - 1}{2} \Big/ ((j-1)(\lceil \frac{2(m+1)}{j} \rceil - 1) - 1) & \text{if } \lceil \frac{2(m+1)}{j} \rceil \text{ is even,} \\ \frac{(j-2)(\lceil \frac{2(m+1)}{j} \rceil - 2) - 1}{2} \Big/ ((j-1)(\lceil \frac{2(m+1)}{j} \rceil - 2) - 1) & \text{if } \lceil \frac{2(m+1)}{j} \rceil \text{ is odd} \end{cases} \tag{2.49}$$

precedes $\frac{j-2}{2(j-1)}$ *in* $\mathcal{F}(\mathbb{B}(2m), m)$*, and the fraction*

$$\begin{cases} \frac{(j-2)(\lceil \frac{2m}{j} \rceil - 1) + 1}{2} \Big/ ((j-1)(\lceil \frac{2m}{j} \rceil - 1) + 1) & \text{if } \lceil \frac{2m}{j} \rceil \text{ is even,} \\ \frac{(j-2)(\lceil \frac{2m}{j} \rceil - 2) + 1}{2} \Big/ ((j-1)(\lceil \frac{2m}{j} \rceil - 2) + 1) & \text{if } \lceil \frac{2m}{j} \rceil \text{ is odd} \end{cases} \tag{2.50}$$

succeeds $\frac{j-2}{2(j-1)}$ *in* $\mathcal{F}(\mathbb{B}(2m), m)$.

(iii) *If* $\frac{j}{2(j-1)} \in \mathcal{F}^{\geq\frac{1}{2}}(\mathbb{B}(2m), m)$ *for some* j*, then the fraction*

$$\begin{cases} \frac{j(\lceil \frac{2m}{j} \rceil - 1) + 1}{2} \Big/ ((j-1)(\lceil \frac{2m}{j} \rceil - 1) + 1) & \text{if } \lceil \frac{2m}{j} \rceil \text{ is even,} \\ \frac{j(\lceil \frac{2m}{j} \rceil - 2) + 1}{2} \Big/ ((j-1)(\lceil \frac{2m}{j} \rceil - 2) + 1) & \text{if } \lceil \frac{2m}{j} \rceil \text{ is odd} \end{cases}$$

precedes $\frac{j}{2(j-1)}$ *in* $\mathcal{F}(\mathbb{B}(2m), m)$*, and the fraction*

$$\begin{cases} \frac{j(\lceil \frac{2(m+1)}{j} \rceil - 1) - 1}{2} \Big/ ((j-1)(\lceil \frac{2(m+1)}{j} \rceil - 1) - 1) & \text{if } \lceil \frac{2(m+1)}{j} \rceil \text{ is even,} \\ \frac{j(\lceil \frac{2(m+1)}{j} \rceil - 2) - 1}{2} \Big/ ((j-1)(\lceil \frac{2(m+1)}{j} \rceil - 2) - 1) & \text{if } \lceil \frac{2(m+1)}{j} \rceil \text{ is odd} \end{cases}$$

succeeds $\frac{j}{2(j-1)}$ *in* $\mathcal{F}(\mathbb{B}(2m), m)$.

(iv) *If* $\frac{j}{j+2} \in \mathcal{F}^{\geq\frac{1}{2}}(\mathbb{B}(2m), m)$ *for some* j, *then the fraction*

$$\begin{cases} \dfrac{j(\lceil\frac{2(m+1)}{j}\rceil - 1) - 1}{2} \Big/ \dfrac{(j+2)(\lceil\frac{2(m+1)}{j}\rceil - 1) - 1}{2} & \text{if } \lceil\frac{2(m+1)}{j}\rceil \text{ is even,} \\ \dfrac{j(\lceil\frac{2(m+1)}{j}\rceil - 2) - 1}{2} \Big/ \dfrac{(j+2)(\lceil\frac{2(m+1)}{j}\rceil - 2) - 1}{2} & \text{if } \lceil\frac{2(m+1)}{j}\rceil \text{ is odd} \end{cases}$$

precedes $\frac{j}{j+2}$ *in* $\mathcal{F}(\mathbb{B}(2m), m)$, *and the fraction*

$$\begin{cases} \dfrac{j(\lceil\frac{2m}{j}\rceil - 1) + 1}{2} \Big/ \dfrac{(j+2)(\lceil\frac{2m}{j}\rceil - 1) + 1}{2} & \text{if } \lceil\frac{2m}{j}\rceil \text{ is even,} \\ \dfrac{j(\lceil\frac{2m}{j}\rceil - 2) + 1}{2} \Big/ \dfrac{(j+2)(\lceil\frac{2m}{j}\rceil - 2) + 1}{2} & \text{if } \lceil\frac{2m}{j}\rceil \text{ is odd} \end{cases}$$

succeeds $\frac{j}{j+2}$ *in* $\mathcal{F}(\mathbb{B}(2m), m)$.

Proof. (i) Recall that, according to Corollary 2.5, the fraction $\frac{2}{j} \in \mathcal{F}_m$ is the image of $\frac{2}{j+2}$ under the order-preserving bijection (2.12). According to Corollary 2.21 (i), the fractions given in (2.39) and (2.40) are the neighbors of the fraction $\frac{2}{j}$ in $\mathcal{F}_m$. As a consequence, their images (2.47) and (2.48) under the order-preserving bijection (2.13) are the neighbors of the fraction $\frac{2}{j+2}$ in $\mathcal{F}(\mathbb{B}(2m), m)$.

To verify assertion (iv), note that the fraction $\frac{2}{j+2} \in \mathcal{F}^{\leq\frac{1}{2}}(\mathbb{B}(2m), m)$ is the image of $\frac{j}{j+2} \in \mathcal{F}^{\geq\frac{1}{2}}(\mathbb{B}(2m), m)$ under the order-reversing bijection $\mathcal{F}(\mathbb{B}(2m), m) \to \mathcal{F}(\mathbb{B}(2m), m)$ defined in (1.8). Thus, assertion (iv) follows directly from assertion (i).

(ii) According to Corollary 2.5, the fraction $\frac{j-2}{j} \in \mathcal{F}_m$ is the image of $\frac{j-2}{2(j-1)}$ under the order-preserving bijection (2.12). According to Corollary 2.21 (ii), the fractions given in (2.41) and (2.42) are the neighbors of the fraction $\frac{j-2}{j}$ in $\mathcal{F}_m$. As a consequence, their images (2.49) and (2.50) under the order-preserving bijection (2.13) are the neighbors of the fraction $\frac{j-2}{2(j-1)}$ in $\mathcal{F}(\mathbb{B}(2m), m)$.

To verify assertion (iii), note that the fraction $\frac{j-2}{2(j-1)} \in \mathcal{F}^{\leq\frac{1}{2}}(\mathbb{B}(2m), m)$ is the image of the fraction $\frac{j}{2(j-1)} \in \mathcal{F}^{\geq\frac{1}{2}}(\mathbb{B}(2m), m)$ under the order-reversing bijection $\mathcal{F}(\mathbb{B}(2m), m) \to \mathcal{F}(\mathbb{B}(2m), m)$ defined in (1.8). Thus, assertion (iii) follows directly from assertion (ii). □

2.2.7.3 The neighbors of $\frac{1}{3}$

Which fractions are the neighbors of the fraction $\frac{1}{3}$ in the Farey subsequences $\mathcal{F}(\mathbb{B}(2m), m)$? The following observation can be made with the help of assertions (i) and (ii) of Corollary 2.22 and of Remark 2.10.

Remark 2.24 (see also Table 2.7). If $m > 1$, then it follows from Remark 2.10, thanks to the order-preserving bijection (2.13), that the fraction

$$\frac{\lfloor (m-1)/2 \rfloor}{3\lfloor (m-1)/2 \rfloor + 1} = \frac{\lceil m/2 \rceil - 1}{3\lceil m/2 \rceil - 2}$$

Table 2.7. The neighbors of the fraction $\frac{1}{3}$ in Farey subsequences; see also Table 2.6.

Sequence	The predecessor of $\frac{1}{3}$		See
$\mathcal{F}(\mathbb{B}(2m), m)$, $m > 1$	$\frac{\lfloor (m-1)/2\rfloor}{3\lfloor (m-1)/2\rfloor+1}$		Remark 2.24
$\mathcal{F}(\mathbb{B}(n), m)$, $n < 2m$, $(n-m) > 1$	$\frac{\lfloor (n-m-1)/2\rfloor}{3\lfloor (n-m-1)/2\rfloor+1}$		Remark 2.42 (i)
$\mathcal{F}(\mathbb{B}(n), m)$, $n > 2m$, $m > 1$	$\frac{m}{3m+1}$	if $n-3m \geq 1$	Remark 2.42 (ii)
	$\frac{\lfloor (n-m-1)/2\rfloor}{3\lfloor (n-m-1)/2\rfloor+1}$	if $n-3m \leq 1$	

Sequence	The successor of $\frac{1}{3}$		See
$\mathcal{F}(\mathbb{B}(2m), m)$, $m > 1$	$\frac{\lceil m/2\rceil}{3\lceil m/2\rceil-1}$		Remark 2.24
$\mathcal{F}(\mathbb{B}(n), m)$, $n < 2m$, $(n-m) > 1$	$\frac{\lceil (n-m)/2\rceil}{3\lceil (n-m)/2\rceil-1}$		Remark 2.42 (i)
$\mathcal{F}(\mathbb{B}(n), m)$, $n > 2m$, $m > 1$	$\frac{\lceil (n-m)/2\rceil}{3\lceil (n-m)/2\rceil-1}$	if $3m-n \geq 1$	Remark 2.42 (ii)
	$\frac{m}{2m-1}$	if $3m-n \leq 1$	

precedes $\frac{1}{3}$ in $\mathcal{F}(\mathbb{B}(2m), m)$, and the fraction

$$\frac{\lceil m/2\rceil}{3\lceil m/2\rceil - 1} = \frac{\lfloor (m-1)/2\rfloor + 1}{3\lfloor (m-1)/2\rfloor + 2}$$

succeeds $\frac{1}{3}$ in $\mathcal{F}(\mathbb{B}(2m), m)$.

Thus, if m is even, then the fractions

$$\frac{(m-2)/2}{(3m-4)/2} < \frac{1}{3} < \frac{m/2}{(3m-2)/2}$$

are consecutive in $\mathcal{F}(\mathbb{B}(2m), m)$. If m is odd, then the fractions

$$\frac{(m-1)/2}{(3m-1)/2} < \frac{1}{3} < \frac{(m+1)/2}{(3m+1)/2}$$

are consecutive in $\mathcal{F}(\mathbb{B}(2m), m)$.

2.2.7.4 The neighbors of $\frac{2}{3}$

Now that we have found in the previous subsection the neighbors of the fraction $\frac{1}{3}$ in the Farey subsequences $\mathcal{F}(\mathbb{B}(2m), m)$, it is natural to ask which fractions are the neighbors of the fraction $\frac{2}{3}$. Let us apply the order-reversing bijection $\mathcal{F}(\mathbb{B}(2m), m) \to \mathcal{F}(\mathbb{B}(2m), m)$ defined in (1.8) to the observations made in Remark 2.24.

Remark 2.25 (see also Table 2.8). If $m > 1$, then the fraction

$$\frac{2\lceil m/2\rceil - 1}{3\lceil m/2\rceil - 1} = \frac{2\lfloor (m-1)/2\rfloor + 1}{3\lfloor (m-1)/2\rfloor + 2}$$

precedes $\frac{2}{3}$ in $\mathcal{F}(\mathbb{B}(2m), m)$, and the fraction

$$\frac{2\lfloor (m-1)/2\rfloor + 1}{3\lfloor (m-1)/2\rfloor + 1} = \frac{2\lceil m/2\rceil - 1}{3\lceil m/2\rceil - 2}$$

succeeds $\frac{2}{3}$ in $\mathcal{F}(\mathbb{B}(2m), m)$.

Table 2.8. The neighbors of the fraction $\frac{2}{3}$ in Farey subsequences; see also Table 2.6.

Sequence	The predecessor of $\frac{2}{3}$	See
$\mathcal{F}(\mathbb{B}(2m), m)$, $m > 1$	$\frac{2\lceil m/2\rceil-1}{3\lceil m/2\rceil-1}$	Remark 2.25
$\mathcal{F}(\mathbb{B}(n), m)$, $n < 2m$, $(n-m) > 1$	$\frac{2\lceil m/2\rceil-1}{3\lceil m/2\rceil-1}$ if $2n-3m \geq 1$ $\frac{n-m-1}{2(n-m)-1}$ if $2n-3m \leq 1$	Remark 2.43 (i)
$\mathcal{F}(\mathbb{B}(n), m)$, $n > 2m$, $m > 1$	$\frac{2\lceil m/2\rceil-1}{3\lceil m/2\rceil-1}$	Remark 2.43 (ii)

Sequence	The successor of $\frac{2}{3}$	See
$\mathcal{F}(\mathbb{B}(2m), m)$, $m > 1$	$\frac{2\lfloor (m-1)/2\rfloor+1}{3\lfloor (m-1)/2\rfloor+1}$	Remark 2.25
$\mathcal{F}(\mathbb{B}(n), m)$, $n < 2m$, $(n-m) > 1$	$\frac{2(n-m)+1}{3(n-m)+1}$ if $3m-2n \geq 1$ $\frac{2\lfloor (m-1)/2\rfloor+1}{3\lfloor (m-1)/2\rfloor+1}$ if $3m-2n \leq 1$	Remark 2.43 (i)
$\mathcal{F}(\mathbb{B}(n), m)$, $n > 2m$, $m > 1$	$\frac{2\lfloor (m-1)/2\rfloor+1}{3\lfloor (m-1)/2\rfloor+1}$	Remark 2.43 (ii)

Thus, if m is even, then the fractions

$$\frac{m-1}{(3m-2)/2} < \frac{2}{3} < \frac{m-1}{(3m-4)/2}$$

are consecutive in $\mathcal{F}(\mathbb{B}(2m), m)$. If m is odd, then the fractions

$$\frac{m}{(3m+1)/2} < \frac{2}{3} < \frac{m}{(3m-1)/2}$$

are consecutive in $\mathcal{F}(\mathbb{B}(2m), m)$.

2.2.8 More on neighboring fractions in $\mathcal{F}_m^\ell$ and $\mathcal{G}_m^\ell$

In this subsection we will describe explicitly the neighbors of fractions of the form $\frac{1}{j}$, $\frac{j-1}{j}$, $\frac{2}{j}$ and $\frac{j-2}{j}$ in the Farey subsequences $\mathcal{F}_m^\ell$ and $\mathcal{G}_m^\ell$, with the help of the formulas provided by Lemmas 2.13 and 2.15. Recall that we have already found in Section 2.2.6 the neighbors of these fractions in the Farey sequences $\mathcal{F}_m$. See Table 2.6 on the selected pairs of neighboring fractions that are considered in this book.

2.2.8.1 The neighbors of $\frac{1}{j}$

Let us find the neighbors of a fraction $\frac{1}{j}$ in the Farey subsequences $\mathcal{F}_m^\ell$ and $\mathcal{G}_m^\ell$.

Corollary 2.26. (i) *Consider a fraction* $\frac{1}{j} \in \mathcal{F}_m^\ell$, *where* $0 < \ell < m$.

(a) *If* $m - j\ell \geq 1$, *then the fraction*

$$\frac{\ell}{j\ell+1} \tag{2.51}$$

precedes the fraction $\frac{1}{j}$ *in* $\mathcal{F}_m^\ell$.

(b) *If* $m - j\ell \le 1$, *then the fraction*

$$\frac{\lceil \frac{m}{j} \rceil - 1}{j(\lceil \frac{m}{j} \rceil - 1) + 1} \tag{2.52}$$

precedes the fraction $\frac{1}{j}$ *in* $\mathcal{F}_m^\ell$.

(ii) *Consider a fraction* $\frac{1}{j} \in \mathcal{F}_m^\ell$, $j > 1$, *where* $0 < \ell < m$.

(a) *If* $j\ell - m \ge 1$, *then the fraction*

$$\frac{\lceil \frac{m+2}{j} \rceil - 1}{j(\lceil \frac{m+2}{j} \rceil - 1) - 1}$$

succeeds the fraction $\frac{1}{j}$ *in* $\mathcal{F}_m^\ell$.

(b) *If* $j\ell - m \le 1$, *then the fraction* $\frac{\ell}{j\ell - 1}$ *succeeds the fraction* $\frac{1}{j}$ *in* $\mathcal{F}_m^\ell$.

Proof. The assertions of the corollary follow directly from the respective assertions of Lemma 2.13. Indeed, the search intervals for the integers a in that lemma turn, for fractions $\frac{h}{k}$ of the form $\frac{1}{j}$, into singleton sets. □

Corollary 2.27. (i) *Consider a fraction* $\frac{1}{j} \in \mathcal{G}_m^\ell$, *where* $0 < \ell < m$.

(a) *If* $m - j\ell \ge 1$, *then the fraction*

$$\frac{\lceil \frac{m}{j} \rceil - 1}{j(\lceil \frac{m}{j} \rceil - 1) + 1}$$

precedes the fraction $\frac{1}{j}$ *in* $\mathcal{G}_m^\ell$.

(b) *If* $m - j\ell \le 1$, *then the fraction*

$$\frac{\lceil \frac{m-\ell}{j-1} \rceil - 1}{j(\lceil \frac{m-\ell}{j-1} \rceil - 1) + 1}$$

precedes the fraction $\frac{1}{j}$ *in* $\mathcal{G}_m^\ell$.

(ii) *Consider a fraction* $\frac{1}{j} \in \mathcal{G}_m^\ell$, $j > 1$, *where* $0 < \ell < m$.

(a) *If* $j\ell - m \ge 1$, *then the fraction*

$$\frac{\lceil \frac{m-\ell+2}{j-1} \rceil - 1}{j(\lceil \frac{m-\ell+2}{j-1} \rceil - 1) - 1}$$

succeeds the fraction $\frac{1}{j}$ *in* $\mathcal{G}_m^\ell$.

(b) *If* $j\ell - m \le 1$, *then the fraction*

$$\frac{\lceil \frac{m+2}{j} \rceil - 1}{j(\lceil \frac{m+2}{j} \rceil - 1) - 1}$$

succeeds the fraction $\frac{1}{j}$ *in* $\mathcal{G}_m^\ell$.

Proof. The corollary follows directly from Lemma 2.15, since the search intervals for the integers a in that lemma turn, for fractions $\frac{h}{k}$ of the form $\frac{1}{j}$, into singleton sets. □

2.2.8.2 The neighbors of $\frac{j-1}{j}$

To find the neighbors of a fraction $\frac{j-1}{j}$ in the Farey subsequences $\mathcal{G}_m^{\ell}$ and $\mathcal{F}_m^{\ell}$, we will now apply the order-reversing bijections $\mathcal{G}_m^{\ell} \leftarrow \mathcal{F}_m^{m-\ell}$ and $\mathcal{F}_m^{\ell} \leftarrow \mathcal{G}_m^{m-\ell}$, defined in (1.8), to the observations made in Corollaries 2.26 and 2.27.

Corollary 2.28. (i) *Consider a fraction* $\frac{j-1}{j} \in \mathcal{G}_m^{\ell}$, $j > 1$, *where* $0 < \ell < m$.

(a) *If* $j(m-\ell) - m \geq 1$, *then the fraction*

$$\frac{(j-1)(\lceil \frac{m+2}{j} \rceil - 1) - 1}{j(\lceil \frac{m+2}{j} \rceil - 1) - 1} \tag{2.53}$$

precedes the fraction $\frac{j-1}{j}$ *in* $\mathcal{G}_m^{\ell}$.

(b) *If* $j(m-\ell) - m \leq 1$, *then the fraction*

$$\frac{(j-1)(m-\ell) - 1}{j(m-\ell) - 1}$$

precedes the fraction $\frac{j-1}{j}$ *in* $\mathcal{G}_m^{\ell}$.

(ii) *Consider a fraction* $\frac{j-1}{j} \in \mathcal{G}_m^{\ell}$, *where* $0 < \ell < m$.

(a) *If* $m - j(m-\ell) \geq 1$, *then the fraction*

$$\frac{(j-1)(m-\ell) + 1}{j(m-\ell) + 1}$$

succeeds the fraction $\frac{j-1}{j}$ *in* $\mathcal{G}_m^{\ell}$.

(b) *If* $m - j(m-\ell) \leq 1$, *then the fraction*

$$\frac{(j-1)(\lceil \frac{m}{j} \rceil - 1) + 1}{j(\lceil \frac{m}{j} \rceil - 1) + 1}$$

succeeds the fraction $\frac{j-1}{j}$ *in* $\mathcal{G}_m^{\ell}$.

Proof. Assertions (i) and (ii) can be derived from Corollary 2.26 (ii) and Corollary 2.26 (i), respectively.

For instance, to verify assertion (i) (a), note that the fraction $\frac{1}{j} \in \mathcal{F}_m^{m-\ell}$ is the image of the fraction $\frac{j-1}{j} \in \mathcal{G}_m^{\ell}$ under the order-reversing bijection (1.8). According to Corollary 2.26 (ii) (a), the fraction $\frac{\lceil (m+2)/j \rceil - 1}{j(\lceil (m+2)/j \rceil - 1) - 1}$ succeeds $\frac{1}{j}$ in $\mathcal{F}_m^{m-\ell}$. As a consequence, its image (2.53) under the order-reversing bijection (1.8) precedes $\frac{j-1}{j}$ in $\mathcal{G}_m^{\ell}$. □

Corollary 2.29. (i) *Consider a fraction* $\frac{j-1}{j} \in \mathcal{F}_m^{\ell}$, $j > 1$, *where* $0 < \ell < m$.

(a) *If* $j(m-\ell) - m \geq 1$, *then the fraction*

$$\frac{(j-1)(\lceil \frac{\ell+2}{j-1} \rceil - 1) - 1}{j(\lceil \frac{\ell+2}{j-1} \rceil - 1) - 1} \tag{2.54}$$

precedes the fraction $\frac{j-1}{j}$ *in* $\mathcal{F}_m^{\ell}$.

(b) *If* $j(m-\ell)-m \le 1$, *then the fraction*

$$\frac{(j-1)(\lceil \frac{m+2}{j} \rceil - 1) - 1}{j(\lceil \frac{m+2}{j} \rceil - 1) - 1}$$

precedes the fraction $\frac{j-1}{j}$ *in* $\mathcal{F}_m^\ell$.

(ii) *Consider a fraction* $\frac{j-1}{j} \in \mathcal{F}_m^\ell$, *where* $0 < \ell < m$.

(a) *If* $m - j(m-\ell) \ge 1$, *then the fraction*

$$\frac{(j-1)(\lceil \frac{m}{j} \rceil - 1) + 1}{j(\lceil \frac{m}{j} \rceil - 1) + 1}$$

succeeds the fraction $\frac{j-1}{j}$ *in* $\mathcal{F}_m^\ell$.

(b) *If* $m - j(m-\ell) \le 1$, *then the fraction*

$$\frac{(j-1)(\lceil \frac{\ell}{j-1} \rceil - 1) + 1}{j(\lceil \frac{\ell}{j-1} \rceil - 1) + 1}$$

succeeds the fraction $\frac{j-1}{j}$ *in* $\mathcal{F}_m^\ell$.

Proof. Assertions (i) and (ii) can be derived from Corollary 2.27 (ii) and Corollary 2.27 (i), respectively.

For instance, to verify assertion (i) (a), note that the fraction $\frac{1}{j} \in \mathcal{G}_m^{m-\ell}$ is the image of the fraction $\frac{j-1}{j} \in \mathcal{F}_m^\ell$ under the order-reversing bijection (1.8). According to Corollary 2.27 (ii) (a), the fraction $\frac{\lceil(\ell+2)/(j-1)\rceil-1}{j(\lceil(\ell+2)/(j-1)\rceil-1)-1}$ succeeds $\frac{1}{j}$ in $\mathcal{G}_m^{m-\ell}$. As a consequence, its image (2.54) under the order-reversing bijection (1.8) precedes $\frac{j-1}{j}$ in $\mathcal{F}_m^\ell$. □

2.2.8.3 The neighbors of $\frac{2}{j}$

To find the neighbors of a fraction $\frac{2}{j}$ in the Farey subsequences $\mathcal{F}_m^\ell$ and $\mathcal{G}_m^\ell$, we again turn to Lemmas 2.13 and 2.15.

Corollary 2.30. *Consider the fraction* $\frac{2}{j} \in \mathcal{F}_m^\ell$ *for some odd* j, *where* $0 < \ell < m$.

(i) (a) *If* $2m - j\ell \ge 1$, *then the fraction*

$$\begin{cases} \ell \Big/ \dfrac{j\ell+1}{2} & \text{if } \ell \text{ is odd,} \\ (\ell-1) \Big/ \dfrac{j(\ell-1)+1}{2} & \text{if } \ell \text{ is even} \end{cases}$$

precedes $\frac{2}{j}$ *in* $\mathcal{F}_m^\ell$.

(b) *If* $2m - j\ell \le 1$, *then the fraction*

$$\begin{cases} (\lceil \frac{2m}{j} \rceil - 1) \Big/ \dfrac{j(\lceil \frac{2m}{j} \rceil - 1)+1}{2} & \text{if } \lceil \frac{2m}{j} \rceil \text{ is even,} \\ (\lceil \frac{2m}{j} \rceil - 2) \Big/ \dfrac{j(\lceil \frac{2m}{j} \rceil - 2)+1}{2} & \text{if } \lceil \frac{2m}{j} \rceil \text{ is odd} \end{cases}$$

precedes $\frac{2}{j}$ *in* $\mathcal{F}_m^\ell$.

(ii) (a) *If $j\ell - 2m \geq 1$, then the fraction*

$$\begin{cases} (\lceil \frac{2(m+1)}{j} \rceil - 1) \Big/ \dfrac{j(\lceil \frac{2(m+1)}{j} \rceil - 1) - 1}{2} & \text{if } \lceil \frac{2(m+1)}{j} \rceil \text{ is even,} \\ (\lceil \frac{2(m+1)}{j} \rceil - 2) \Big/ \dfrac{j(\lceil \frac{2(m+1)}{j} \rceil - 2) - 1}{2} & \text{if } \lceil \frac{2(m+1)}{j} \rceil \text{ is odd} \end{cases}$$

succeeds $\frac{2}{j}$ *in* $\mathcal{F}_m^\ell$.

(b) *If $j\ell - 2m \leq 1$, then the fraction*

$$\begin{cases} \ell \Big/ \dfrac{j\ell - 1}{2} & \text{if } \ell \text{ is odd,} \\ (\ell - 1) \Big/ \dfrac{j(\ell - 1) - 1}{2} & \text{if } \ell \text{ is even} \end{cases}$$

succeeds $\frac{2}{j}$ *in* $\mathcal{F}_m^\ell$.

Proof. The assertions of this corollary follow directly from the respective assertions of Lemma 2.13. For instance, under the hypothesis of assertion (i) (a), Lemma 2.13 (i) (a) implies that the fraction $\frac{2}{j}$ succeeds the fraction $a/\frac{ja+1}{2}$, where

$$ja \equiv -1 \pmod 2, \quad \ell - 1 \leq a \leq \ell. \qquad \square$$

Corollary 2.31. *Consider the fraction $\frac{2}{j} \in \mathcal{G}_m^\ell$ for some odd j, where $0 < \ell < m$.*

(i) (a) *If $2m - j\ell \geq 1$, then the fraction*

$$\begin{cases} (\lceil \frac{2m}{j} \rceil - 1) \Big/ \dfrac{j(\lceil \frac{2m}{j} \rceil - 1) + 1}{2} & \text{if } \lceil \frac{2m}{j} \rceil \text{ is even,} \\ (\lceil \frac{2m}{j} \rceil - 2) \Big/ \dfrac{j(\lceil \frac{2m}{j} \rceil - 2) + 1}{2} & \text{if } \lceil \frac{2m}{j} \rceil \text{ is odd} \end{cases}$$

precedes $\frac{2}{j}$ *in* $\mathcal{G}_m^\ell$.

(b) *If $2m - j\ell \leq 1$, then the fraction*

$$\begin{cases} (\lceil \frac{2(m-\ell)}{j-2} \rceil - 1) \Big/ \dfrac{j(\lceil \frac{2(m-\ell)}{j-2} \rceil - 1) + 1}{2} & \text{if } \lceil \frac{2(m-\ell)}{j-2} \rceil \text{ is even,} \\ (\lceil \frac{2(m-\ell)}{j-2} \rceil - 2) \Big/ \dfrac{j(\lceil \frac{2(m-\ell)}{j-2} \rceil - 2) + 1}{2}, & \text{if } \lceil \frac{2(m-\ell)}{j-2} \rceil \text{ is odd} \end{cases}$$

precedes $\frac{2}{j}$ *in* $\mathcal{G}_m^\ell$.

(ii) (a) *If $j\ell - 2m \geq 1$, then the fraction*

$$\begin{cases} (\lceil \frac{2(m-\ell+1)}{j-2} \rceil - 1) \Big/ \dfrac{j(\lceil \frac{2(m-\ell+1)}{j-2} \rceil - 1) - 1}{2} & \text{if } \lceil \frac{2(m-\ell+1)}{j-2} \rceil \text{ is even,} \\ (\lceil \frac{2(m-\ell+1)}{j-2} \rceil - 2) \Big/ \dfrac{j(\lceil \frac{2(m-\ell+1)}{j-2} \rceil - 2) - 1}{2} & \text{if } \lceil \frac{2(m-\ell+1)}{j-2} \rceil \text{ is odd} \end{cases}$$

succeeds $\frac{2}{j}$ *in* $\mathcal{G}_m^\ell$.

(b) *If* $j\ell - 2m \le 1$, *then the fraction*

$$\begin{cases} \big(\lceil \frac{2(m+1)}{j} \rceil - 1\big) \Big/ \dfrac{j(\lceil \frac{2(m+1)}{j} \rceil - 1) - 1}{2} & \text{if } \lceil \frac{2(m+1)}{j} \rceil \text{ is even,} \\ \big(\lceil \frac{2(m+1)}{j} \rceil - 2\big) \Big/ \dfrac{j(\lceil \frac{2(m+1)}{j} \rceil - 2) - 1}{2} & \text{if } \lceil \frac{2(m+1)}{j} \rceil \text{ is odd} \end{cases}$$

succeeds $\frac{2}{j}$ *in* $\mathcal{G}_m^\ell$.

Proof. The assertions of this corollary follow directly from the respective assertions of Lemma 2.15. For instance, under the hypothesis of assertion (i) (a), Lemma 2.15 (i) (a) implies that the fraction $\frac{2}{j}$ succeeds the fraction $a/\frac{ja+1}{2}$, where

$$ja \equiv -1 \pmod 2), \quad \left\lceil \frac{2m}{j} \right\rceil - 2 \le a \le \left\lceil \frac{2m}{j} \right\rceil - 1. \qquad \square$$

2.2.8.4 The neighbors of $\frac{j-2}{j}$

To find the neighbors of a fraction $\frac{j-2}{j}$ in the Farey subsequences $\mathcal{G}_m^\ell$ and $\mathcal{F}_m^\ell$, we will now apply the order-reversing bijections $\mathcal{G}_m^\ell \leftarrow \mathcal{F}_m^{m-\ell}$ and $\mathcal{F}_m^\ell \leftarrow \mathcal{G}_m^{m-\ell}$, defined in (1.8), to the observations in Corollaries 2.30 and 2.31.

Corollary 2.32. *Consider the fraction* $\frac{j-2}{j} \in \mathcal{G}_m^\ell$ *for some odd* j, *where* $0 < \ell < m$.

(i) (a) *If* $j(m-\ell) - 2m \ge 1$, *then the fraction*

$$\begin{cases} \dfrac{(j-2)(\lceil \frac{2(m+1)}{j} \rceil - 1) - 1}{2} \Big/ \dfrac{j(\lceil \frac{2(m+1)}{j} \rceil - 1) - 1}{2} & \text{if } \lceil \frac{2(m+1)}{j} \rceil \text{ is even,} \\ \dfrac{(j-2)(\lceil \frac{2(m+1)}{j} \rceil - 2) - 1}{2} \Big/ \dfrac{j(\lceil \frac{2(m+1)}{j} \rceil - 2) - 1}{2} & \text{if } \lceil \frac{2(m+1)}{j} \rceil \text{ is odd} \end{cases} \tag{2.55}$$

precedes $\frac{j-2}{j}$ *in* $\mathcal{G}_m^\ell$.

(b) *If* $j(m-\ell) - 2m \le 1$, *then the fraction*

$$\begin{cases} \dfrac{(j-2)(m-\ell) - 1}{2} \Big/ \dfrac{j(m-\ell) - 1}{2} & \text{if } (m-\ell) \text{ is odd,} \\ \dfrac{(j-2)(m-\ell-1) - 1}{2} \Big/ \dfrac{j(m-\ell-1) - 1}{2} & \text{if } (m-\ell) \text{ is even} \end{cases}$$

precedes $\frac{j-2}{j}$ *in* $\mathcal{G}_m^\ell$.

(ii) (a) *If* $2m - j(m-\ell) \ge 1$, *then the fraction*

$$\begin{cases} \dfrac{(j-2)(m-\ell) + 1}{2} \Big/ \dfrac{j(m-\ell) + 1}{2} & \text{if } (m-\ell) \text{ is odd,} \\ \dfrac{(j-2)(m-\ell-1) + 1}{2} \Big/ \dfrac{j(m-\ell-1) + 1}{2} & \text{if } (m-\ell) \text{ is even} \end{cases}$$

succeeds $\frac{j-2}{j}$ *in* $\mathcal{G}_m^\ell$.

(b) *If* $2m - j(m - \ell) \leq 1$, *then the fraction*

$$\begin{cases} \dfrac{(j-2)(\lceil \frac{2m}{j} \rceil - 1) + 1}{2} \Big/ \dfrac{j(\lceil \frac{2m}{j} \rceil - 1) + 1}{2} & \text{if } \lceil \frac{2m}{j} \rceil \text{ is even,} \\ \dfrac{(j-2)(\lceil \frac{2m}{j} \rceil - 2) + 1}{2} \Big/ \dfrac{j(\lceil \frac{2m}{j} \rceil - 2) + 1}{2} & \text{if } \lceil \frac{2m}{j} \rceil \text{ is odd} \end{cases}$$

succeeds $\frac{j-2}{j}$ *in* $\mathcal{G}^{\ell}_{m}$.

Proof. Assertions (i) and (ii) follow from Corollary 2.30 (ii) and Corollary 2.30 (i), respectively.

Note that the fraction $\frac{2}{j} \in \mathcal{F}^{m-\ell}_{m}$ is the image of the fraction $\frac{j-2}{j} \in \mathcal{G}^{\ell}_{m}$ under the order-reversing bijection (1.8).

For instance, to prove assertion (i) (a), note that if $j(m - \ell) - 2m \geq 1$, then, according to Corollary 2.30 (ii) (a), the fraction

$$\begin{cases} (\lceil \frac{2(m+1)}{j} \rceil - 1) \Big/ \dfrac{j(\lceil \frac{2(m+1)}{j} \rceil - 1) - 1}{2} & \text{if} \lceil \frac{2(m+1)}{j} \rceil \text{ is even,} \\ (\lceil \frac{2(m+1)}{j} \rceil - 2) \Big/ \dfrac{j(\lceil \frac{2(m+1)}{j} \rceil - 2) - 1}{2} & \text{if } \lceil \frac{2(m+1)}{j} \rceil \text{ is odd} \end{cases}$$

succeeds $\frac{2}{j}$ in $\mathcal{F}^{m-\ell}_{m}$. As a consequence, its image (2.55) under the order-reversing bijection (1.8) precedes the fraction $\frac{j-2}{j}$ in $\mathcal{G}^{\ell}_{m}$. □

Corollary 2.33. *Consider the fraction* $\frac{j-2}{j} \in \mathcal{F}^{\ell}_{m}$ *for some odd* j, *where* $0 < \ell < m$.

(i) (a) *If* $j(m - \ell) - 2m \geq 1$, *then the fraction*

$$\begin{cases} \dfrac{(j-2)(\lceil \frac{2(\ell+1)}{j-2} \rceil - 1) - 1}{2} \Big/ \dfrac{j(\lceil \frac{2(\ell+1)}{j-2} \rceil - 1) - 1}{2} & \text{if } \lceil \frac{2(\ell+1)}{j-2} \rceil \text{ is even,} \\ \dfrac{(j-2)(\lceil \frac{2(\ell+1)}{j-2} \rceil - 2) - 1}{2} \Big/ \dfrac{j(\lceil \frac{2(\ell+1)}{j-2} \rceil - 2) - 1}{2} & \text{if } \lceil \frac{2(\ell+1)}{j-2} \rceil \text{ is odd} \end{cases} \tag{2.56}$$

precedes $\frac{j-2}{j}$ *in* $\mathcal{F}^{\ell}_{m}$.

(b) *If* $j(m - \ell) - 2m \leq 1$, *then the fraction*

$$\begin{cases} \dfrac{(j-2)(\lceil \frac{2(m+1)}{j} \rceil - 1) - 1}{2} \Big/ \dfrac{j(\lceil \frac{2(m+1)}{j} \rceil - 1) - 1}{2} & \text{if } \lceil \frac{2(m+1)}{j} \rceil \text{ is even,} \\ \dfrac{(j-2)(\lceil \frac{2(m+1)}{j} \rceil - 2) - 1}{2} \Big/ \dfrac{j(\lceil \frac{2(m+1)}{j} \rceil - 2) - 1}{2} & \text{if } \lceil \frac{2(m+1)}{j} \rceil \text{ is odd} \end{cases}$$

precedes $\frac{j-2}{j}$ *in* $\mathcal{F}^{\ell}_{m}$.

(ii) (a) *If* $2m - j(m-\ell) \geq 1$, *then the fraction*

$$\begin{cases} \dfrac{(j-2)(\lceil\frac{2m}{j}\rceil - 1) + 1}{2} \Big/ \dfrac{j(\lceil\frac{2m}{j}\rceil - 1) + 1}{2} & \textit{if } \lceil\frac{2m}{j}\rceil \textit{ is even,} \\ \dfrac{(j-2)(\lceil\frac{2m}{j}\rceil - 2) + 1}{2} \Big/ \dfrac{j(\lceil\frac{2m}{j}\rceil - 2) + 1}{2} & \textit{if } \lceil\frac{2m}{j}\rceil \textit{ is odd} \end{cases}$$

succeeds $\frac{j-2}{j}$ *in* $\mathcal{F}^{\ell}_{m}$.

(b) *If* $2m - j(m-\ell) \leq 1$, *then the fraction*

$$\begin{cases} \dfrac{(j-2)(\lceil\frac{2\ell}{j-2}\rceil - 1) + 1}{2} \Big/ \dfrac{j(\lceil\frac{2\ell}{j-2}\rceil - 1) + 1}{2} & \textit{if } \lceil\frac{2\ell}{j-2}\rceil \textit{ is even,} \\ \dfrac{(j-2)(\lceil\frac{2\ell}{j-2}\rceil - 2) + 1}{2} \Big/ \dfrac{j(\lceil\frac{2\ell}{j-2}\rceil - 2) + 1}{2} & \textit{if } \lceil\frac{2\ell}{j-2}\rceil \textit{ is odd} \end{cases}$$

succeeds $\frac{j-2}{j}$ *in* $\mathcal{F}^{\ell}_{m}$.

Proof. Assertions (i) and (ii) follow from Corollary 2.31 (ii) and Corollary 2.31 (i), respectively.

Note that the fraction $\frac{2}{j} \in \mathcal{G}^{m-\ell}_{m}$ is the image of the fraction $\frac{j-2}{j} \in \mathcal{F}^{\ell}_{m}$ under the order-reversing bijection (1.8).

For instance, to prove assertion (i) (a), note that if $j(m-\ell) - 2m \geq 1$, then, according to Corollary 2.31 (ii) (a), the fraction

$$\begin{cases} (\lceil\frac{2(\ell+1)}{j-2}\rceil - 1) \Big/ \dfrac{j(\lceil\frac{2(\ell+1)}{j-2}\rceil - 1) - 1}{2} & \text{if } \lceil\frac{2(\ell+1)}{j-2}\rceil \text{ is even,} \\ (\lceil\frac{2(\ell+1)}{j-2}\rceil - 2) \Big/ \dfrac{j(\lceil\frac{2(\ell+1)}{j-2}\rceil - 2) - 1}{2} & \text{if } \lceil\frac{2(\ell+1)}{j-2}\rceil \text{ is odd} \end{cases}$$

succeeds $\frac{2}{j}$ in $\mathcal{G}^{m-\ell}_{m}$. As a consequence, its image (2.56) under the order-reversing bijection (1.8) precedes the fraction $\frac{j-2}{j}$ in $\mathcal{F}^{\ell}_{m}$. □

2.2.9 More on neighboring fractions in $\mathcal{F}(\mathbb{B}(n), m)$, $n \neq 2m$

In this subsection, on the one hand, we use the characterization of Farey duality given in Section 2.1.4 to find the neighbors of fractions of the form $\frac{1}{j+1}$, $\frac{j-1}{2j-1}$, $\frac{2}{j+2}$ and $\frac{j-2}{2(j-1)}$ in the left half sequence $\mathcal{F}^{\leq\frac{1}{2}}(\mathbb{B}(n), m)$ of the sequence $\mathcal{F}(\mathbb{B}(n), m)$, with $n \neq 2m$. On the other hand, we apply the order-reversing bijections $\mathcal{F}^{\geq\frac{1}{2}}(\mathbb{B}(n), m) \leftrightarrow \mathcal{F}^{\leq\frac{1}{2}}(\mathbb{B}(n), n-m)$, defined in (1.8), to the obtained results, in order to make similar observations on fractions of the form $\frac{j}{j+1}$, $\frac{j}{2j-1}$, $\frac{j}{j+2}$ and $\frac{j}{2(j-1)}$ in the right half sequence $\mathcal{F}^{\geq\frac{1}{2}}(\mathbb{B}(n), m)$.

2.2.9.1 The neighbors of $\frac{1}{j+1}$ in $\mathcal{F}^{\leq\frac{1}{2}}(\mathbb{B}(n), m)$

Let us find the neighbors of a fraction $\frac{1}{j+1}$ in the left half sequences $\mathcal{F}^{\leq\frac{1}{2}}(\mathbb{B}(n), m)$ of the sequences $\mathcal{F}(\mathbb{B}(n), m)$, with $n \neq 2m$.

Corollary 2.34. *Consider the fraction* $\frac{1}{j+1} \in \mathcal{F}^{\leq\frac{1}{2}}(\mathbb{B}(n), m) - \{\frac{1}{2}\}$ *for some* j, *where* $n \neq 2m$.

(i) (a) *Suppose that* $n < 2m$. *Then the fraction*

$$\frac{\lceil \frac{n-m}{j} \rceil - 1}{(j+1)(\lceil \frac{n-m}{j} \rceil - 1) + 1} \tag{2.57}$$

precedes $\frac{1}{j+1}$ *in* $\mathcal{F}(\mathbb{B}(n), m)$.

(b) *Suppose that* $n > 2m$. *If* $n - (j+1)m \geq 1$, *then the fraction*

$$\frac{m}{(j+1)m + 1} \tag{2.58}$$

precedes $\frac{1}{j+1}$ *in* $\mathcal{F}(\mathbb{B}(n), m)$. *If* $n - (j+1)m \leq 1$, *then the fraction*

$$\frac{\lceil \frac{n-m}{j} \rceil - 1}{(j+1)(\lceil \frac{n-m}{j} \rceil - 1) + 1} \tag{2.59}$$

precedes $\frac{1}{j+1}$ *in* $\mathcal{F}(\mathbb{B}(n), m)$.

(ii) (a) *Suppose that* $n < 2m$. *Then the fraction*

$$\frac{\lceil \frac{n-m+2}{j} \rceil - 1}{(j+1)(\lceil \frac{n-m+2}{j} \rceil - 1) - 1} \tag{2.60}$$

succeeds $\frac{1}{j+1}$ *in* $\mathcal{F}(\mathbb{B}(n), m)$.

(b) *Suppose that* $n > 2m$. *If* $(j+1)m - n \geq 1$, *then the fraction*

$$\frac{\lceil \frac{n-m+2}{j} \rceil - 1}{(j+1)(\lceil \frac{n-m+2}{j} \rceil - 1) - 1} \tag{2.61}$$

succeeds $\frac{1}{j+1}$ *in* $\mathcal{F}(\mathbb{B}(n), m)$. *If* $(j+1)m - n \leq 1$, *then the fraction*

$$\frac{m}{(j+1)m - 1} \tag{2.62}$$

succeeds $\frac{1}{j+1}$ *in* $\mathcal{F}(\mathbb{B}(n), m)$.

Proof. Let us verify assertions (i) (a) and (ii) (a). According to Corollary 2.8 (i), the fraction $\frac{1}{j} \in \mathcal{F}_{n-m}$ is the image of the fraction $\frac{1}{j+1} \in \mathcal{F}^{\leq\frac{1}{2}}(\mathbb{B}(n), m)$ under the order-preserving bijection (2.20). According to Corollary 2.20 (i), the fraction

$$\frac{\lceil \frac{n-m}{j} \rceil - 1}{j(\lceil \frac{n-m}{j} \rceil - 1) + 1}$$

precedes $\frac{1}{j}$ in $\mathcal{F}_{n-m}$; as a consequence, its image (2.57) under the order-preserving bijection (2.21) precedes $\frac{1}{j+1}$ in $\mathcal{F}^{\leq\frac{1}{2}}(\mathbb{B}(n), m)$.

According to Corollary 2.20 (i), the fraction

$$\frac{\lceil \frac{n-m+2}{j} \rceil - 1}{j(\lceil \frac{n-m+2}{j} \rceil - 1) - 1}$$

succeeds $\frac{1}{j}$ in $\mathcal{F}_{n-m}$. As a consequence, its image (2.60) under the order-preserving bijection (2.21) succeeds $\frac{1}{j+1}$ in $\mathcal{F}^{\leq\frac{1}{2}}(\mathbb{B}(n), m)$.

Let us verify assertion (i) (b). According to Corollary 2.8 (ii), the fraction $\frac{1}{j} \in \mathcal{F}^{m}_{n-m}$ is the image of the fraction $\frac{1}{j+1} \in \mathcal{F}^{\leq\frac{1}{2}}(\mathbb{B}(n), m)$ under the order-preserving bijection (2.25).

If $(n-m) - jm = n - (j+1)m \geq 1$, then Corollary 2.26 (i) (a) states that the fraction $\frac{m}{jm+1}$ precedes $\frac{1}{j}$ in $\mathcal{F}^{m}_{n-m}$. As a consequence, its image (2.58) under the order-preserving bijection (2.26) precedes $\frac{1}{j+1}$ in $\mathcal{F}^{\leq\frac{1}{2}}(\mathbb{B}(n), m)$.

If $n - (j+1)m \leq 1$, then Corollary 2.26 (i) (b) states that the fraction

$$\frac{\lceil \frac{n-m}{j} \rceil - 1}{j(\lceil \frac{n-m}{j} \rceil - 1) + 1}$$

precedes $\frac{1}{j}$ in $\mathcal{F}^{m}_{n-m}$. As a consequence, its image (2.59) under the order-preserving bijection (2.26) precedes $\frac{1}{j+1}$ in $\mathcal{F}^{\leq\frac{1}{2}}(\mathbb{B}(n), m)$.

Let us verify assertion (ii) (b).

If $jm - (n-m) = (j+1)m - n \geq 1$, then Corollary 2.26 (ii) (a) states that the fraction

$$\frac{\lceil \frac{n-m+2}{j} \rceil - 1}{j(\lceil \frac{n-m+2}{j} \rceil - 1) - 1}$$

succeeds $\frac{1}{j}$ in $\mathcal{F}^{m}_{n-m}$. As a consequence, its image (2.61) under the order-preserving bijection (2.26) succeeds $\frac{1}{j+1}$ in $\mathcal{F}^{\leq\frac{1}{2}}(\mathbb{B}(n), m)$.

If $(j+1)m - n \leq 1$, then Corollary 2.26 (ii) (b) states that the fraction $\frac{m}{jm-1}$ succeeds $\frac{1}{j}$ in $\mathcal{F}^{m}_{n-m}$. As a consequence, its image (2.62) under the order-preserving bijection (2.26) succeeds $\frac{1}{j+1}$ in $\mathcal{F}^{\leq\frac{1}{2}}(\mathbb{B}(n), m)$. □

2.2.9.2 The neighbors of $\frac{j}{j+1}$ in $\mathcal{F}^{\geq\frac{1}{2}}(\mathbb{B}(n), m)$

In order to describe the neighbors of a fraction $\frac{j}{j+1}$ in the right half sequences $\mathcal{F}^{\geq\frac{1}{2}}(\mathbb{B}(n), m)$ of the sequences $\mathcal{F}(\mathbb{B}(n), m)$, with $n \neq 2m$, we will apply the order-reversing bijections $\mathcal{F}^{\geq\frac{1}{2}}(\mathbb{B}(n), m) \leftrightarrow \mathcal{F}^{\leq\frac{1}{2}}(\mathbb{B}(n), n-m)$, defined in (1.8), to the observations, made in Corollary 2.34, on the neighbors of the fraction $\frac{1}{j+1}$ in $\mathcal{F}^{\leq\frac{1}{2}}(\mathbb{B}(n), n-m)$.

Corollary 2.35. *Consider the fraction $\frac{j}{j+1} \in \mathcal{F}^{\geq\frac{1}{2}}(\mathbb{B}(n), m) - \{\frac{1}{2}\}$ for some j, where $n \neq 2m$.*

(i) (a) *Suppose that $n < 2m$. If $jn - (j+1)m \geq 1$, then the fraction*

$$\frac{j(\lceil \frac{m+2}{j} \rceil - 1) - 1}{(j+1)(\lceil \frac{m+2}{j} \rceil - 1) - 1} \tag{2.63}$$

precedes $\frac{j}{j+1}$ in $\mathcal{F}(\mathbb{B}(n), m)$. If $jn - (j+1)m \leq 1$, then the fraction

$$\frac{j(n-m) - 1}{(j+1)(n-m) - 1} \tag{2.64}$$

precedes $\frac{j}{j+1}$ in $\mathcal{F}(\mathbb{B}(n), m)$.

(b) *Suppose that* $n > 2m$. *Then the fraction*

$$\frac{j(\lceil \frac{m+2}{j} \rceil - 1) - 1}{(j+1)(\lceil \frac{m+2}{j} \rceil - 1) - 1}$$

precedes $\frac{j}{j+1}$ *in* $\mathcal{F}(\mathbb{B}(n), m)$.

(ii) (a) *Suppose that* $n < 2m$. *If* $(j+1)m - jn \geq 1$, *then the fraction*

$$\frac{j(n-m)+1}{(j+1)(n-m)+1}$$

succeeds $\frac{j}{j+1}$ *in* $\mathcal{F}(\mathbb{B}(n), m)$. *If* $(j+1)m - jn \leq 1$, *then the fraction*

$$\frac{j(\lceil \frac{m}{j} \rceil - 1) + 1}{(j+1)(\lceil \frac{m}{j} \rceil - 1) + 1}$$

succeeds $\frac{j}{j+1}$ *in* $\mathcal{F}(\mathbb{B}(n), m)$.

(b) *Suppose that* $n > 2m$. *Then the fraction*

$$\frac{j(\lceil \frac{m}{j} \rceil - 1) + 1}{(j+1)(\lceil \frac{m}{j} \rceil - 1) + 1}$$

succeeds $\frac{j}{j+1}$ *in* $\mathcal{F}(\mathbb{B}(n), m)$.

Proof. Note that the fraction $\frac{1}{j+1} \in \mathcal{F}^{\leq \frac{1}{2}}(\mathbb{B}(n), n-m)$ is the image of the fraction $\frac{j}{j+1} \in \mathcal{F}^{\geq \frac{1}{2}}(\mathbb{B}(n), m)$ under the order-reversing bijection (1.8). Thus, Corollary 2.35 is derived directly from Corollary 2.34 with the help of the order-reversing bijections $\mathcal{F}^{\geq \frac{1}{2}}(\mathbb{B}(n), m) \leftrightarrow \mathcal{F}^{\leq \frac{1}{2}}(\mathbb{B}(n), n-m)$ defined in (1.8).

For instance, suppose that $n < 2m$. Let us verify assertion (i) (a). Note that $n - 2(n-m) = 2m - n > 0$, that is, $n > 2(n-m)$.

If $(j+1)(n-m) - n = jn - (j+1)m \geq 1$, then, according to Corollary 2.34 (ii) (b), the fraction

$$\frac{\lceil \frac{m+2}{j} \rceil - 1}{(j+1)(\lceil \frac{m+2}{j} \rceil - 1) - 1}$$

succeeds $\frac{1}{j+1}$ in $\mathcal{F}^{\leq \frac{1}{2}}(\mathbb{B}(n), n-m)$. As a consequence, its image (2.63) under the order-reversing bijection (1.8) precedes $\frac{j}{j+1}$ in $\mathcal{F}^{\geq \frac{1}{2}}(\mathbb{B}(n), m)$.

If $jn - (j+1)m \leq 1$, then, according to Corollary 2.34 (ii) (b), the fraction

$$\frac{n-m}{(j+1)(n-m)-1}$$

succeeds $\frac{1}{j+1}$ in $\mathcal{F}^{\leq \frac{1}{2}}(\mathbb{B}(n), n-m)$. As a consequence, its image (2.64) under the order-reversing bijection (1.8) precedes $\frac{j}{j+1}$ in $\mathcal{F}^{\geq \frac{1}{2}}(\mathbb{B}(n), m)$. □

2.2.9.3 The neighbors of $\frac{j-1}{2j-1}$ in $\mathcal{F}^{\leq\frac{1}{2}}(\mathbb{B}(n), m)$

Here we will describe the neighbors of a fraction $\frac{j-1}{2j-1}$ in the left half sequences $\mathcal{F}^{\leq\frac{1}{2}}(\mathbb{B}(n), m)$ of the sequences $\mathcal{F}(\mathbb{B}(n), m)$, with $n \neq 2m$.

Corollary 2.36. *Consider the fraction* $\frac{j-1}{2j-1} \in \mathcal{F}^{\leq\frac{1}{2}}(\mathbb{B}(n), m)$ *for some* j, *where* $n \neq 2m$.

(i) (a) *Suppose that* $n < 2m$. *If* $j > 1$, *then the fraction*

$$\frac{(j-1)(\lceil\frac{n-m+2}{j}\rceil - 1) - 1}{(2j-1)(\lceil\frac{n-m+2}{j}\rceil - 1) - 2} \tag{2.65}$$

precedes $\frac{j-1}{2j-1}$ *in* $\mathcal{F}(\mathbb{B}(n), m)$.

(b) *Suppose that* $n > 2m$. *If* $j > 1$ *and* $(j-1)n - (2j-1)m \geq 1$, *then the fraction*

$$\frac{(j-1)(\lceil\frac{m+2}{j-1}\rceil - 1) - 1}{(2j-1)(\lceil\frac{m+2}{j-1}\rceil - 1) - 2} \tag{2.66}$$

precedes $\frac{j-1}{2j-1}$ *in* $\mathcal{F}(\mathbb{B}(n), m)$. *If* $j > 1$ *and* $(j-1)n - (2j-1)m \leq 1$, *then the fraction*

$$\frac{(j-1)(\lceil\frac{n-m+2}{j}\rceil - 1) - 1}{(2j-1)(\lceil\frac{n-m+2}{j}\rceil - 1) - 2} \tag{2.67}$$

precedes $\frac{j-1}{2j-1}$ *in* $\mathcal{F}(\mathbb{B}(n), m)$.

(ii) (a) *Suppose that* $n < 2m$. *Then the fraction*

$$\frac{(j-1)(\lceil\frac{n-m}{j}\rceil - 1) + 1}{(2j-1)(\lceil\frac{n-m}{j}\rceil - 1) + 2} \tag{2.68}$$

succeeds $\frac{j-1}{2j-1}$ *in* $\mathcal{F}(\mathbb{B}(n), m)$.

(b) *Suppose that* $n > 2m$. *If* $(2j-1)m - (j-1)n \geq 1$, *then the fraction*

$$\frac{(j-1)(\lceil\frac{n-m}{j}\rceil - 1) + 1}{(2j-1)(\lceil\frac{n-m}{j}\rceil - 1) + 2} \tag{2.69}$$

succeeds $\frac{j-1}{2j-1}$ *in* $\mathcal{F}(\mathbb{B}(n), m)$. *If* $(2j-1)m - (j-1)n \leq 1$, *then the fraction*

$$\frac{(j-1)(\lceil\frac{m}{j-1}\rceil - 1) + 1}{(2j-1)(\lceil\frac{m}{j-1}\rceil - 1) + 2} \tag{2.70}$$

succeeds $\frac{j-1}{2j-1}$ *in* $\mathcal{F}(\mathbb{B}(n), m)$.

Proof. Let us verify assertions (i) (a) and (ii) (a). According to Corollary 2.8 (i), the fraction $\frac{j-1}{j} \in \mathcal{F}_{n-m}$ is the image of the fraction $\frac{j-1}{2j-1} \in \mathcal{F}^{\leq\frac{1}{2}}(\mathbb{B}(n), m)$ under the order-preserving bijection (2.20). According to Corollary 2.20 (ii), the fraction

$$\frac{(j-1)(\lceil\frac{n-m+2}{j}\rceil - 1) - 1}{j(\lceil\frac{n-m+2}{j}\rceil - 1) - 1}$$

precedes $\frac{j-1}{j}$ in $\mathcal{F}_{n-m}$. As a consequence, its image (2.65) under the order-preserving bijection (2.21) precedes $\frac{j-1}{2j-1}$ in $\mathcal{F}^{\leq\frac{1}{2}}(\mathbb{B}(n), m)$.

According to Corollary 2.20 (ii), the fraction

$$\frac{(j-1)(\lceil\frac{n-m}{j}\rceil-1)+1}{j(\lceil\frac{n-m}{j}\rceil-1)+1}$$

succeeds $\frac{j-1}{j}$ in $\mathcal{F}_{n-m}$. As a consequence, its image (2.68) under the order-preserving bijection (2.21) succeeds $\frac{j-1}{2j-1}$ in $\mathcal{F}^{\leq\frac{1}{2}}(\mathbb{B}(n),m)$.

Let us verify assertion (i) (b). According to Corollary 2.8 (ii), the fraction $\frac{j-1}{j}\in\mathcal{F}^{m}_{n-m}$ is the image of the fraction $\frac{j-1}{2j-1}\in\mathcal{F}^{\leq\frac{1}{2}}(\mathbb{B}(n),m)$ under the order-preserving bijection (2.25).

If $j((n-m)-m)-(n-m)=(j-1)n-(2j-1)m\geq 1$, then Corollary 2.29 (i) (a) states that the fraction

$$\frac{(j-1)(\lceil\frac{m+2}{j-1}\rceil-1)-1}{j(\lceil\frac{m+2}{j-1}\rceil-1)-1}$$

precedes $\frac{j-1}{j}$ in $\mathcal{F}^{m}_{n-m}$. As a consequence, its image (2.66) under the order-preserving bijection (2.26) precedes $\frac{j-1}{2j-1}$ in $\mathcal{F}^{\leq\frac{1}{2}}(\mathbb{B}(n),m)$.

If $(j-1)n-(2j-1)m\leq 1$, then Corollary 2.29 (i) (b) states that the fraction

$$\frac{(j-1)(\lceil\frac{n-m+2}{j}\rceil-1)-1}{j(\lceil\frac{n-m+2}{j}\rceil-1)-1}$$

precedes $\frac{j-1}{j}$ in $\mathcal{F}^{m}_{n-m}$. As a consequence, its image (2.67) under the order-preserving bijection (2.26) precedes $\frac{j-1}{2j-1}$ in $\mathcal{F}^{\leq\frac{1}{2}}(\mathbb{B}(n),m)$.

Let us verify assertion (ii) (b).

If $(n-m)-j((n-m)-m)=(2j-1)m-(j-1)n\geq 1$, then Corollary 2.29 (ii) (a) states that the fraction

$$\frac{(j-1)(\lceil\frac{n-m}{j}\rceil-1)+1}{j(\lceil\frac{n-m}{j}\rceil-1)+1}$$

succeeds $\frac{j-1}{j}$ in $\mathcal{F}^{m}_{n-m}$. As a consequence, its image (2.69) under the order-preserving bijection (2.26) succeeds $\frac{j-1}{2j-1}$ in $\mathcal{F}^{\leq\frac{1}{2}}(\mathbb{B}(n),m)$.

If $(2j-1)m-(j-1)n\leq 1$, then Corollary 2.29 (ii) (b) states that the fraction

$$\frac{(j-1)(\lceil\frac{m}{j-1}\rceil-1)+1}{j(\lceil\frac{m}{j-1}\rceil-1)+1}$$

succeeds $\frac{j-1}{j}$ in $\mathcal{F}^{m}_{n-m}$. As a consequence, its image (2.70) under the order-preserving bijection (2.26) succeeds $\frac{j-1}{2j-1}$ in $\mathcal{F}^{\leq\frac{1}{2}}(\mathbb{B}(n),m)$. □

2.2.9.4 The neighbors of $\frac{j}{2j-1}$ in $\mathcal{F}^{\geq\frac{1}{2}}(\mathbb{B}(n),m)$

In order to describe the neighbors of a fraction $\frac{j}{2j-1}$ in the right half sequences $\mathcal{F}^{\geq\frac{1}{2}}(\mathbb{B}(n),m)$ of the sequences $\mathcal{F}(\mathbb{B}(n),m)$, with $n\neq 2m$, we will apply the order-reversing bijections $\mathcal{F}^{\geq\frac{1}{2}}(\mathbb{B}(n),m)\leftrightarrow\mathcal{F}^{\leq\frac{1}{2}}(\mathbb{B}(n),n-m)$, defined in (1.8), to the observations, made in Corollary 2.36, on the neighbors of the fraction $\frac{j-1}{2j-1}$ in $\mathcal{F}^{\leq\frac{1}{2}}(\mathbb{B}(n),n-m)$.

Corollary 2.37. *Consider the fraction* $\frac{j}{2j-1} \in \mathcal{F}^{\geq\frac{1}{2}}(\mathbb{B}(n), m) - \{\frac{1}{1}\}$ *for some* j, *where* $n \neq 2m$.

(i) (a) *Suppose that* $n < 2m$. *If* $jn - (2j-1)m \geq 1$, *then the fraction*

$$\frac{j(\lceil\frac{m}{j}\rceil - 1) + 1}{(2j-1)(\lceil\frac{m}{j}\rceil - 1) + 2} \tag{2.71}$$

precedes $\frac{j}{2j-1}$ *in* $\mathcal{F}(\mathbb{B}(n), m)$. *If* $jn - (2j-1)m \leq 1$, *then the fraction*

$$\frac{j(\lceil\frac{n-m}{j-1}\rceil - 1) + 1}{(2j-1)(\lceil\frac{n-m}{j-1}\rceil - 1) + 2} \tag{2.72}$$

precedes $\frac{j}{2j-1}$ *in* $\mathcal{F}(\mathbb{B}(n), m)$.

(b) *Suppose that* $n > 2m$. *Then the fraction*

$$\frac{j(\lceil\frac{m}{j}\rceil - 1) + 1}{(2j-1)(\lceil\frac{m}{j}\rceil - 1) + 2}$$

precedes $\frac{j}{2j-1}$ *in* $\mathcal{F}(\mathbb{B}(n), m)$.

(ii) (a) *Suppose that* $n < 2m$. *If* $(2j-1)m - jn \geq 1$, *then the fraction*

$$\frac{j(\lceil\frac{n-m+2}{j-1}\rceil - 1) - 1}{(2j-1)(\lceil\frac{n-m+2}{j-1}\rceil - 1) - 2}$$

succeeds $\frac{j}{2j-1}$ *in* $\mathcal{F}(\mathbb{B}(n), m)$. *If* $(2j-1)m - jn \leq 1$, *then the fraction*

$$\frac{j(\lceil\frac{m+2}{j}\rceil - 1) - 1}{(2j-1)(\lceil\frac{m+2}{j}\rceil - 1) - 2}$$

succeeds $\frac{j}{2j-1}$ *in* $\mathcal{F}(\mathbb{B}(n), m)$.

(b) *Suppose that* $n > 2m$. *Then the fraction*

$$\frac{j(\lceil\frac{m+2}{j}\rceil - 1) - 1}{(2j-1)(\lceil\frac{m+2}{j}\rceil - 1) - 2}$$

succeeds $\frac{j}{2j-1}$ *in* $\mathcal{F}(\mathbb{B}(n), m)$.

Proof. Note that the fraction $\frac{j-1}{2j-1} \in \mathcal{F}^{\leq\frac{1}{2}}(\mathbb{B}(n), n-m)$ is the image of the fraction $\frac{j}{2j-1} \in \mathcal{F}^{\geq\frac{1}{2}}(\mathbb{B}(n), m)$ under the order-reversing bijection (1.8). Thus, Corollary 2.37 is derived directly from Corollary 2.36, with the help of the order-reversing bijections $\mathcal{F}^{\geq\frac{1}{2}}(\mathbb{B}(n), m) \leftrightarrow \mathcal{F}^{\leq\frac{1}{2}}(\mathbb{B}(n), n-m)$ defined in (1.8).

For instance, suppose that $n < 2m$; let us verify assertion (i)(a). Note that $n - 2(n-m) = 2m - n > 0$, that is, $n > 2(n-m)$.

If $(2j-1)(n-m)-(j-1)n = jn-(2j-1)m \geq 1$, then, according to Corollary 2.36 (ii)(b), the fraction

$$\frac{(j-1)(\lceil\frac{m}{j}\rceil - 1) + 1}{(2j-1)(\lceil\frac{m}{j}\rceil - 1) + 2}$$

succeeds $\frac{j-1}{2j-1}$ in $\mathcal{F}^{\leq\frac{1}{2}}(\mathbb{B}(n), n-m)$. As a consequence, its image (2.71) under the order-reversing bijection (1.8) precedes $\frac{j}{2j-1}$ in $\mathcal{F}^{\geq\frac{1}{2}}(\mathbb{B}(n), m)$.

If $jn-(j+1)m \leq 1$, then, according to Corollary 2.36 (ii) (b), the fraction

$$\frac{(j-1)(\lceil\frac{n-m}{j-1}\rceil - 1) + 1}{(2j-1)(\lceil\frac{n-m}{j-1}\rceil - 1) + 2}$$

succeeds $\frac{j-1}{2j-1}$ in $\mathcal{F}^{\leq\frac{1}{2}}(\mathbb{B}(n), n-m)$. As a consequence, its image (2.72) under the order-reversing bijection (1.8) precedes $\frac{j}{2j-1}$ in $\mathcal{F}^{\geq\frac{1}{2}}(\mathbb{B}(n), m)$. □

2.2.9.5 The neighbors of $\frac{2}{j+2}$ in $\mathcal{F}^{\leq\frac{1}{2}}(\mathbb{B}(n), m)$

Let us describe the neighbors of a fraction $\frac{2}{j+2}$ in the left half sequences $\mathcal{F}^{\leq\frac{1}{2}}(\mathbb{B}(n), m)$ of the sequences $\mathcal{F}(\mathbb{B}(n), m)$, with $n \neq 2m$.

Corollary 2.38. *Consider the fraction $\frac{2}{j+2} \in \mathcal{F}^{\leq\frac{1}{2}}(\mathbb{B}(n), m)$ for some odd j, where $n \neq 2m$.*

(i) (a) *Suppose that $n < 2m$. Then the fraction*

$$\begin{cases} (\lceil\frac{2(n-m)}{j}\rceil - 1)\Big/\dfrac{(j+2)(\lceil\frac{2(n-m)}{j}\rceil - 1) + 1}{2} & \text{if } \lceil\frac{2(n-m)}{j}\rceil \text{ is even,} \\ (\lceil\frac{2(n-m)}{j}\rceil - 2)\Big/\dfrac{(j+2)(\lceil\frac{2(n-m)}{j}\rceil - 2) + 1}{2} & \text{if } \lceil\frac{2(n-m)}{j}\rceil \text{ is odd} \end{cases} \tag{2.73}$$

precedes $\frac{2}{j+2}$ in $\mathcal{F}(\mathbb{B}(n), m)$.

(b) *Suppose that $n > 2m$. If $2n-(j+2)m \geq 1$, then the fraction*

$$\begin{cases} m\Big/\dfrac{(j+2)m + 1}{2} & \text{if } m \text{ is odd,} \\ (m-1)\Big/\dfrac{(j+2)(m-1) + 1}{2} & \text{if } m \text{ is even} \end{cases} \tag{2.74}$$

precedes $\frac{2}{j+2}$ in $\mathcal{F}(\mathbb{B}(n), m)$. If $2n-(j+2)m \leq 1$, then the fraction

$$\begin{cases} (\lceil\frac{2(n-m)}{j}\rceil - 1)\Big/\dfrac{(j+2)(\lceil\frac{2(n-m)}{j}\rceil - 1) + 1}{2} & \text{if } \lceil\frac{2(n-m)}{j}\rceil \text{ is even,} \\ (\lceil\frac{2(n-m)}{j}\rceil - 2)\Big/\dfrac{(j+2)(\lceil\frac{2(n-m)}{j}\rceil - 2) + 1}{2} & \text{if } \lceil\frac{2(n-m)}{j}\rceil \text{ is odd} \end{cases} \tag{2.75}$$

precedes $\frac{2}{j+2}$ in $\mathcal{F}(\mathbb{B}(n), m)$.

(ii) (a) *Suppose that $n < 2m$. Then the fraction*

$$\begin{cases} (\lceil\frac{2(n-m+1)}{j}\rceil - 1)\Big/\dfrac{(j+2)(\lceil\frac{2(n-m+1)}{j}\rceil - 1) - 1}{2} & \text{if } \lceil\frac{2(n-m+1)}{j}\rceil \text{ is even,} \\ (\lceil\frac{2(n-m+1)}{j}\rceil - 2)\Big/\dfrac{(j+2)(\lceil\frac{2(n-m+1)}{j}\rceil - 2) - 1}{2} & \text{if } \lceil\frac{2(n-m+1)}{j}\rceil \text{ is odd} \end{cases} \tag{2.76}$$

succeeds $\frac{2}{j+2}$ in $\mathcal{F}(\mathbb{B}(n), m)$.

(b) *Suppose that $n > 2m$. If $(j+2)m - 2n \geq 1$, then the fraction*

$$\begin{cases} \left(\lceil \frac{2(n-m+1)}{j} \rceil - 1\right) \Big/ \dfrac{(j+2)(\lceil \frac{2(n-m+1)}{j} \rceil - 1) - 1}{2} & \text{if } \lceil \frac{2(n-m+1)}{j} \rceil \text{ is even,} \\ \left(\lceil \frac{2(n-m+1)}{j} \rceil - 2\right) \Big/ \dfrac{(j+2)(\lceil \frac{2(n-m+1)}{j} \rceil - 2) - 1}{2} & \text{if } \lceil \frac{2(n-m+1)}{j} \rceil \text{ is odd} \end{cases} \tag{2.77}$$

succeeds $\frac{2}{j+2}$ in $\mathcal{F}(\mathbb{B}(n), m)$. If $(j+2)m - 2n \leq 1$, then the fraction

$$\begin{cases} m \Big/ \dfrac{(j+2)m - 1}{2} & \text{if } m \text{ is odd,} \\ (m-1) \Big/ \dfrac{(j+2)(m-1) - 1}{2} & \text{if } m \text{ is even} \end{cases} \tag{2.78}$$

succeeds $\frac{2}{j+2}$ in $\mathcal{F}(\mathbb{B}(n), m)$.

Proof. Let us verify assertions (i) (a) and (ii) (a). According to Corollary 2.8 (i), the fraction $\frac{2}{j} \in \mathcal{F}_{n-m}$ is the image of the fraction $\frac{2}{j+2} \in \mathcal{F}^{\leq\frac{1}{2}}(\mathbb{B}(n), m)$ under the order-preserving bijection (2.20). According to Corollary 2.21 (i), the fraction

$$\begin{cases} \left(\lceil \frac{2(n-m)}{j} \rceil - 1\right) \Big/ \dfrac{j(\lceil \frac{2(n-m)}{j} \rceil - 1) + 1}{2} & \text{if } \lceil \frac{2(n-m)}{j} \rceil \text{ is even,} \\ \left(\lceil \frac{2(n-m)}{j} \rceil - 2\right) \Big/ \dfrac{j(\lceil \frac{2(n-m)}{j} \rceil - 2) + 1}{2} & \text{if } \lceil \frac{2(n-m)}{j} \rceil \text{ is odd} \end{cases}$$

precedes $\frac{2}{j}$ in $\mathcal{F}_{n-m}$. As a consequence, its image (2.73) under the order-preserving bijection (2.21) precedes $\frac{2}{j+2}$ in $\mathcal{F}^{\leq\frac{1}{2}}(\mathbb{B}(n), m)$.

According to Corollary 2.21 (i), the fraction

$$\begin{cases} \left(\lceil \frac{2(n-m+1)}{j} \rceil - 1\right) \Big/ \dfrac{j(\lceil \frac{2(n-m+1)}{j} \rceil - 1) - 1}{2} & \text{if } \lceil \frac{2(n-m+1)}{j} \rceil \text{ is even,} \\ \left(\lceil \frac{2(n-m+1)}{j} \rceil - 2\right) \Big/ \dfrac{j(\lceil \frac{2(n-m+1)}{j} \rceil - 2) - 1}{2} & \text{if } \lceil \frac{2(n-m+1)}{j} \rceil \text{ is odd} \end{cases}$$

succeeds $\frac{2}{j}$ in $\mathcal{F}_{n-m}$. As a consequence, its image (2.76) under the order-preserving bijection (2.21) succeeds $\frac{2}{j+2}$ in $\mathcal{F}^{\leq\frac{1}{2}}(\mathbb{B}(n), m)$.

Let us verify assertion (i) (b). According to Corollary 2.8 (ii), the fraction $\frac{2}{j} \in \mathcal{F}^m_{n-m}$ is the image of the fraction $\frac{2}{j+2} \in \mathcal{F}^{\leq\frac{1}{2}}(\mathbb{B}(n), m)$ under the order-preserving bijection (2.25).

If $2(n-m) - jm = 2n - (j+2)m \geq 1$, then Corollary 2.30 (i) (a) states that the fraction

$$\begin{cases} m \Big/ \dfrac{jm + 1}{2} & \text{if } m \text{ is odd,} \\ (m-1) \Big/ \dfrac{j(m-1) + 1}{2} & \text{if } m \text{ is even} \end{cases}$$

precedes $\frac{2}{j}$ in $\mathcal{F}^m_{n-m}$. As a consequence, its image (2.74) under the order-preserving bijection (2.26) precedes $\frac{2}{j+2}$ in $\mathcal{F}^{\leq\frac{1}{2}}(\mathbb{B}(n), m)$.

If $2n - (j+2)m \leq 1$, then Corollary 2.30 (i) (b) states that the fraction

$$\begin{cases} (\lceil \frac{2(n-m)}{j} \rceil - 1) \Big/ \dfrac{j(\lceil \frac{2(n-m)}{j} \rceil - 1) + 1}{2} & \text{if } \lceil \frac{2(n-m)}{j} \rceil \text{ is even,} \\ (\lceil \frac{2(n-m)}{j} \rceil - 2) \Big/ \dfrac{j(\lceil \frac{2(n-m)}{j} \rceil - 2) + 1}{2} & \text{if } \lceil \frac{2(n-m)}{j} \rceil \text{ is odd} \end{cases}$$

precedes $\frac{2}{j}$ in $\mathcal{F}^m_{n-m}$. As a consequence, its image (2.75) under the order-preserving bijection (2.26) precedes $\frac{2}{j+2}$ in $\mathcal{F}^{\leq \frac{1}{2}}(\mathbb{B}(n), m)$.

Let us verify assertion (ii) (b).

If $jm - 2(n-m) = (j+2)m - 2n \geq 1$, then Corollary 2.30 (ii) (a) states that the fraction

$$\begin{cases} (\lceil \frac{2(n-m+1)}{j} \rceil - 1) \Big/ \dfrac{j(\lceil \frac{2(n-m+1)}{j} \rceil - 1) - 1}{2} & \text{if } \lceil \frac{2(n-m+1)}{j} \rceil \text{ is even,} \\ (\lceil \frac{2(n-m+1)}{j} \rceil - 2) \Big/ \dfrac{j(\lceil \frac{2(n-m+1)}{j} \rceil - 2) - 1}{2} & \text{if } \lceil \frac{2(n-m+1)}{j} \rceil \text{ is odd} \end{cases}$$

succeeds $\frac{2}{j}$ in $\mathcal{F}^m_{n-m}$. As a consequence, its image (2.77) under the order-preserving bijection (2.26) succeeds $\frac{2}{j+2}$ in $\mathcal{F}^{\leq \frac{1}{2}}(\mathbb{B}(n), m)$.

If $(j+2)m - 2n \leq 1$, then Corollary 2.30 (ii) (b) states that the fraction

$$\begin{cases} m \Big/ \dfrac{jm - 1}{2} & \text{if } m \text{ is odd,} \\ (m-1) \Big/ \dfrac{j(m-1) - 1}{2} & \text{if } m \text{ is even} \end{cases}$$

succeeds $\frac{2}{j}$ in $\mathcal{F}^m_{n-m}$. As a consequence, its image (2.78) under the order-preserving bijection (2.26) succeeds $\frac{2}{j+2}$ in $\mathcal{F}^{\leq \frac{1}{2}}(\mathbb{B}(n), m)$. □

2.2.9.6 The neighbors of $\frac{j}{j+2}$ in $\mathcal{F}^{\geq \frac{1}{2}}(\mathbb{B}(n), m)$

In order to describe the neighbors of a fraction $\frac{j}{j+2}$ in the right half sequences $\mathcal{F}^{\geq \frac{1}{2}}(\mathbb{B}(n), m)$ of the sequences $\mathcal{F}(\mathbb{B}(n), m)$, with $n \neq 2m$, we will apply the order-reversing bijections $\mathcal{F}^{\geq \frac{1}{2}}(\mathbb{B}(n), m) \leftrightarrow \mathcal{F}^{\leq \frac{1}{2}}(\mathbb{B}(n), n-m)$, defined in (1.8), to the observations, made in Corollary 2.38, on the neighbors of the fraction $\frac{2}{j+2}$ in $\mathcal{F}^{\leq \frac{1}{2}}(\mathbb{B}(n), n-m)$.

Corollary 2.39. *Consider the fraction $\frac{j}{j+2} \in \mathcal{F}^{\geq \frac{1}{2}}(\mathbb{B}(n), m)$ for some odd j, where $n \neq 2m$.*

(i) (a) *Suppose that $n < 2m$. If $jn - (j+2)m \geq 1$, then the fraction*

$$\begin{cases} \dfrac{j(\lceil \frac{2(m+1)}{j} \rceil - 1) - 1}{2} \Big/ \dfrac{(j+2)(\lceil \frac{2(m+1)}{j} \rceil - 1) - 1}{2} & \text{if } \lceil \frac{2(m+1)}{j} \rceil \text{ is even,} \\ \dfrac{j(\lceil \frac{2(m+1)}{j} \rceil - 2) - 1}{2} \Big/ \dfrac{(j+2)(\lceil \frac{2(m+1)}{j} \rceil - 2) - 1}{2} & \text{if } \lceil \frac{2(m+1)}{j} \rceil \text{ is odd} \end{cases} \tag{2.79}$$

precedes $\frac{j}{j+2}$ in $\mathcal{F}(\mathbb{B}(n), m)$.

If $jn-(j+2)m\le 1$, then the fraction

$$\begin{cases} \frac{j(n-m)-1}{2}\Big/\frac{(j+2)(n-m)-1}{2} & \text{if } (n-m) \text{ is odd,}\\ \frac{j(n-m-1)-1}{2}\Big/\frac{(j+2)(n-m-1)-1}{2} & \text{if } (n-m) \text{ is even} \end{cases} \tag{2.80}$$

precedes $\frac{j}{j+2}$ in $\mathcal{F}(\mathbb{B}(n),m)$.

(b) *Suppose that $n>2m$. Then the fraction*

$$\begin{cases} \frac{j(\lceil\frac{2(m+1)}{j}\rceil-1)-1}{2}\Big/\frac{(j+2)(\lceil\frac{2(m+1)}{j}\rceil-1)-1}{2} & \text{if } \lceil\frac{2(m+1)}{j}\rceil \text{ is even,}\\ \frac{j(\lceil\frac{2(m+1)}{j}\rceil-2)-1}{2}\Big/\frac{(j+2)(\lceil\frac{2(m+1)}{j}\rceil-2)-1}{2} & \text{if } \lceil\frac{2(m+1)}{j}\rceil \text{ is odd} \end{cases}$$

precedes $\frac{j}{j+2}$ in $\mathcal{F}(\mathbb{B}(n),m)$.

(ii) (a) *Suppose that $n<2m$. If $(j+2)m-jn\ge 1$, then the fraction*

$$\begin{cases} \frac{j(n-m)+1}{2}\Big/\frac{(j+2)(n-m)+1}{2} & \text{if } (n-m) \text{ is odd,}\\ \frac{j(n-m-1)+1}{2}\Big/\frac{(j+2)(n-m-1)+1}{2} & \text{if } (n-m) \text{ is even} \end{cases}$$

succeeds $\frac{j}{j+2}$ in $\mathcal{F}(\mathbb{B}(n),m)$. If $(j+2)m-jn\le 1$, then the fraction

$$\begin{cases} \frac{j(\lceil\frac{2m}{j}\rceil-1)+1}{2}\Big/\frac{(j+2)(\lceil\frac{2m}{j}\rceil-1)+1}{2} & \text{if } \lceil\frac{2m}{j}\rceil \text{ is even,}\\ \frac{j(\lceil\frac{2m}{j}\rceil-2)+1}{2}\Big/\frac{(j+2)(\lceil\frac{2m}{j}\rceil-2)+1}{2} & \text{if } \lceil\frac{2m}{j}\rceil \text{ is odd} \end{cases}$$

succeeds $\frac{j}{j+2}$ in $\mathcal{F}(\mathbb{B}(n),m)$.

(b) *Suppose that $n>2m$. Then the fraction*

$$\begin{cases} \frac{j(\lceil\frac{2m}{j}\rceil-1)+1}{2}\Big/\frac{(j+2)(\lceil\frac{2m}{j}\rceil-1)+1}{2} & \text{if } \lceil\frac{2m}{j}\rceil \text{ is even,}\\ \frac{j(\lceil\frac{2m}{j}\rceil-2)+1}{2}\Big/\frac{(j+2)(\lceil\frac{2m}{j}\rceil-2)+1}{2} & \text{if } \lceil\frac{2m}{j}\rceil \text{ is odd} \end{cases}$$

succeeds $\frac{j}{j+2}$ in $\mathcal{F}(\mathbb{B}(n),m)$.

Proof. Note that the fraction $\frac{2}{j+2}\in\mathcal{F}^{\le\frac{1}{2}}(\mathbb{B}(n),n-m)$ is the image of the fraction $\frac{j}{j+2}\in\mathcal{F}^{\ge\frac{1}{2}}(\mathbb{B}(n),m)$ under the order-reversing bijection (1.8). Thus, Corollary 2.39 is derived directly from Corollary 2.38, with the help of the order-reversing bijections $\mathcal{F}^{\ge\frac{1}{2}}(\mathbb{B}(n),m)\leftrightarrow\mathcal{F}^{\le\frac{1}{2}}(\mathbb{B}(n),n-m)$ defined in (1.8).

For instance, suppose that $n<2m$. Let us verify assertion (i) (a). Note that $n-2(n-m)=2m-n>0$, that is, $n>2(n-m)$.

If $(j+2)(n-m)-2n = jn-(j+2)m \geq 1$, then, according to Corollary 2.38 (ii) (b), the fraction

$$\begin{cases} (\lceil \frac{2(m+1)}{j} \rceil - 1) \Big/ \dfrac{(j+2)(\lceil \frac{2(m+1)}{j} \rceil - 1) - 1}{2} & \text{if } \lceil \frac{2(m+1)}{j} \rceil \text{ is even,} \\ (\lceil \frac{2(m+1)}{j} \rceil - 2) \Big/ \dfrac{(j+2)(\lceil \frac{2(m+1)}{j} \rceil - 2) - 1}{2} & \text{if } \lceil \frac{2(m+1)}{j} \rceil \text{ is odd} \end{cases}$$

succeeds $\frac{2}{j+2}$ in $\mathcal{F}^{\leq\frac{1}{2}}(\mathbb{B}(n), n-m)$. As a consequence, its image (2.79) under the order-reversing bijection (1.8) precedes $\frac{j}{j+2}$ in $\mathcal{F}^{\geq\frac{1}{2}}(\mathbb{B}(n), m)$.

If $jn-(j+2)m \leq 1$, then, according to Corollary 2.38 (ii) (b), the fraction

$$\begin{cases} (n-m) \Big/ \dfrac{(j+2)(n-m)-1}{2} & \text{if } (n-m) \text{ is odd,} \\ (n-m-1) \Big/ \dfrac{(j+2)(n-m-1)-1}{2} & \text{if } (n-m) \text{ is even} \end{cases}$$

succeeds $\frac{2}{j+2}$ in $\mathcal{F}^{\leq\frac{1}{2}}(\mathbb{B}(n), n-m)$. As a consequence, its image (2.80) under the order-reversing bijection (1.8) precedes $\frac{j}{j+2}$ in $\mathcal{F}^{\geq\frac{1}{2}}(\mathbb{B}(n), m)$. □

2.2.9.7 The neighbors of $\frac{j-2}{2(j-1)}$ in $\mathcal{F}^{\leq\frac{1}{2}}(\mathbb{B}(n), m)$

Let us find the neighbors of a fraction $\frac{j-2}{2(j-1)}$ in the left half sequences $\mathcal{F}^{\leq\frac{1}{2}}(\mathbb{B}(n), m)$ of the sequences $\mathcal{F}(\mathbb{B}(n), m)$, with $n \neq 2m$.

Corollary 2.40. *Consider the fraction $\frac{j-2}{2(j-1)} \in \mathcal{F}^{\leq\frac{1}{2}}(\mathbb{B}(n), m)$ for some odd j, where $n \neq 2m$.*

(i) (a) *Suppose that $n < 2m$. Then the fraction*

$$\begin{cases} \dfrac{(j-2)(\lceil \frac{2(n-m+1)}{j} \rceil - 1) - 1}{2} \Big/ ((j-1)(\lceil \frac{2(n-m+1)}{j} \rceil - 1) - 1) \\ \text{if } \lceil \frac{2(n-m+1)}{j} \rceil \text{ is even,} \\ \dfrac{(j-2)(\lceil \frac{2(n-m+1)}{j} \rceil - 2) - 1}{2} \Big/ ((j-1)(\lceil \frac{2(n-m+1)}{j} \rceil - 2) - 1) \\ \text{if } \lceil \frac{2(n-m+1)}{j} \rceil \text{ is odd} \end{cases} \tag{2.81}$$

precedes $\frac{j-2}{2(j-1)}$ in $\mathcal{F}(\mathbb{B}(n), m)$.

(b) *Suppose that $n > 2m$. If $(j-2)n - 2(j-1)m \geq 1$, then the fraction*

$$\begin{cases} \dfrac{(j-2)(\lceil \frac{2(m+1)}{j-2} \rceil - 1) - 1}{2} \Big/ ((j-1)(\lceil \frac{2(m+1)}{j-2} \rceil - 1) - 1) \\ \text{if } \lceil \frac{2(m+1)}{j-2} \rceil \text{ is even,} \\ \dfrac{(j-2)(\lceil \frac{2(m+1)}{j-2} \rceil - 2) - 1}{2} \Big/ ((j-1)(\lceil \frac{2(m+1)}{j-2} \rceil - 2) - 1) \\ \text{if } \lceil \frac{2(m+1)}{j-2} \rceil \text{ is odd} \end{cases} \tag{2.82}$$

precedes $\frac{j-2}{2(j-1)}$ in $\mathcal{F}(\mathbb{B}(n), m)$.

If $(j-2)n - 2(j-1)m \le 1$, *then the fraction*

$$\begin{cases} \dfrac{(j-2)(\lceil\frac{2(n-m+1)}{j}\rceil - 1) - 1}{2} \Big/ ((j-1)(\lceil\frac{2(n-m+1)}{j}\rceil - 1) - 1) \\ \text{if } \lceil\frac{2(n-m+1)}{j}\rceil \text{ is even,} \\ \dfrac{(j-2)(\lceil\frac{2(n-m+1)}{j}\rceil - 2) - 1}{2} \Big/ ((j-1)(\lceil\frac{2(n-m+1)}{j}\rceil - 2) - 1) \\ \text{if } \lceil\frac{2(n-m+1)}{j}\rceil \text{ is odd} \end{cases} \tag{2.83}$$

precedes $\frac{j-2}{2(j-1)}$ *in* $\mathcal{F}(\mathbb{B}(n), m)$.

(ii) (a) *Suppose that* $n < 2m$. *Then the fraction*

$$\begin{cases} \dfrac{(j-2)(\lceil\frac{2(n-m)}{j}\rceil - 1) + 1}{2} \Big/ ((j-1)(\lceil\frac{2(n-m)}{j}\rceil - 1) + 1) \\ \text{if } \lceil\frac{2(n-m)}{j}\rceil \text{ is even,} \\ \dfrac{(j-2)(\lceil\frac{2(n-m)}{j}\rceil - 2) + 1}{2} \Big/ ((j-1)(\lceil\frac{2(n-m)}{j}\rceil - 2) + 1) \\ \text{if } \lceil\frac{2(n-m)}{j}\rceil \text{ is odd} \end{cases} \tag{2.84}$$

succeeds $\frac{j-2}{2(j-1)}$ *in* $\mathcal{F}(\mathbb{B}(n), m)$.

(b) *Suppose that* $n > 2m$. *If* $2(j-1)m - (j-2)n \ge 1$, *then the fraction*

$$\begin{cases} \dfrac{(j-2)(\lceil\frac{2(n-m)}{j}\rceil - 1) + 1}{2} \Big/ ((j-1)(\lceil\frac{2(n-m)}{j}\rceil - 1) + 1) \\ \text{if } \lceil\frac{2(n-m)}{j}\rceil \text{ is even,} \\ \dfrac{(j-2)(\lceil\frac{2(n-m)}{j}\rceil - 2) + 1}{2} \Big/ ((j-1)(\lceil\frac{2(n-m)}{j}\rceil - 2) + 1) \\ \text{if } \lceil\frac{2(n-m)}{j}\rceil \text{ is odd} \end{cases} \tag{2.85}$$

succeeds $\frac{j-2}{2(j-1)}$ *in* $\mathcal{F}(\mathbb{B}(n), m)$. *If* $2(j-1)m - (j-2)n \le 1$, *then the fraction*

$$\begin{cases} \dfrac{(j-2)(\lceil\frac{2m}{j-2}\rceil - 1) + 1}{2} \Big/ ((j-1)(\lceil\frac{2m}{j-2}\rceil - 1) + 1) & \text{if } \lceil\frac{2m}{j-2}\rceil \text{ is even,} \\ \dfrac{(j-2)(\lceil\frac{2m}{j-2}\rceil - 2) + 1}{2} \Big/ ((j-1)(\lceil\frac{2m}{j-2}\rceil - 2) + 1) & \text{if } \lceil\frac{2m}{j-2}\rceil \text{ is odd} \end{cases} \tag{2.86}$$

succeeds $\frac{j-2}{2(j-1)}$ *in* $\mathcal{F}(\mathbb{B}(n), m)$.

Proof. Let us verify assertions (i) (a) and (ii) (a). According to Corollary 2.8 (i), the fraction $\frac{j-2}{j} \in \mathcal{F}_{n-m}$ is the image of the fraction $\frac{j-2}{2(j-1)} \in \mathcal{F}^{\le\frac{1}{2}}(\mathbb{B}(n), m)$ under the order-preserving bijection (2.20). According to Corollary 2.21 (ii), the fraction

$$\begin{cases} \dfrac{(j-2)(\lceil\frac{2(n-m+1)}{j}\rceil - 1) - 1}{2} \Big/ \dfrac{j(\lceil\frac{2(n-m+1)}{j}\rceil - 1) - 1}{2} & \text{if } \lceil\frac{2(n-m+1)}{j}\rceil \text{ is even,} \\ \dfrac{(j-2)(\lceil\frac{2(n-m+1)}{j}\rceil - 2) - 1}{2} \Big/ \dfrac{j(\lceil\frac{2(n-m+1)}{j}\rceil - 2) - 1}{2} & \text{if } \lceil\frac{2(n-m+1)}{j}\rceil \text{ is odd} \end{cases}$$

precedes $\frac{j-2}{j}$ in $\mathcal{F}_{n-m}$. As a consequence, its image (2.81) under the order-preserving bijection (2.21) precedes $\frac{j-2}{2(j-1)}$ in $\mathcal{F}^{\leq\frac{1}{2}}(\mathbb{B}(n), m)$.

According to Corollary 2.21 (ii), the fraction

$$\begin{cases} \dfrac{(j-2)(\lceil\frac{2(n-m)}{j}\rceil-1)+1}{2}\Big/\dfrac{j(\lceil\frac{2(n-m)}{j}\rceil-1)+1}{2} & \text{if } \lceil\frac{2(n-m)}{j}\rceil \text{ is even,} \\ \dfrac{(j-2)(\lceil\frac{2(n-m)}{j}\rceil-2)+1}{2}\Big/\dfrac{j(\lceil\frac{2(n-m)}{j}\rceil-2)+1}{2} & \text{if } \lceil\frac{2(n-m)}{j}\rceil \text{ is odd} \end{cases}$$

succeeds $\frac{j-2}{j}$ in $\mathcal{F}_{n-m}$. As a consequence, its image (2.84) under the order-preserving bijection (2.21) succeeds $\frac{j-2}{2(j-1)}$ in $\mathcal{F}^{\leq\frac{1}{2}}(\mathbb{B}(n), m)$.

Let us verify assertion (i) (b). According to Corollary 2.8 (ii), the fraction $\frac{j-2}{j} \in \mathcal{F}^m_{n-m}$ is the image of the fraction $\frac{j-2}{2(j-1)} \in \mathcal{F}^{\leq\frac{1}{2}}(\mathbb{B}(n), m)$ under the order-preserving bijection (2.25).

If $j((n-m)-m)-2(n-m) = (j-2)n-2(j-1)m \geq 1$, then Corollary 2.33 (i) (a) states that the fraction

$$\begin{cases} \dfrac{(j-2)(\lceil\frac{2(m+1)}{j-2}\rceil-1)-1}{2}\Big/\dfrac{j(\lceil\frac{2(m+1)}{j-2}\rceil-1)-1}{2} & \text{if } \lceil\frac{2(m+1)}{j-2}\rceil \text{ is even,} \\ \dfrac{(j-2)(\lceil\frac{2(m+1)}{j-2}\rceil-2)-1}{2}\Big/\dfrac{j(\lceil\frac{2(m+1)}{j-2}\rceil-2)-1}{2} & \text{if } \lceil\frac{2(m+1)}{j-2}\rceil \text{ is odd} \end{cases}$$

precedes $\frac{j-2}{j}$ in $\mathcal{F}^m_{n-m}$. As a consequence, its image (2.82) under the order-preserving bijection (2.26) precedes $\frac{j-2}{2(j-1)}$ in $\mathcal{F}^{\leq\frac{1}{2}}(\mathbb{B}(n), m)$.

If $(j-2)n-2(j-1)m \leq 1$, then Corollary 2.33 (i) (b) states that the fraction

$$\begin{cases} \dfrac{(j-2)(\lceil\frac{2(n-m+1)}{j}\rceil-1)-1}{2}\Big/\dfrac{j(\lceil\frac{2(n-m+1)}{j}\rceil-1)-1}{2} & \text{if } \lceil\frac{2(n-m+1)}{j}\rceil \text{ is even,} \\ \dfrac{(j-2)(\lceil\frac{2(n-m+1)}{j}\rceil-2)-1}{2}\Big/\dfrac{j(\lceil\frac{2(n-m+1)}{j}\rceil-2)-1}{2} & \text{if } \lceil\frac{2(n-m+1)}{j}\rceil \text{ is odd} \end{cases}$$

precedes $\frac{j-2}{j}$ in $\mathcal{F}^m_{n-m}$. As a consequence, its image (2.83) under the order-preserving bijection (2.26) precedes $\frac{j-2}{2(j-1)}$ in $\mathcal{F}^{\leq\frac{1}{2}}(\mathbb{B}(n), m)$.

Let us verify assertion (ii) (b).

If $2(n-m)-j((n-m)-m) = 2(j-1)m-(j-2)n \geq 1$, then Corollary 2.33 (ii) (a) states that the fraction

$$\begin{cases} \dfrac{(j-2)(\lceil\frac{2(n-m)}{j}\rceil-1)+1}{2}\Big/\dfrac{j(\lceil\frac{2(n-m)}{j}\rceil-1)+1}{2} & \text{if } \lceil\frac{2(n-m)}{j}\rceil \text{ is even,} \\ \dfrac{(j-2)(\lceil\frac{2(n-m)}{j}\rceil-2)+1}{2}\Big/\dfrac{j(\lceil\frac{2(n-m)}{j}\rceil-2)+1}{2} & \text{if } \lceil\frac{2(n-m)}{j}\rceil \text{ is odd} \end{cases}$$

succeeds $\frac{j-2}{j}$ in $\mathcal{F}^m_{n-m}$. As a consequence, its image (2.85) under the order-preserving bijection (2.26) succeeds $\frac{j-2}{2(j-1)}$ in $\mathcal{F}^{\leq\frac{1}{2}}(\mathbb{B}(n), m)$.

If $2(j-1)m-(j-2)n \le 1$, then Corollary 2.33 (ii) (b) states that the fraction

$$\begin{cases} \dfrac{(j-2)(\lceil \frac{2m}{j-2} \rceil - 1) + 1}{2} \Big/ \dfrac{j(\lceil \frac{2m}{j-2} \rceil - 1) + 1}{2} & \text{if } \lceil \frac{2m}{j-2} \rceil \text{ is even,} \\ \dfrac{(j-2)(\lceil \frac{2m}{j-2} \rceil - 2) + 1}{2} \Big/ \dfrac{j(\lceil \frac{2m}{j-2} \rceil - 2) + 1}{2} & \text{if } \lceil \frac{2m}{j-2} \rceil \text{ is odd} \end{cases}$$

succeeds $\frac{j-2}{j}$ in $\mathcal{F}^m_{n-m}$. As a consequence, its image (2.86) under the order-preserving bijection (2.26) succeeds $\frac{j-2}{2(j-1)}$ in $\mathcal{F}^{\le\frac{1}{2}}(\mathbb{B}(n), m)$. □

2.2.9.8 The neighbors of $\frac{j}{2(j-1)}$ in $\mathcal{F}^{\ge\frac{1}{2}}(\mathbb{B}(n), m)$

In order to describe the neighbors of a fraction $\frac{j}{2(j-1)}$ in the right half sequences $\mathcal{F}^{\ge\frac{1}{2}}(\mathbb{B}(n), m)$ of the sequences $\mathcal{F}(\mathbb{B}(n), m)$, with $n \neq 2m$, we will apply the order-reversing bijections $\mathcal{F}^{\ge\frac{1}{2}}(\mathbb{B}(n), m) \leftrightarrow \mathcal{F}^{\le\frac{1}{2}}(\mathbb{B}(n), n-m)$ defined in (1.8) to the observations, made in Corollary 2.40, on the neighbors of the fraction $\frac{j-2}{2(j-1)}$ in $\mathcal{F}^{\le\frac{1}{2}}(\mathbb{B}(n), n-m)$.

Corollary 2.41. *Consider the fraction* $\frac{j}{2(j-1)} \in \mathcal{F}^{\ge\frac{1}{2}}(\mathbb{B}(n), m) - \{\frac{1}{2}\}$ *for some odd* j*, where* $n \neq 2m$.

(i) (a) *Suppose that* $n < 2m$. *If* $jn - 2(j-1)m \ge 1$, *then the fraction*

$$\begin{cases} \dfrac{j(\lceil \frac{2m}{j} \rceil - 1) + 1}{2} \Big/ ((j-1)(\lceil \frac{2m}{j} \rceil - 1) + 1) & \text{if } \lceil \frac{2m}{j} \rceil \text{ is even,} \\ \dfrac{j(\lceil \frac{2m}{j} \rceil - 2) + 1}{2} \Big/ ((j-1)(\lceil \frac{2m}{j} \rceil - 2) + 1) & \text{if } \lceil \frac{2m}{j} \rceil \text{ is odd} \end{cases} \tag{2.87}$$

precedes $\frac{j}{2(j-1)}$ *in* $\mathcal{F}(\mathbb{B}(n), m)$. *If* $jn - 2(j-1)m \le 1$, *then the fraction*

$$\begin{cases} \dfrac{j(\lceil \frac{2(n-m)}{j-2} \rceil - 1) + 1}{2} \Big/ ((j-1)(\lceil \frac{2(n-m)}{j-2} \rceil - 1) + 1) & \text{if } \lceil \frac{2(n-m)}{j-2} \rceil \text{ is even,} \\ \dfrac{j(\lceil \frac{2(n-m)}{j-2} \rceil - 2) + 1}{2} \Big/ ((j-1)(\lceil \frac{2(n-m)}{j-2} \rceil - 2) + 1) & \text{if } \lceil \frac{2(n-m)}{j-2} \rceil \text{ is odd} \end{cases} \tag{2.88}$$

precedes $\frac{j}{2(j-1)}$ *in* $\mathcal{F}(\mathbb{B}(n), m)$.

(b) *Suppose that* $n > 2m$. *Then the fraction*

$$\begin{cases} \dfrac{j(\lceil \frac{2m}{j} \rceil - 1) + 1}{2} \Big/ ((j-1)(\lceil \frac{2m}{j} \rceil - 1) + 1) & \text{if } \lceil \frac{2m}{j} \rceil \text{ is even,} \\ \dfrac{j(\lceil \frac{2m}{j} \rceil - 2) + 1}{2} \Big/ ((j-1)(\lceil \frac{2m}{j} \rceil - 2) + 1) & \text{if } \lceil \frac{2m}{j} \rceil \text{ is odd} \end{cases}$$

precedes $\frac{j}{2(j-1)}$ *in* $\mathcal{F}(\mathbb{B}(n), m)$.

(ii) (a) *Suppose that* $n < 2m$. *If* $2(j-1)m - jn \ge 1$, *then the fraction*

$$\begin{cases} \dfrac{j(\lceil \frac{2(n-m+1)}{j-2} \rceil - 1) - 1}{2} \Big/ ((j-1)(\lceil \frac{2(n-m+1)}{j-2} \rceil - 1) - 1) & \text{if } \lceil \frac{2(n-m+1)}{j-2} \rceil \text{ is even,} \\ \dfrac{j(\lceil \frac{2(n-m+1)}{j-2} \rceil - 2) - 1}{2} \Big/ ((j-1)(\lceil \frac{2(n-m+1)}{j-2} \rceil - 2) - 1) & \text{if } \lceil \frac{2(n-m+1)}{j-2} \rceil \text{ is odd} \end{cases}$$

succeeds $\frac{j}{2(j-1)}$ *in* $\mathcal{F}(\mathbb{B}(n), m)$.

If $2(j-1)m - jn \le 1$, then the fraction

$$\begin{cases} \dfrac{j(\lceil \frac{2(m+1)}{j} \rceil - 1) - 1}{2} \Big/ ((j-1)(\lceil \frac{2(m+1)}{j} \rceil - 1) - 1) & \text{if } \lceil \frac{2(m+1)}{j} \rceil \text{ is even,} \\ \dfrac{j(\lceil \frac{2(m+1)}{j} \rceil - 2) - 1}{2} \Big/ ((j-1)(\lceil \frac{2(m+1)}{j} \rceil - 2) - 1) & \text{if } \lceil \frac{2(m+1)}{j} \rceil \text{ is odd} \end{cases}$$

succeeds $\frac{j}{2(j-1)}$ in $\mathcal{F}(\mathbb{B}(n), m)$.

(b) *Suppose that $n > 2m$. Then the fraction*

$$\begin{cases} \dfrac{j(\lceil \frac{2(m+1)}{j} \rceil - 1) - 1}{2} \Big/ ((j-1)(\lceil \frac{2(m+1)}{j} \rceil - 1) - 1) & \text{if } \lceil \frac{2(m+1)}{j} \rceil \text{ is even,} \\ \dfrac{j(\lceil \frac{2(m+1)}{j} \rceil - 2) - 1}{2} \Big/ ((j-1)(\lceil \frac{2(m+1)}{j} \rceil - 2) - 1) & \text{if } \lceil \frac{2(m+1)}{j} \rceil \text{ is odd} \end{cases}$$

succeeds $\frac{j}{2(j-1)}$ in $\mathcal{F}(\mathbb{B}(n), m)$.

Proof. Note that the fraction $\frac{j-2}{2(j-1)} \in \mathcal{F}^{\le\frac{1}{2}}(\mathbb{B}(n), n-m)$ is the image of the fraction $\frac{j}{2(j-1)} \in \mathcal{F}^{\ge\frac{1}{2}}(\mathbb{B}(n), m)$ under the order-reversing bijection (1.8). Thus, Corollary 2.41 is derived directly from Corollary 2.40, with the help of the order-reversing bijections $\mathcal{F}^{\ge\frac{1}{2}}(\mathbb{B}(n), m) \leftrightarrow \mathcal{F}^{\le\frac{1}{2}}(\mathbb{B}(n), n-m)$ defined in (1.8).

For instance, suppose that $n < 2m$. Let us verify assertion (i) (a). Note that $n - 2(n-m) = 2m - n > 0$, that is, $n > 2(n-m)$.

If $2(j-1)(n-m) - (j-2)n = jn - 2(j-1)m \ge 1$, then, according to Corollary 2.40 (ii) (b), the fraction

$$\begin{cases} \dfrac{(j-2)(\lceil \frac{2m}{j} \rceil - 1) + 1}{2} \Big/ ((j-1)(\lceil \frac{2m}{j} \rceil - 1) + 1) & \text{if } \lceil \frac{2m}{j} \rceil \text{ is even,} \\ \dfrac{(j-2)(\lceil \frac{2m}{j} \rceil - 2) + 1}{2} \Big/ ((j-1)(\lceil \frac{2m}{j} \rceil - 2) + 1) & \text{if } \lceil \frac{2m}{j} \rceil \text{ is odd} \end{cases}$$

succeeds $\frac{j-2}{2(j-1)}$ in $\mathcal{F}^{\le\frac{1}{2}}(\mathbb{B}(n), n-m)$. As a consequence, its image (2.87) under the order-reversing bijection (1.8) precedes $\frac{j}{2(j-1)}$ in $\mathcal{F}^{\ge\frac{1}{2}}(\mathbb{B}(n), m)$.

If $jn - 2(j-1)m \le 1$, then, according to Corollary 2.40 (ii) (b), the fraction

$$\begin{cases} \dfrac{(j-2)(\lceil \frac{2(n-m)}{j-2} \rceil - 1) + 1}{2} \Big/ ((j-1)(\lceil \frac{2(n-m)}{j-2} \rceil - 1) + 1) & \text{if } \lceil \frac{2(n-m)}{j-2} \rceil \text{ is even,} \\ \dfrac{(j-2)(\lceil \frac{2(n-m)}{j-2} \rceil - 2) + 1}{2} \Big/ ((j-1)(\lceil \frac{2(n-m)}{j-2} \rceil - 2) + 1) & \text{if } \lceil \frac{2(n-m)}{j-2} \rceil \text{ is odd} \end{cases}$$

succeeds $\frac{j-2}{2(j-1)}$ in $\mathcal{F}^{\le\frac{1}{2}}(\mathbb{B}(n), n-m)$. As a consequence, its image (2.88) under the order-reversing bijection (1.8) precedes $\frac{j}{2(j-1)}$ in $\mathcal{F}^{\ge\frac{1}{2}}(\mathbb{B}(n), m)$. □

2.2.9.9 The neighbors of $\frac{1}{3}$

We will now find the neighbors of the fraction $\frac{1}{3}$ in the Farey subsequences $\mathcal{F}^{\le\frac{1}{2}}(\mathbb{B}(n), m)$, with $n \ne 2m$.

Remark 2.42 (see also Table 2.7). (i) Consider a Farey subsequence $\mathcal{F}(\mathbb{B}(n), m)$ such that $n < 2m$ and $n - m > 1$.

From Remark 2.10 and the order-preserving bijection $\mathcal{F}_{n-m} \to \mathcal{F}^{\leq\frac{1}{2}}(\mathbb{B}(n), m)$ given in (2.21), it follows that the fraction

$$\frac{\lfloor (n-m-1)/2 \rfloor}{3\lfloor (n-m-1)/2 \rfloor + 1} = \frac{\lceil (n-m)/2 \rceil - 1}{3\lceil (n-m)/2 \rceil - 2}$$

precedes $\frac{1}{3}$ in $\mathcal{F}^{\leq\frac{1}{2}}(\mathbb{B}(n), m)$, and the fraction

$$\frac{\lceil (n-m)/2 \rceil}{3\lceil (n-m)/2 \rceil - 1} = \frac{\lfloor (n-m-1)/2 \rfloor + 1}{3\lfloor (n-m-1)/2 \rfloor + 2}$$

succeeds $\frac{1}{3}$ in $\mathcal{F}^{\leq\frac{1}{2}}(\mathbb{B}(n), m)$.

Thus, if $(n - m)$ is even, then the fractions

$$\frac{(n-m-2)/2}{(3(n-m)-4)/2} < \frac{1}{3} < \frac{(n-m)/2}{(3(n-m)-2)/2}$$

are consecutive in $\mathcal{F}^{\leq\frac{1}{2}}(\mathbb{B}(n), m)$. If $(n - m)$ is odd, then the fractions

$$\frac{(n-m-1)/2}{(3(n-m)-1)/2} < \frac{1}{3} < \frac{(n-m+1)/2}{(3(n-m)+1)/2}$$

are consecutive in $\mathcal{F}^{\leq\frac{1}{2}}(\mathbb{B}(n), m)$.

(ii) Consider a Farey subsequence $\mathcal{F}(\mathbb{B}(n), m)$ such that $n > 2m$ and $m > 1$.

Let us apply the order-preserving bijection $\mathcal{F}^{m}_{n-m} \to \mathcal{F}^{\leq\frac{1}{2}}(\mathbb{B}(n), m)$ given in (2.26) to Remark 2.14.

If $(n - m) - 2m = n - 3m \geq 1$, then $\frac{m}{3m+1}$ precedes $\frac{1}{3}$ in $\mathcal{F}^{\leq\frac{1}{2}}(\mathbb{B}(n), m)$. If $n - 3m \leq 1$, then

$$\frac{\lfloor (n-m-1)/2 \rfloor}{3\lfloor (n-m-1)/2 \rfloor + 1} = \frac{\lceil (n-m)/2 \rceil - 1}{3\lceil (n-m)/2 \rceil - 2}$$

precedes $\frac{1}{3}$ in $\mathcal{F}^{\leq\frac{1}{2}}(\mathbb{B}(n), m)$.

Thus, if $n - 3m \leq 1$ and $(n - m)$ is even, then

$$\frac{(n-m-2)/2}{(3(n-m)-4)/2}$$

precedes $\frac{1}{3}$ in $\mathcal{F}^{\leq\frac{1}{2}}(\mathbb{B}(n), m)$. If $n - 3m \leq 1$ and $(n - m)$ is odd, then

$$\frac{(n-m-1)/2}{(3(n-m)-1)/2}$$

precedes $\frac{1}{3}$ in $\mathcal{F}^{\leq\frac{1}{2}}(\mathbb{B}(n), m)$.

If $2m - (n - m) = 3m - n \geq 1$, then

$$\frac{\lceil (n-m)/2 \rceil}{3\lceil (n-m)/2 \rceil - 1} = \frac{\lfloor (n-m-1)/2 \rfloor + 1}{3\lfloor (n-m-1)/2 \rfloor + 2}$$

succeeds $\frac{1}{3}$ in $\mathcal{F}^{\leq\frac{1}{2}}(\mathbb{B}(n), m)$.

Thus, if $3m - n \geq 1$ and $(n - m)$ is even, then

$$\frac{(n - m)/2}{(3(n - m) - 2)/2}$$

succeeds $\frac{1}{3}$ in $\mathcal{F}^{\leq\frac{1}{2}}(\mathbb{B}(n), m)$. If $3m - n \geq 1$ and $(n - m)$ is odd, then

$$\frac{(n - m + 1)/2}{(3(n - m) - 1)/2}$$

succeeds $\frac{1}{3}$ in $\mathcal{F}^{\leq\frac{1}{2}}(\mathbb{B}(n), m)$.

If $3m - n \leq 1$, then $\frac{m}{2m-1}$ succeeds $\frac{1}{3}$ in $\mathcal{F}^{\leq\frac{1}{2}}(\mathbb{B}(n), m)$.

2.2.9.10 The neighbors of $\frac{2}{3}$

In order to find the neighbors of the fraction $\frac{2}{3}$ in the Farey subsequences $\mathcal{F}(\mathbb{B}(n), m)$, with $n \neq 2m$, let us apply the order-reversing bijections $\mathcal{F}(\mathbb{B}(n), m) \leftrightarrow \mathcal{F}(\mathbb{B}(n), n - m)$ defined in (1.8) to the observations, made in Remark 2.42, on the neighbors of the fraction $\frac{1}{3}$.

Remark 2.43 (see also Table 2.8). (i) Consider a Farey subsequence $\mathcal{F}(\mathbb{B}(n), m)$ such that $n < 2m$ and $n - m > 1$.

If $2n - 3m \geq 1$, then

$$\frac{2\lceil m/2\rceil - 1}{3\lceil m/2\rceil - 1} = \frac{2\lfloor (m - 1)/2\rfloor + 1}{3\lfloor (m - 1)/2\rfloor + 2}$$

precedes $\frac{2}{3}$ in $\mathcal{F}^{\geq\frac{1}{2}}(\mathbb{B}(n), m)$.

Thus, if $2n - 3m \geq 1$ and m is even, then

$$\frac{m - 1}{(3m - 2)/2}$$

precedes $\frac{2}{3}$ in $\mathcal{F}^{\geq\frac{1}{2}}(\mathbb{B}(n), m)$. If $2n - 3m \geq 1$ and m is odd, then

$$\frac{m - 1}{(3m - 1)/2}$$

precedes $\frac{2}{3}$ in $\mathcal{F}^{\geq\frac{1}{2}}(\mathbb{B}(n), m)$.

If $2n - 3m \leq 1$, then

$$\frac{n - m - 1}{2(n - m) - 1}$$

precedes $\frac{2}{3}$ in $\mathcal{F}^{\geq\frac{1}{2}}(\mathbb{B}(n), m)$.

If $3m - 2n \geq 1$, then

$$\frac{2(n - m) + 1}{3(n - m) + 1}$$

succeeds $\frac{2}{3}$ in $\mathcal{F}^{\geq\frac{1}{2}}(\mathbb{B}(n), m)$. If $3m - 2n \leq 1$, then

$$\frac{2\lfloor (m - 1)/2\rfloor + 1}{3\lfloor (m - 1)/2\rfloor + 1} = \frac{2\lceil m/2\rceil - 1}{3\lceil m/2\rceil - 2}$$

succeeds $\frac{2}{3}$ in $\mathcal{F}^{\geq\frac{1}{2}}(\mathbb{B}(n), m)$.

Thus, if $3m - 2n \le 1$ and m is even, then

$$\frac{m-1}{(3m-4)/2}$$

succeeds $\frac{2}{3}$ in $\mathcal{F}^{\ge\frac{1}{2}}(\mathbb{B}(n), m)$. If $3m - 2n \le 1$ and m is odd, then

$$\frac{m}{(3m-1)/2}$$

succeeds $\frac{2}{3}$ in $\mathcal{F}^{\ge\frac{1}{2}}(\mathbb{B}(n), m)$.

(ii) Consider a Farey subsequence $\mathcal{F}(\mathbb{B}(n), m)$ such that $n > 2m$ and $m > 1$. The fraction

$$\frac{2\lceil m/2\rceil - 1}{3\lceil m/2\rceil - 1} = \frac{2\lfloor (m-1)/2\rfloor + 1}{3\lfloor (m-1)/2\rfloor + 2}$$

precedes $\frac{2}{3}$ in $\mathcal{F}^{\ge\frac{1}{2}}(\mathbb{B}(n), m)$, and the fraction

$$\frac{2\lfloor (m-1)/2\rfloor + 1}{3\lfloor (m-1)/2\rfloor + 1} = \frac{2\lceil m/2\rceil - 1}{3\lceil m/2\rceil - 2}$$

succeeds $\frac{2}{3}$ in $\mathcal{F}^{\ge\frac{1}{2}}(\mathbb{B}(n), m)$.

Thus, if m is even, then the fractions

$$\frac{m-1}{(3m-2)/2} < \frac{2}{3} < \frac{m-1}{(3m-4)/2}$$

are consecutive in $\mathcal{F}^{\ge\frac{1}{2}}(\mathbb{B}(n), m)$. If m is odd, then the fractions

$$\frac{m}{(3m+1)/2} < \frac{2}{3} < \frac{m}{(3m-1)/2}$$

are consecutive in $\mathcal{F}^{\ge\frac{1}{2}}(\mathbb{B}(n), m)$.

2.3 Triples of consecutive fractions. II

In this section we pursue the investigation, initiated in Section 1.5.5, of recurrent relations that describe, by means of duality, triples of consecutive fractions in the Farey subsequences $\mathcal{F}(\mathbb{B}(n), m)$; see Table 2.9. Recall that the triples of consecutive fractions whose central fraction is $\frac{1}{2}$ are presented in Table 2.5.

2.3.1 Triples of consecutive fractions in $\mathcal{F}^{\le\frac{1}{2}}(\mathbb{B}(2m), m)$ and $\mathcal{F}^{\ge\frac{1}{2}}(\mathbb{B}(2m), m)$

Let us find recurrent relations that describe triples of consecutive fractions in the left and right half sequences of the sequences $\mathcal{F}(\mathbb{B}(2m), m)$; cf. Proposition 1.28.

Table 2.9. Triples of consecutive fractions $\frac{h_j}{k_j} < \frac{h_{j+1}}{k_{j+1}} < \frac{h_{j+2}}{k_{j+2}}$ in Farey (sub)sequences; see also Table 1.6.

Sequence and Formulas	**See**
$\mathcal{F}^{\leq \frac{1}{2}}(\mathbb{B}(2m), m)$	
$h_j = \lfloor \frac{m+k_{j+2}-h_{j+2}}{k_{j+1}-h_{j+1}} \rfloor h_{j+1} - h_{j+2},\ k_j = \lfloor \frac{m+k_{j+2}-h_{j+2}}{k_{j+1}-h_{j+1}} \rfloor k_{j+1} - k_{j+2}$ $h_{j+2} = \lfloor \frac{m+k_j-h_j}{k_{j+1}-h_{j+1}} \rfloor h_{j+1} - h_j,\ k_{j+2} = \lfloor \frac{m+k_j-h_j}{k_{j+1}-h_{j+1}} \rfloor k_{j+1} - k_j$	Corollary 2.44 (i)
$\mathcal{F}^{\geq \frac{1}{2}}(\mathbb{B}(2m), m)$	
$h_j = \lfloor \frac{m+h_{j+2}}{h_{j+1}} \rfloor h_{j+1} - h_{j+2},\ k_j = \lfloor \frac{m+h_{j+2}}{h_{j+1}} \rfloor k_{j+1} - k_{j+2}$ $h_{j+2} = \lfloor \frac{m+h_j}{h_{j+1}} \rfloor h_{j+1} - h_j,\ k_{j+2} = \lfloor \frac{m+h_j}{h_{j+1}} \rfloor k_{j+1} - k_j$	Corollary 2.44 (ii)
$\mathcal{F}^{\leq \frac{1}{2}}(\mathbb{B}(n), m),\ n < 2m$	
$h_j = \lfloor \frac{(n-m)+(k_{j+2}-h_{j+2})}{k_{j+1}-h_{j+1}} \rfloor h_{j+1} - h_{j+2},\ k_j = \lfloor \frac{(n-m)+(k_{j+2}-h_{j+2})}{k_{j+1}-h_{j+1}} \rfloor k_{j+1} - k_{j+2}$ $h_{j+2} = \lfloor \frac{(n-m)+(k_j-h_j)}{k_{j+1}-h_{j+1}} \rfloor h_{j+1} - h_j,\ k_{j+2} = \lfloor \frac{(n-m)+(k_j-h_j)}{k_{j+1}-h_{j+1}} \rfloor k_{j+1} - k_j$	Corollary 2.45 (i)
$\mathcal{F}^{\geq \frac{1}{2}}(\mathbb{B}(n), m),\ n < 2m$	
$h_j = \lfloor \frac{m+h_{j+2}}{h_{j+1}} \rfloor h_{j+1} - h_{j+2},\ k_j = \lfloor \frac{m+h_{j+2}}{h_{j+1}} \rfloor k_{j+1} - k_{j+2}$ if $h_{j+1}n - k_{j+1}m \geq 1$ $h_j = \lfloor \frac{(n-m)+(k_{j+2}-h_{j+2})}{k_{j+1}-h_{j+1}} \rfloor h_{j+1} - h_{j+2},\ k_j = \lfloor \frac{(n-m)+(k_{j+2}-h_{j+2})}{k_{j+1}-h_{j+1}} \rfloor k_{j+1} - k_{j+2}$ if $h_{j+1}n - k_{j+1}m \leq 1$	Corollary 2.45 (ii) (a)
$h_{j+2} = \lfloor \frac{(n-m)+(k_j-h_j)}{k_{j+1}-h_{j+1}} \rfloor h_{j+1} - h_j,\ k_{j+2} = \lfloor \frac{(n-m)+(k_j-h_j)}{k_{j+1}-h_{j+1}} \rfloor k_{j+1} - k_j$ if $k_{j+1}m - h_{j+1}n \geq 1$ $h_{j+2} = \lfloor \frac{m+h_j}{h_{j+1}} \rfloor h_{j+1} - h_j,\ k_{j+2} = \lfloor \frac{m+h_j}{h_{j+1}} \rfloor k_{j+1} - k_j$ if $k_{j+1}m - h_{j+1}n \leq 1$	Corollary 2.45 (ii) (b)
$\mathcal{F}^{\leq \frac{1}{2}}(\mathbb{B}(n), m),\ n > 2m$	
$h_j = \lfloor \frac{m+h_{j+2}}{h_{j+1}} \rfloor h_{j+1} - h_{j+2},\ k_j = \lfloor \frac{m+h_{j+2}}{h_{j+1}} \rfloor k_{j+1} - k_{j+2}$ if $h_{j+1}n - k_{j+1}m \geq 1$ $h_j = \lfloor \frac{(n-m)+(k_{j+2}-h_{j+2})}{k_{j+1}-h_{j+1}} \rfloor h_{j+1} - h_{j+2},\ k_j = \lfloor \frac{(n-m)+(k_{j+2}-h_{j+2})}{k_{j+1}-h_{j+1}} \rfloor k_{j+1} - k_{j+2}$ if $h_{j+1}n - k_{j+1}m \leq 1$	Corollary 2.46 (i) (a)
$h_{j+2} = \lfloor \frac{(n-m)+(k_j-h_j)}{k_{j+1}-h_{j+1}} \rfloor h_{j+1} - h_j,\ k_{j+2} = \lfloor \frac{(n-m)+(k_j-h_j)}{k_{j+1}-h_{j+1}} \rfloor k_{j+1} - k_j$ if $k_{j+1}m - h_{j+1}n \geq 1$ $h_{j+2} = \lfloor \frac{m+h_j}{h_{j+1}} \rfloor h_{j+1} - h_j,\ k_{j+2} = \lfloor \frac{m+h_j}{h_{j+1}} \rfloor k_{j+1} - k_j$ if $k_{j+1}m - h_{j+1}n \leq 1$	Corollary 2.46 (i) (b)
$\mathcal{F}^{\geq \frac{1}{2}}(\mathbb{B}(n), m),\ n > 2m$	
$h_j = \lfloor \frac{m+h_{j+2}}{h_{j+1}} \rfloor h_{j+1} - h_{j+2},\ k_j = \lfloor \frac{m+h_{j+2}}{h_{j+1}} \rfloor k_{j+1} - k_{j+2},$ $h_{j+2} = \lfloor \frac{m+h_j}{h_{j+1}} \rfloor h_{j+1} - h_j,\ k_{j+2} = \lfloor \frac{m+h_j}{h_{j+1}} \rfloor k_{j+1} - k_j$	Corollary 2.46 (ii)

Corollary 2.44. (i) *If* $\frac{h_j}{k_j} < \frac{h_{j+1}}{k_{j+1}} < \frac{h_{j+2}}{k_{j+2}}$ *are three consecutive fractions in the sequence* $\mathcal{F}^{\leq\frac{1}{2}}(\mathbb{B}(2m), m)$, *then*

$$h_j = \left\lfloor \frac{m + k_{j+2} - h_{j+2}}{k_{j+1} - h_{j+1}} \right\rfloor h_{j+1} - h_{j+2}, \tag{2.89}$$

$$k_j = \left\lfloor \frac{m + k_{j+2} - h_{j+2}}{k_{j+1} - h_{j+1}} \right\rfloor k_{j+1} - k_{j+2}, \tag{2.90}$$

$$h_{j+2} = \left\lfloor \frac{m + k_j - h_j}{k_{j+1} - h_{j+1}} \right\rfloor h_{j+1} - h_j, \tag{2.91}$$

$$k_{j+2} = \left\lfloor \frac{m + k_j - h_j}{k_{j+1} - h_{j+1}} \right\rfloor k_{j+1} - k_j. \tag{2.92}$$

(ii) *If* $\frac{h_j}{k_j} < \frac{h_{j+1}}{k_{j+1}} < \frac{h_{j+2}}{k_{j+2}}$ *are three consecutive fractions in the sequence* $\mathcal{F}^{\geq\frac{1}{2}}(\mathbb{B}(2m), m)$, *then*

$$h_j = \left\lfloor \frac{m + h_{j+2}}{h_{j+1}} \right\rfloor h_{j+1} - h_{j+2},$$

$$k_j = \left\lfloor \frac{m + h_{j+2}}{h_{j+1}} \right\rfloor k_{j+1} - k_{j+2},$$

$$h_{j+2} = \left\lfloor \frac{m + h_j}{h_{j+1}} \right\rfloor h_{j+1} - h_j,$$

$$k_{j+2} = \left\lfloor \frac{m + h_j}{h_{j+1}} \right\rfloor k_{j+1} - k_j.$$

Proof. (i) According to Corollary 2.5, the three consecutive fractions

$$\frac{h_j}{k_j - h_j} < \frac{h_{j+1}}{k_{j+1} - h_{j+1}} < \frac{h_{j+2}}{k_{j+2} - h_{j+2}} \tag{2.93}$$

in $\mathcal{F}_m$ are the images of the fractions $\frac{h_j}{k_j} < \frac{h_{j+1}}{k_{j+1}} < \frac{h_{j+2}}{k_{j+2}}$ from $\mathcal{F}^{\leq\frac{1}{2}}(\mathbb{B}(2m), m)$ under the order-preserving bijection (2.12). It follows from Proposition 1.25 that the integers determining the fractions in (2.93) obey (2.89) and the equality

$$k_j - h_j = \left\lfloor \frac{m + k_{j+2} - h_{j+2}}{k_{j+1} - h_{j+1}} \right\rfloor (k_{j+1} - h_{j+1}) - (k_{j+2} - h_{j+2});$$

hence, (2.90) follows.

Equations (2.91) and (2.92) are proved analogously to the proof of (2.89) and (2.90).

To prove assertion (ii), it suffices to note that the three consecutive fractions

$$\frac{k_{j+2} - h_{j+2}}{k_{j+2}} < \frac{k_{j+1} - h_{j+1}}{k_{j+1}} < \frac{k_j - h_j}{k_j}$$

in $\mathcal{F}^{\leq\frac{1}{2}}(\mathbb{B}(2m), m)$ are the images of the fractions $\frac{h_j}{k_j} < \frac{h_{j+1}}{k_{j+1}} < \frac{h_{j+2}}{k_{j+2}}$ from $\mathcal{F}^{\geq\frac{1}{2}}(\mathbb{B}(2m), m)$ under the order-reversing bijection defined in (1.8). Now, assertion (ii) can be derived from assertion (i). □

2.3.2 Triples of consecutive fractions in $\mathcal{F}^{\leq\frac{1}{2}}(\mathbb{B}(n), m)$ and $\mathcal{F}^{\geq\frac{1}{2}}(\mathbb{B}(n), m)$, $n \neq 2m$

In this subsection we will present recurrent relations that describe triples of consecutive fractions in the left and right half sequences of the sequences $\mathcal{F}(\mathbb{B}(n), m)$, with $n \neq 2m$; cf. Proposition 1.28.

Corollary 2.45. *Let $\mathcal{F}(\mathbb{B}(n), m)$ be a Farey subsequence such that $n < 2m$.*

(i) *Let*

$$\frac{h_j}{k_j} < \frac{h_{j+1}}{k_{j+1}} < \frac{h_{j+2}}{k_{j+2}} \tag{2.94}$$

be three consecutive fractions in the subsequence $\mathcal{F}^{\leq\frac{1}{2}}(\mathbb{B}(n), m)$. Then we have

$$\begin{aligned} h_j &= \left\lfloor \frac{(n-m) + (k_{j+2} - h_{j+2})}{k_{j+1} - h_{j+1}} \right\rfloor h_{j+1} - h_{j+2}, \\ k_j &= \left\lfloor \frac{(n-m) + (k_{j+2} - h_{j+2})}{k_{j+1} - h_{j+1}} \right\rfloor k_{j+1} - k_{j+2}, \\ h_{j+2} &= \left\lfloor \frac{(n-m) + (k_j - h_j)}{k_{j+1} - h_{j+1}} \right\rfloor h_{j+1} - h_j, \\ k_{j+2} &= \left\lfloor \frac{(n-m) + (k_j - h_j)}{k_{j+1} - h_{j+1}} \right\rfloor k_{j+1} - k_j. \end{aligned}$$

(ii) *Let*

$$\frac{h_j}{k_j} < \frac{h_{j+1}}{k_{j+1}} < \frac{h_{j+2}}{k_{j+2}} \tag{2.95}$$

be three consecutive fractions in the subsequence $\mathcal{F}^{\geq\frac{1}{2}}(\mathbb{B}(n), m)$.

(a) *If $h_{j+1}n - k_{j+1}m \geq 1$, then*

$$h_j = \left\lfloor \frac{m + h_{j+2}}{h_{j+1}} \right\rfloor h_{j+1} - h_{j+2}, \quad k_j = \left\lfloor \frac{m + h_{j+2}}{h_{j+1}} \right\rfloor k_{j+1} - k_{j+2}.$$

If $h_{j+1}n - k_{j+1}m \leq 1$, then

$$\begin{aligned} h_j &= \left\lfloor \frac{(n-m) + (k_{j+2} - h_{j+2})}{k_{j+1} - h_{j+1}} \right\rfloor h_{j+1} - h_{j+2}, \\ k_j &= \left\lfloor \frac{(n-m) + (k_{j+2} - h_{j+2})}{k_{j+1} - h_{j+1}} \right\rfloor k_{j+1} - k_{j+2}. \end{aligned}$$

(b) *If $k_{j+1}m - h_{j+1}n \geq 1$, then*

$$\begin{aligned} h_{j+2} &= \left\lfloor \frac{(n-m) + (k_j - h_j)}{k_{j+1} - h_{j+1}} \right\rfloor h_{j+1} - h_j, \\ k_{j+2} &= \left\lfloor \frac{(n-m) + (k_j - h_j)}{k_{j+1} - h_{j+1}} \right\rfloor k_{j+1} - k_j. \end{aligned}$$

If $k_{j+1}m - h_{j+1}n \leq 1$, then

$$h_{j+2} = \left\lfloor \frac{m + h_j}{h_{j+1}} \right\rfloor h_{j+1} - h_j, \quad k_{j+2} = \left\lfloor \frac{m + h_j}{h_{j+1}} \right\rfloor k_{j+1} - k_j.$$

Proof. (i) According to Corollary 2.8 (i), the three consecutive fractions

$$\frac{h_j}{k_j - h_j} < \frac{h_{j+1}}{k_{j+1} - h_{j+1}} < \frac{h_{j+2}}{k_{j+2} - h_{j+2}}$$

in $\mathcal{F}_{n-m}$ are the images of the fractions in (2.94) under the order-preserving bijection (2.20). Hence, assertion (i) follows from Proposition 1.25.

(ii) Note that, according to Corollary 2.8 (i), the three consecutive fractions

$$\frac{k_{j+2} - h_{j+2}}{h_{j+2}} < \frac{k_{j+1} - h_{j+1}}{h_{j+1}} < \frac{k_j - h_j}{h_j}$$

in $\mathcal{F}_m^{n-m}$ are the images of the fractions in (2.95) under the order-reversing bijection (2.23). Assertion (ii) (a) now follows from Proposition 1.26 (ii) (b), and assertion (ii) (b) follows from Proposition 1.26 (ii) (a). □

Corollary 2.46. *Let $\mathcal{F}(\mathbb{B}(n), m)$ be a Farey subsequence such that $n > 2m$.*

(i) *Let*

$$\frac{h_j}{k_j} < \frac{h_{j+1}}{k_{j+1}} < \frac{h_{j+2}}{k_{j+2}} \tag{2.96}$$

be three consecutive fractions in the subsequence $\mathcal{F}^{\leq\frac{1}{2}}(\mathbb{B}(n), m)$.

(a) *If $h_{j+1}n - k_{j+1}m \geq 1$, then*

$$h_j = \left\lfloor \frac{m + h_{j+2}}{h_{j+1}} \right\rfloor h_{j+1} - h_{j+2},$$

$$k_j = \left\lfloor \frac{m + h_{j+2}}{h_{j+1}} \right\rfloor k_{j+1} - k_{j+2}.$$

If $h_{j+1}n - k_{j+1}m \leq 1$, then

$$h_j = \left\lfloor \frac{(n - m) + (k_{j+2} - h_{j+2})}{k_{j+1} - h_{j+1}} \right\rfloor h_{j+1} - h_{j+2},$$

$$k_j = \left\lfloor \frac{(n - m) + (k_{j+2} - h_{j+2})}{k_{j+1} - h_{j+1}} \right\rfloor k_{j+1} - k_{j+2}.$$

(b) *If $k_{j+1}m - h_{j+1}n \geq 1$, then*

$$h_{j+2} = \left\lfloor \frac{(n - m) + (k_j - h_j)}{k_{j+1} - h_{j+1}} \right\rfloor h_{j+1} - h_j,$$

$$k_{j+2} = \left\lfloor \frac{(n - m) + (k_j - h_j)}{k_{j+1} - h_{j+1}} \right\rfloor k_{j+1} - k_j.$$

If $k_{j+1}m - h_{j+1}n \leq 1$, then

$$h_{j+2} = \left\lfloor \frac{m + h_j}{h_{j+1}} \right\rfloor h_{j+1} - h_j,$$

$$k_{j+2} = \left\lfloor \frac{m + h_j}{h_{j+1}} \right\rfloor k_{j+1} - k_j.$$

(ii) *Let*

$$\frac{h_j}{k_j} < \frac{h_{j+1}}{k_{j+1}} < \frac{h_{j+2}}{k_{j+2}} \tag{2.97}$$

be three consecutive fractions in the subsequence $\mathcal{F}^{\geq\frac{1}{2}}(\mathbb{B}(n), m)$. *Then we have*

$$h_j = \left\lfloor \frac{m + h_{j+2}}{h_{j+1}} \right\rfloor h_{j+1} - h_{j+2},$$

$$k_j = \left\lfloor \frac{m + h_{j+2}}{h_{j+1}} \right\rfloor k_{j+1} - k_{j+2},$$

$$h_{j+2} = \left\lfloor \frac{m + h_j}{h_{j+1}} \right\rfloor h_{j+1} - h_j,$$

$$k_{j+2} = \left\lfloor \frac{m + h_j}{h_{j+1}} \right\rfloor k_{j+1} - k_j.$$

Proof. (i) Note that, according to Corollary 2.8 (ii), the three consecutive fractions

$$\frac{h_j}{k_j - h_j} < \frac{h_{j+1}}{k_{j+1} - h_{j+1}} < \frac{h_{j+2}}{k_{j+2} - h_{j+2}}$$

in $\mathcal{F}_{n-m}^{m}$ are the images of the fractions in (2.96) under the order-preserving bijection (2.25). Assertion (i) (a) now follows from Proposition 1.26 (ii) (a), and assertion (i) (b) follows from Proposition 1.26 (ii) (b).

(ii) According to Corollary 2.8 (ii), the three consecutive fractions

$$\frac{k_{j+2} - h_{j+2}}{h_{j+2}} < \frac{k_{j+1} - h_{j+1}}{h_{j+1}} < \frac{k_j - h_j}{h_j}$$

in $\mathcal{F}_m$ are the images of the fractions in (2.97) under the order-reversing bijection (2.29). Hence, assertion (ii) follows from Proposition 1.25. □

2.4 The number of fractions in Farey (sub)sequences. II

In Section 1.6 we counted the number of fractions in the Farey (sub)sequences $\mathcal{F}_n$, $\mathcal{F}_n^m$ and $\mathcal{G}_n^m$. In this section we first count, by means of duality, the number of fractions in the Farey subsequences $\mathcal{F}(\mathbb{B}(n), m)$, and then we find several other related quantities; see Table 1.7.

2.4.1 The number of fractions in $\mathcal{F}(\mathbb{B}(n), m)$

The monotone bijections given in (2.2) and (2.8) of Theorem 2.1 allow us to write down the formula

$$|\mathcal{F}(\mathbb{B}(n), m)| = |\mathcal{F}_{n-m}^{m}| + |\mathcal{F}_{m}^{n-m}| - 1 = |\mathcal{F}_{n-m}^{\min\{m,n-m\}}| + |\mathcal{F}_{m}^{\min\{m,n-m\}}| - 1$$

for the number of fractions in any sequence $\mathcal{F}(\mathbb{B}(n), m)$:

$$|\mathcal{F}(\mathbb{B}(n), m)| = |\mathcal{F}^{\le\frac{1}{2}}(\mathbb{B}(n), m)| + |\mathcal{F}^{\ge\frac{1}{2}}(\mathbb{B}(n), m)| - 1.$$

Recall that (see Proposition 1.29)

$$|\mathcal{F}_q^p| = \frac{3}{2} + \sum_{d\ge1} \mu(d)\left\lfloor \frac{p}{d} \right\rfloor\left(\left\lfloor \frac{q}{d} \right\rfloor - \frac{1}{2}\left\lfloor \frac{p}{d} \right\rfloor\right).$$

Thus, we have

$$\begin{aligned}|\mathcal{F}(\mathbb{B}(n), m)| = {}&\frac{3}{2} + \sum_{d\ge1} \mu(d)\left\lfloor \frac{\min\{m, n-m\}}{d} \right\rfloor\left(\left\lfloor \frac{n-m}{d} \right\rfloor - \frac{1}{2}\left\lfloor \frac{\min\{m, n-m\}}{d} \right\rfloor\right)\\ &+ \frac{3}{2} + \sum_{d\ge1} \mu(d)\left\lfloor \frac{\min\{m, n-m\}}{d} \right\rfloor\left(\left\lfloor \frac{m}{d} \right\rfloor - \frac{1}{2}\left\lfloor \frac{\min\{m, n-m\}}{d} \right\rfloor\right) - 1\end{aligned}$$

or

$$|\mathcal{F}(\mathbb{B}(n), m)| = 2 + \sum_{d\ge1} \mu(d)\left\lfloor \frac{\min\{m, n-m\}}{d} \right\rfloor\left(\left\lfloor \frac{n-m}{d} \right\rfloor + \left\lfloor \frac{m}{d} \right\rfloor - \left\lfloor \frac{\min\{m, n-m\}}{d} \right\rfloor\right).$$

We come to the following conclusion.

Proposition 2.47. *The number of fractions in the Farey subsequence $\mathcal{F}(\mathbb{B}(n), m)$ is*

$$|\mathcal{F}(\mathbb{B}(n), m)| = 2 + \sum_{d\ge1} \mu(d)\left\lfloor \frac{m}{d} \right\rfloor\left\lfloor \frac{n-m}{d} \right\rfloor.$$

In particular,

$$|\mathcal{F}(\mathbb{B}(2m), m)| = 2 + \sum_{d=1}^{m} \mu(d)\left\lfloor \frac{m}{d} \right\rfloor^2.$$

Remark 2.48. For all positive integers m, we have (cf. (1.117))

$$\sum_{d=1}^{m} \mu(d)\left\lfloor \frac{m}{d} \right\rfloor^2 = |\mathcal{F}(\mathbb{B}(2m), m)| - 2 = 2|\mathcal{F}_m| - 3. \tag{2.98}$$

2.4.2 More on the number of fractions

Let us count the number of fractions in several subsequences of the Farey sequences; the simple observation made in (1.7) should be recalled here. The computations made in this subsection and other related results are collected in Table 1.7.

For positive integers m and n, where $0 < m < n$, Propositions 1.29 and 2.47 yield

$$|\mathcal{F}_n^m| - |\mathcal{F}(\mathbb{B}(n), m)| = \underbrace{\frac{3}{2} + \sum_{d=1}^{m} \mu(d) \left\lfloor \frac{m}{d} \right\rfloor \left(\left\lfloor \frac{n}{d} \right\rfloor - \frac{1}{2} \left\lfloor \frac{m}{d} \right\rfloor \right)}_{|\mathcal{F}_n^m|} - \left(2 + \sum_{d=1}^{m} \mu(d) \left\lfloor \frac{m}{d} \right\rfloor \left\lfloor \frac{n-m}{d} \right\rfloor \right)$$

$$= -\frac{1}{2} + \sum_{d=1}^{m} \mu(d) \left\lfloor \frac{m}{d} \right\rfloor \left(\left\lfloor \frac{n}{d} \right\rfloor - \frac{1}{2} \left\lfloor \frac{m}{d} \right\rfloor - \left\lfloor \frac{n-m}{d} \right\rfloor \right).$$

Similarly, Corollary 1.31 and Proposition 2.47 yield

$$|\mathcal{G}_n^m| - |\mathcal{F}(\mathbb{B}(n), m)| = \underbrace{\frac{3}{2} + \sum_{d=1}^{n-m} \mu(d) \left\lfloor \frac{n-m}{d} \right\rfloor \left(\left\lfloor \frac{n}{d} \right\rfloor - \frac{1}{2} \left\lfloor \frac{n-m}{d} \right\rfloor \right)}_{|\mathcal{G}_n^m|}$$

$$- \left(2 + \sum_{d=1}^{n-m} \mu(d) \left\lfloor \frac{m}{d} \right\rfloor \left\lfloor \frac{n-m}{d} \right\rfloor \right)$$

$$= -\frac{1}{2} + \sum_{d=1}^{n-m} \mu(d) \left\lfloor \frac{n-m}{d} \right\rfloor \left(\left\lfloor \frac{n}{d} \right\rfloor - \frac{1}{2} \left\lfloor \frac{n-m}{d} \right\rfloor - \left\lfloor \frac{m}{d} \right\rfloor \right).$$

In particular,

$$|\mathcal{F}_{2m}^m| - |\mathcal{F}(\mathbb{B}(2m), m)| = |\mathcal{G}_{2m}^m| - |\mathcal{F}(\mathbb{B}(2m), m)|$$

$$= -\frac{1}{2} + \sum_{d=1}^{m} \mu(d) \left\lfloor \frac{m}{d} \right\rfloor \left(\left\lfloor \frac{2m}{d} \right\rfloor - \frac{3}{2} \left\lfloor \frac{m}{d} \right\rfloor \right).$$

As a consequence,

$$|\mathcal{F}_n| - |\mathcal{F}(\mathbb{B}(n), m)| = (|\mathcal{F}_n^m| - |\mathcal{F}(\mathbb{B}(n), m)|) + (|\mathcal{G}_n^m| - |\mathcal{F}(\mathbb{B}(n), m)|)$$

$$= -4 + \sum_{d \geq 1} \mu(d) \left\lfloor \frac{n}{d} \right\rfloor \left(\left\lfloor \frac{m}{d} \right\rfloor + \left\lfloor \frac{n-m}{d} \right\rfloor \right)$$

and

$$|\mathcal{F}_{2m}| - |\mathcal{F}(\mathbb{B}(2m), m)| = 2(|\mathcal{F}_{2m}^m| - |\mathcal{F}(\mathbb{B}(2m), m)|)$$

$$= -4 + 2 \sum_{d=1}^{m} \mu(d) \left\lfloor \frac{2m}{d} \right\rfloor \left\lfloor \frac{m}{d} \right\rfloor.$$

2.5 The position of a fraction in a Farey (sub)sequence. II

In Section 1.7 we gave answers to the questions of how to find the indices t of fractions $\frac{h}{k} =: f_t$ in the Farey (sub)sequences $\mathcal{F}_n$, $\mathcal{F}_n^m$ and $\mathcal{G}_n^m$; recall that we always index the fractions starting with zero. In this section we use, by means of duality, those results for computing the indices of fractions in the Farey subsequences $\mathcal{F}(\mathbb{B}(n), m)$.

2.5.1 The index of $\frac{1}{2}$ in $\mathcal{F}(\mathbb{B}(n), m)$

Let us begin with a simple application of the Farey duality to the problem of finding the index t of the fraction $\frac{1}{2} =: f_t$ in the sequences $\mathcal{F}(\mathbb{B}(n), m)$.

Remark 2.49. If $f_t := \frac{1}{2} \in \mathcal{F}(\mathbb{B}(2m), m)$, then the bijections given in (2.12) and (2.13) of Corollary 2.5 guarantee that $t = |\mathcal{F}_m| - 1$, that is, according to Corollary 1.30, we have

$$t = \frac{1}{2} + \frac{1}{2}\sum_{d=1}^{m} \mu(d)\left\lfloor \frac{m}{d} \right\rfloor^2 .$$

If $f_t := \frac{1}{2} \in \mathcal{F}(\mathbb{B}(n), m)$, where $n < 2m$, then the bijections given in (2.20) and (2.21) of Corollary 2.8 (i) guarantee that $t = |\mathcal{F}_{n-m}| - 1$, that is, according to Corollary 1.30, we have

$$t = \frac{1}{2} + \frac{1}{2}\sum_{d=1}^{n-m} \mu(d)\left\lfloor \frac{n-m}{d} \right\rfloor^2 .$$

If $f_t := \frac{1}{2} \in \mathcal{F}(\mathbb{B}(n), m)$, where $n > 2m$, then the bijections given in (2.25) and (2.26) of Corollary 2.8 (ii) guarantee that $t = |\mathcal{F}^{m}_{n-m}| - 1$, that is, according to Proposition 1.29, we have

$$t = \frac{1}{2} + \sum_{d \geq 1} \mu(d)\left\lfloor \frac{m}{d} \right\rfloor\left(\left\lfloor \frac{n-m}{d} \right\rfloor - \frac{1}{2}\left\lfloor \frac{m}{d} \right\rfloor\right).$$

2.5.2 The indices of fractions in $\mathcal{F}(\mathbb{B}(2m), m)$

In order to find the indices t of fractions $\frac{h}{k} =: f_t$ in the Farey subsequences $\mathcal{F}(\mathbb{B}(2m), m)$, in this subsection we will revisit, via duality, the results of Section 1.7.1.

Proposition 2.50. *Consider a Farey subsequence $\mathcal{F}(\mathbb{B}(2m), m)$.*
(i) *If $f_t := \frac{h}{k} \in \mathcal{F}^{\leq\frac{1}{2}}(\mathbb{B}(2m), m) - \{\frac{0}{1}, \frac{1}{2}\}$, then*

$$t = \sum_{d=1}^{m} \mu(d) \sum_{j=1}^{\lfloor m/d \rfloor} \left\lfloor \frac{hj}{k-h} \right\rfloor = \sum_{j=2}^{m} M\left(\frac{m}{j}\right)\left\lfloor \frac{hj}{k-h} \right\rfloor .$$

(ii) *If $f_t := \frac{h}{k} \in \mathcal{F}^{\geq\frac{1}{2}}(\mathbb{B}(2m), m) - \{\frac{1}{2}, \frac{1}{1}\}$, then*

$$\begin{aligned} t &= 1 + \sum_{d \geq 1} \mu(d)\left(\left\lfloor \frac{m}{d} \right\rfloor^2 - \sum_{j=1}^{\lfloor m/d \rfloor} \left\lfloor \frac{(k-h)j}{h} \right\rfloor\right) \\ &= 1 + \sum_{d \geq 1} \mu(d)\left\lfloor \frac{m}{d} \right\rfloor^2 - \sum_{j=2}^{m} M\left(\frac{m}{j}\right)\left\lfloor \frac{(k-h)j}{h} \right\rfloor . \end{aligned}$$

Proof. (i) According to Corollary 2.5, the fraction $e_s := \frac{h}{k-h} \in \mathcal{F}_m$ is the image of f_t under the order-preserving bijection (2.12); we have $t = s$. Since $\frac{0}{1} < f_t < \frac{1}{2}$, we have $m > 1$. Hence, the assertion follows from Corollary 1.33.

(ii) According to Corollary 2.5, the fraction $e_s := \frac{k-h}{h} \in \mathcal{F}_m$ is the image of f_t under the order-reversing bijection (2.18); note that $t = |\mathcal{F}(\mathbb{B}(2m), m)| - s - 1$. Since $\frac{1}{2} < f_t < \frac{1}{1}$, we have $m > 1$. Hence, the assertion follows from Proposition 2.47 and Corollary 1.33. □

2.5.3 The indices of fractions in $\mathcal{F}(\mathbb{B}(n), m)$, $n \neq 2m$

In this subsection we will find the indices t of fractions $\frac{h}{k} =: f_t$ in the Farey subsequences $\mathcal{F}(\mathbb{B}(n), m)$, with $n \neq 2m$. The results of Section 1.7.1 will be applied, via the Farey duality, to the left and right half sequences of the sequences $\mathcal{F}(\mathbb{B}(n), m)$.

Proposition 2.51. *Consider a Farey subsequence $\mathcal{F}(\mathbb{B}(n), m)$, where $n \neq 2m$.*

(i) *Suppose that $n < 2m$.*

(a) *If $f_t := \frac{h}{k} \in \mathcal{F}^{\leq\frac{1}{2}}(\mathbb{B}(n), m) - \{\frac{0}{1}, \frac{1}{2}\}$, then*

$$t = \sum_{d=1}^{n-m} \mu(d) \sum_{j=1}^{\lfloor (n-m)/d \rfloor} \left\lfloor \frac{hj}{k-h} \right\rfloor = \sum_{j=2}^{n-m} M\left(\frac{n-m}{j}\right) \left\lfloor \frac{hj}{k-h} \right\rfloor .$$

(b) *If $f_t := \frac{h}{k} \in \mathcal{F}^{\geq\frac{1}{2}}(\mathbb{B}(n), m) - \{\frac{1}{2}, \frac{1}{1}\}$, then*

$$t = 1 + \sum_{d\geq 1} \mu(d) \left(\left\lfloor \frac{m}{d} \right\rfloor \left\lfloor \frac{n-m}{d} \right\rfloor - \sum_{j=1}^{\lfloor m/d \rfloor} \left\lfloor \min\left\{ \frac{n-m}{d}, \frac{(k-h)j}{h} \right\} \right\rfloor \right).$$

(ii) *Suppose that $n > 2m$.*

(a) *If $f_t := \frac{h}{k} \in \mathcal{F}^{\leq\frac{1}{2}}(\mathbb{B}(n), m) - \{\frac{0}{1}, \frac{1}{2}\}$, then*

$$t = \sum_{d\geq 1} \mu(d) \sum_{j=1}^{\lfloor (n-m)/d \rfloor} \left\lfloor \min\left\{ \frac{m}{d}, \frac{hj}{k-h} \right\} \right\rfloor .$$

(b) *If $f_t := \frac{h}{k} \in \mathcal{F}^{\geq\frac{1}{2}}(\mathbb{B}(n), m) - \{\frac{1}{2}, \frac{1}{1}\}$, then*

$$\begin{aligned} t &= 1 + \sum_{d\geq 1} \mu(d) \left(\left\lfloor \frac{m}{d} \right\rfloor \left\lfloor \frac{n-m}{d} \right\rfloor - \sum_{j=1}^{\lfloor m/d \rfloor} \left\lfloor \frac{(k-h)j}{h} \right\rfloor \right) \\ &= 1 + \sum_{d\geq 1} \mu(d) \left\lfloor \frac{m}{d} \right\rfloor \left\lfloor \frac{n-m}{d} \right\rfloor - \sum_{j=2}^{m} M\left(\frac{m}{j}\right) \left\lfloor \frac{(k-h)j}{h} \right\rfloor . \end{aligned}$$

Proof. (i) (a) According to Corollary 2.8 (i), the fraction $e_s := \frac{h}{k-h} \in \mathcal{F}_{n-m}$ is the image of f_t under the order-preserving bijection (2.20); we have $t = s$. Since $\frac{0}{1} < f_t < \frac{1}{2}$, we have $n - m > 1$. Hence, the assertion follows from Corollary 1.33.

(i) (b) According to Corollary 2.8 (i), the fraction $e_s := \frac{k-h}{h} \in \mathcal{F}_m^{n-m}$ is the image of f_t under the order-reversing bijection (2.23); note that $t = |\mathcal{F}(\mathbb{B}(n), m)| - s - 1$. Hence, the assertion follows from Propositions 2.47 and 1.32.

(ii) (a) According to Corollary 2.8 (ii), the fraction $e_s := \frac{h}{k-h} \in \mathcal{F}^m_{n-m}$ is the image of f_t under the order-preserving bijection (2.25); we have $t = s$. The assertion follows from Proposition 1.32.

(ii) (b) According to Corollary 2.8 (ii), the fraction $e_s := \frac{k-h}{h} \in \mathcal{F}_m$ is the image of f_t under the order-reversing bijection (2.29); note that $t = |\mathcal{F}(\mathbb{B}(n), m)| - s - 1$. Since $\frac{1}{2} < f_t < \frac{1}{1}$, we have $m > 1$. Hence, the assertion follows from Proposition 2.47 and Corollary 1.33. □

2.6 The rank problem. II

Recall that the rank problem is the search in a Farey (sub)sequence for the index of a fraction that is nearest to a given real number r, $0 \le r \le 1$; see Section 1.8. This section is concerned with the rank problem for the Farey subsequences $\mathcal{F}(\mathbb{B}(n), m)$.

2.6.1 The rank problem for $\mathcal{F}(\mathbb{B}(2m), m)$

When we deal with the rank problem for the Farey subsequences $\mathcal{F}(\mathbb{B}(2m), m)$, we use the Farey map $T(x)$ considered in Section 2.1.2, and the observation made in Remark 1.35 (ii).

Remark 2.52 (cf. Proposition 2.50). (i) Let $r \in \mathbb{R}$, $0 < r < \frac{1}{2}$. If

$$f_t = \max\{\tfrac{h}{k} \in \mathcal{F}(\mathbb{B}(2m), m) : \tfrac{h}{k} \le r\},$$

then

$$t = \sum_{d \in [1,m]} \mu(d) \sum_{j=1}^{\lfloor m/d \rfloor} \left\lfloor \frac{jr}{1-r} \right\rfloor = \sum_{j=2}^{m} M\left(\frac{m}{j}\right) \left\lfloor \frac{jr}{1-r} \right\rfloor .$$

The real number $\frac{r}{1-r}$ is the image $T_0(r)$ of r under the Farey map, and here we have actually found, with the help of the order-preserving bijection (2.12) and Remark 1.35 (ii), the index s of the fraction $e_s \in \mathcal{F}_m$ such that $e_s = \max\{\frac{h}{k-h} \in \mathcal{F}_m : \frac{h}{k-h} \le \frac{r}{1-r}\}$ and $t = s$.

(ii) Let $r \in \mathbb{R}$, $\frac{1}{2} < r < 1$. If

$$f_t = \min\{\tfrac{h}{k} \in \mathcal{F}(\mathbb{B}(2m), m) : r \le \tfrac{h}{k}\},$$

then

$$t = 1 + \sum_{d \ge 1} \mu(d) \left(\left\lfloor \frac{m}{d} \right\rfloor^2 - \sum_{j=1}^{\lfloor m/d \rfloor} \left\lfloor \frac{j(1-r)}{r} \right\rfloor \right) = 1 + \sum_{d \ge 1} \mu(d) \left\lfloor \frac{m}{d} \right\rfloor^2 - \sum_{j=2}^{m} M\left(\frac{m}{j}\right) \left\lfloor \frac{j(1-r)}{r} \right\rfloor .$$

Indeed, the real number $\frac{1-r}{r}$ is the image $T_1(r)$ of r under the Farey map. The order-reversing bijection (2.18) and Remark 1.35 (ii) allow us to find the index s of the fraction $e_s \in \mathcal{F}_m$ such that $e_s = \max\{\frac{k-h}{h} \in \mathcal{F}_m : \frac{k-h}{h} \le \frac{1-r}{r}\}$. It suffices to note that $t = |\mathcal{F}(\mathbb{B}(2m), m)| - s - 1$, and use Proposition 2.47.

2.6.2 The rank problem for $\mathcal{F}(\mathbb{B}(n), m)$, $n \neq 2m$

In this subsection we will use the Farey duality and the Farey map for solving the rank problem for the sequences $\mathcal{F}(\mathbb{B}(n), m)$, with $n \neq 2m$.

Remark 2.53 (cf. Proposition 2.51). Consider a Farey subsequence $\mathcal{F}(\mathbb{B}(n), m)$, where $n \neq 2m$.

(i) Suppose that $n < 2m$.

(a) Let $r \in \mathbb{R}$, $0 < r < \frac{1}{2}$. If $f_t = \max\{\frac{h}{k} \in \mathcal{F}(\mathbb{B}(n), m) : \frac{h}{k} \leq r\}$, then

$$t = \sum_{d=1}^{n-m} \mu(d) \sum_{j=1}^{\lfloor (n-m)/d \rfloor} \left\lfloor \frac{jr}{1-r} \right\rfloor = \sum_{j=2}^{n-m} M\left(\frac{n-m}{j}\right) \left\lfloor \frac{jr}{1-r} \right\rfloor .$$

Indeed, the real number $\frac{r}{1-r}$ is the image $T_0(r)$ of r under the Farey map, and here we have found, with the help of the order-preserving bijection (2.20) and Remark 1.35 (ii), the index s of the fraction $e_s \in \mathcal{F}_{n-m}$ such that $e_s = \max\{\frac{h}{k-h} \in \mathcal{F}_{n-m} : \frac{h}{k-h} \leq \frac{r}{1-r}\}$, and $t = s$.

(b) Let $r \in \mathbb{R}$, $\frac{1}{2} < r < 1$. If $f_t = \min\{\frac{h}{k} \in \mathcal{F}(\mathbb{B}(n), m) : r \leq \frac{h}{k}\}$, then

$$t = 1 + \sum_{d \geq 1} \mu(d) \left(\left\lfloor \frac{m}{d} \right\rfloor \left\lfloor \frac{n-m}{d} \right\rfloor - \sum_{j=1}^{\lfloor m/d \rfloor} \left\lfloor \min\left\{ \frac{n-m}{d}, \frac{j(1-r)}{r} \right\} \right\rfloor \right).$$

Indeed, the real number $\frac{1-r}{r}$ is the image $T_1(r)$ of r under the Farey map. The order-reversing bijection (2.23) and Remark 1.35 (i) allow us to find the index s of the fraction $e_s \in \mathcal{F}_m^{n-m}$ such that $e_s = \max\{\frac{k-h}{h} \in \mathcal{F}_m^{n-m} : \frac{k-h}{h} \leq \frac{1-r}{r}\}$. It suffices to note that $t = |\mathcal{F}(\mathbb{B}(n), m)| - s - 1$, and use Proposition 2.47 in order to determine the cardinality $|\mathcal{F}(\mathbb{B}(n), m)|$.

(ii) Suppose that $n > 2m$.

(a) Let $r \in \mathbb{R}$, $0 < r < \frac{1}{2}$. If $f_t = \max\{\frac{h}{k} \in \mathcal{F}(\mathbb{B}(n), m) : \frac{h}{k} \leq r\}$, then

$$t = \sum_{d \geq 1} \mu(d) \sum_{j=1}^{\lfloor (n-m)/d \rfloor} \left\lfloor \min\left\{ \frac{m}{d}, \frac{jr}{1-r} \right\} \right\rfloor .$$

The real number $\frac{r}{1-r}$ is the image $T_0(r)$ of r under the Farey map, and we have found, with the help of the order-preserving bijection (2.25) and Remark 1.35 (i), the index s of the fraction $e_s \in \mathcal{F}_{n-m}^m$ such that $e_s = \max\{\frac{h}{k-h} \in \mathcal{F}_{n-m}^m : \frac{h}{k-h} \leq \frac{r}{1-r}\}$ and $t = s$.

(b) Let $r \in \mathbb{R}$, $\frac{1}{2} < r < 1$. If $f_t = \min\{\frac{h}{k} \in \mathcal{F}(\mathbb{B}(n), m) : r \leq \frac{h}{k}\}$, then

$$\begin{aligned} t &= 1 + \sum_{d \geq 1} \mu(d) \left(\left\lfloor \frac{m}{d} \right\rfloor \left\lfloor \frac{n-m}{d} \right\rfloor - \sum_{j=1}^{\lfloor m/d \rfloor} \left\lfloor \frac{j(1-r)}{r} \right\rfloor \right) \\ &= 1 + \sum_{d \geq 1} \mu(d) \left\lfloor \frac{m}{d} \right\rfloor \left\lfloor \frac{n-m}{d} \right\rfloor - \sum_{j=2}^{m} M\left(\frac{m}{j}\right) \left\lfloor \frac{j(1-r)}{r} \right\rfloor . \end{aligned}$$

The real number $\frac{1-r}{r}$ is the image $T_1(r)$ of r under the Farey map. The order-reversing bijection (2.29) and Remark 1.35 (ii) allow us to find the index s of the fraction $e_s \in \mathcal{F}_m$ such that $e_s = \max\{\frac{k-h}{h} \in \mathcal{F}_m : \frac{k-h}{h} \le \frac{1-r}{r}\}$. It suffices to note that $t = |\mathcal{F}(\mathbb{B}(n), m)| - s - 1$, and use Proposition 2.47 for finding the quantity $|\mathcal{F}(\mathbb{B}(n), m)|$.

2.7 Well-structured subsequences of consecutive fractions

A few well-structured Farey subsequences were presented in Section 1.2.2. We will now give further instances of this kind by describing subsequences, with simple structure, of consecutive fractions in the Farey sequences, among which, perhaps the most interesting well-structured subsequences are neighborhoods of the fraction $\frac{1}{2}$ in the sequences $\mathcal{F}(\mathbb{B}(2m), m)$.

2.7.1 Well-structured subsequences within $\mathcal{F}_m$, $\mathcal{F}_m^\ell$ and $\mathcal{G}_m^\ell$

Let us mention some subsequences, with simple structure, of consecutive fractions in the Farey (sub)sequences $\mathcal{F}_m$, $\mathcal{F}_m^\ell$ and $\mathcal{G}_m^\ell$.

Remark 2.54. (i) The expression (2.38) in Corollary 2.20 (i) implies that the fractions

$$\frac{0}{1} < \frac{1}{m} < \frac{1}{m-1} < \frac{1}{m-2} < \cdots < \frac{1}{\lceil m/2 \rceil} \tag{2.99}$$

are consecutive in $\mathcal{F}_m$. As a consequence, the fractions

$$\frac{\lceil m/2 \rceil - 1}{\lceil m/2 \rceil} < \cdots < \frac{m-3}{m-2} < \frac{m-2}{m-1} < \frac{m-1}{m} < \frac{1}{1}$$

are consecutive in $\mathcal{F}_m$, thanks to the order-reversing bijection (1.8).

(ii) Similarly, from the expressions (2.51) and (2.52) in Corollary 2.26 (i), it follows that the fractions

$$\frac{0}{1} < \frac{1}{m} < \frac{1}{m-1} < \frac{1}{m-2} < \cdots < \frac{1}{\lceil m/2 \rceil}$$

are consecutive in any sequence $\mathcal{F}_m^\ell$, $0 < \ell < m$. As a consequence, the fractions

$$\frac{\lceil m/2 \rceil - 1}{\lceil m/2 \rceil} < \cdots < \frac{m-3}{m-2} < \frac{m-2}{m-1} < \frac{m-1}{m} < \frac{1}{1}$$

are consecutive in any sequence $\mathcal{G}_m^\ell$, $0 < \ell < m$, thanks to the order-reversing bijections $\mathcal{F}_m^\ell \leftrightarrow \mathcal{G}_m^{m-\ell}$ defined in (1.8).

2.7.2 Well-structured subsequences within $\mathcal{F}(\mathbb{B}(2m), m)$

We will now describe three well-structured subsequences of consecutive fractions in the sequences $\mathcal{F}(\mathbb{B}(2m), m)$, and show that perhaps the most remarkable subsequences are certain neighborhoods of the fraction $\frac{1}{2}$.

Remark 2.55 (see also Remark 2.56). (i) The fractions

$$\tfrac{0}{1} < \tfrac{1}{m+1} < \tfrac{1}{m} < \tfrac{1}{m-1} < \cdots < \tfrac{1}{\lceil m/2\rceil+1}$$

are consecutive in $\mathcal{F}^{\le\frac{1}{2}}(\mathbb{B}(2m), m)$.

(ii) The fractions

$$\tfrac{\lceil m/2\rceil-1}{2\lceil m/2\rceil-1} < \cdots < \tfrac{m-3}{2m-5} < \tfrac{m-2}{2m-3} < \tfrac{m-1}{2m-1} < \tfrac{1}{2}$$

are consecutive in $\mathcal{F}^{\le\frac{1}{2}}(\mathbb{B}(2m), m)$.

(iii) The fractions

$$\tfrac{1}{2} < \tfrac{m}{2m-1} < \tfrac{m-1}{2m-3} < \tfrac{m-2}{2m-5} < \cdots < \tfrac{\lceil m/2\rceil}{2\lceil m/2\rceil-1}$$

are consecutive in $\mathcal{F}^{\ge\frac{1}{2}}(\mathbb{B}(2m), m)$.

(iv) The fractions

$$\tfrac{\lceil m/2\rceil}{\lceil m/2\rceil+1} < \cdots < \tfrac{m-2}{m-1} < \tfrac{m-1}{m} < \tfrac{m}{m+1} < \tfrac{1}{1}$$

are consecutive in $\mathcal{F}^{\ge\frac{1}{2}}(\mathbb{B}(2m), m)$.

Indeed, since the fractions that compose the sequence (2.99) are consecutive in $\mathcal{F}_m$, as noted in Remark 2.54 (i), we arrive at the conclusions (i), (ii), (iii) and (iv) with the help of the monotone bijections (2.13), (2.17), (2.15) and (2.19), respectively.

2.7.3 Well-structured subsequences within $\mathcal{F}(\mathbb{B}(n), m)$, $n \ne 2m$

In this subsection we describe well-structured subsequences of consecutive fractions in the sequences $\mathcal{F}(\mathbb{B}(n), m)$, with $n \ne 2m$. See Section 2.7.2 about similar observations on the sequences $\mathcal{F}(\mathbb{B}(2m), m)$.

Remark 2.56 (cf. Remark 2.55). Consider a Farey subsequence $\mathcal{F}(\mathbb{B}(n), m)$, where $n \ne 2m$.

(i) Suppose that $n < 2m$.

(a) Remark 2.54 (i) and the order-preserving bijection (2.21) in Corollary 2.8 (i) guarantee that the fractions

$$\tfrac{0}{1} < \tfrac{1}{(n-m)+1} < \tfrac{1}{n-m} < \tfrac{1}{(n-m)-1} < \cdots < \tfrac{1}{\lceil (n-m)/2\rceil+1}$$

are consecutive in $\mathcal{F}^{\le\frac{1}{2}}(\mathbb{B}(n), m)$.

(b) Remark 2.54 (i) and the order-reversing bijection (2.22) guarantee that the fractions

$$\tfrac{\lceil (n-m)/2\rceil-1}{2\lceil (n-m)/2\rceil-1} < \cdots < \tfrac{(n-m)-3}{2(n-m)-5} < \tfrac{(n-m)-2}{2(n-m)-3} < \tfrac{(n-m)-1}{2(n-m)-1} < \tfrac{1}{2}$$

are consecutive in $\mathcal{F}^{\le\frac{1}{2}}(\mathbb{B}(n), m)$.

(c) Remark 2.54 (ii) and the order-reversing bijection (2.24) guarantee that the fractions

$$\tfrac{\lceil m/2\rceil}{\lceil m/2\rceil+1} < \cdots < \tfrac{m-2}{m-1} < \tfrac{m-1}{m} < \tfrac{m}{m+1} < \tfrac{1}{1}$$

are consecutive in $\mathcal{F}^{\ge\frac{1}{2}}(\mathbb{B}(n), m)$.

(ii) Suppose that $n > 2m$.

(a) Remark 2.54 (ii) and the order-preserving bijection (2.26) guarantee that the fractions

$$\tfrac{0}{1} < \tfrac{1}{(n-m)+1} < \tfrac{1}{n-m} < \tfrac{1}{(n-m)-1} < \cdots < \tfrac{1}{\lceil (n-m)/2 \rceil + 1}$$

are consecutive in $\mathcal{F}^{\leq\frac{1}{2}}(\mathbb{B}(n), m)$.

(b) Remark 2.54 (i) and the order-preserving bijection (2.28) guarantee that the fractions

$$\tfrac{1}{2} < \tfrac{m}{2m-1} < \tfrac{m-1}{2m-3} < \tfrac{m-2}{2m-5} < \cdots < \tfrac{\lceil m/2 \rceil}{2\lceil m/2 \rceil - 1}$$

are consecutive in $\mathcal{F}^{\geq\frac{1}{2}}(\mathbb{B}(n), m)$.

(c) Remark 2.54 (i) and the order-reversing bijection (2.30) guarantee that the fractions

$$\tfrac{\lceil m/2 \rceil}{\lceil m/2 \rceil + 1} < \cdots < \tfrac{m-2}{m-1} < \tfrac{m-1}{m} < \tfrac{m}{m+1} < \tfrac{1}{1}$$

are consecutive in $\mathcal{F}^{\geq\frac{1}{2}}(\mathbb{B}(n), m)$.

Notes

The Farey map, defined by (2.9), was introduced in the works [57, 58, 85]; see also [47, 97, 136].

The recent book [88] provides much information on the Farey map and its properties.

From the point of view of number theory, the Farey map encodes the *continued fraction* algorithm. This map is a simple model of the physical phenomenon of *intermittency*, see, e.g., [19, 53, 134].

The inverse branches $T_0^{-1}(x)$ and $T_1^{-1}(x)$ of the Farey map, given in (2.10), can also be described as follows (see [82, p. 1523]):

$$T_0^{-1}(x) = \frac{1}{2} - \frac{1}{2} \cdot \frac{1-x}{1+x} \quad \text{and} \quad T_1^{-1}(x) = \frac{1}{2} + \frac{1}{2} \cdot \frac{1-x}{1+x}.$$

See, e.g., [13, 19, 23, 35, 49, 53, 59, 60, 77, 78, 83, 89, 120, 135] on various aspects of the Farey map, Farey rows, continued fractions and *circle maps*.

The modified Farey map, defined by (2.11), is considered, e.g., in [36, 84]; see also [89, Section 2.2].

The recent book [132] is an encyclopedia of square matrices of order 2.

All the 2×2 matrices appearing in Theorem 2.1 and Corollary 2.5 are elements of the general linear group $\mathrm{GL}(2, \mathbb{Z})$.

The *general linear group* $\mathrm{GL}(2, \mathbb{Z})$ is defined to be the group of 2×2 matrices $\boldsymbol{G}$, with integer entries, such that $\det \boldsymbol{G} \in \{-1, 1\}$.

The (*homogeneous*) *modular group* (or *special linear group*) $\mathrm{SL}(2,\mathbb{Z})$ is defined to be the subgroup of matrices $\boldsymbol{S} \in \mathrm{GL}(2,\mathbb{Z})$ such that $\det \boldsymbol{S} = 1$.

The four matrices $\left[\begin{smallmatrix} a & b \\ c & d \end{smallmatrix}\right] \in \mathrm{SL}(2,\mathbb{Z})$ that determine the order-preserving maps in Theorem 2.1 and Corollary 2.5 are matrices with *trace* $a + d = 2$. Such matrices are *parabolic* elements of the modular group [7, p. 95].

Let us use the terminology and results of [138, Chapter 1] and [7, Section 5.1], and consider several distinguished elements of $\mathrm{SL}(2,\mathbb{Z})$:

$$\boldsymbol{U} := \left[\begin{smallmatrix} 1 & 1 \\ 0 & 1 \end{smallmatrix}\right], \quad \boldsymbol{V} := \left[\begin{smallmatrix} 0 & -1 \\ 1 & 0 \end{smallmatrix}\right],$$
$$\boldsymbol{W} := \left[\begin{smallmatrix} 1 & 0 \\ 1 & 1 \end{smallmatrix}\right], \quad \boldsymbol{Q} := \left[\begin{smallmatrix} -1 & -1 \\ 1 & 0 \end{smallmatrix}\right].$$

The group $\mathrm{SL}(2,\mathbb{Z})$ is *generated* by the pair of its elements $\{\boldsymbol{U}, \boldsymbol{V}\}$, see [7, Proposition 5.8]. The following result [7, Theorem 5.9] concerns canonical forms of elements of the modular group:

Every $\boldsymbol{T} \in \mathrm{SL}(2,\mathbb{Z})$ can be uniquely written in the form

$$\boldsymbol{T} = (-1)^r \boldsymbol{Q}^{q_0} \boldsymbol{V} \boldsymbol{Q}^{q_1} \boldsymbol{V} \cdots \boldsymbol{V} \boldsymbol{Q}^{q_n},$$

where $r \in \{0, 1\}$, $q_i \in \{0, 1, 2\}$, and $q_i > 0$ for $0 < i < n$.

The matrix $\boldsymbol{W}$ appears in the definition of the order-preserving bijection in (2.2). Let $\boldsymbol{Z} \in \mathrm{SL}(2,\mathbb{Z})$ denote the matrix determining the order-preserving bijection in (2.4):

$$\boldsymbol{Z} := \left[\begin{smallmatrix} 0 & 1 \\ -1 & 2 \end{smallmatrix}\right].$$

This matrix is *conjugate* to $\boldsymbol{U}$. Indeed, we have

$$\boldsymbol{Z} = \left[\begin{smallmatrix} 1 & 0 \\ -1 & 1 \end{smallmatrix}\right]^{-1} \boldsymbol{U} \left[\begin{smallmatrix} 1 & 0 \\ -1 & 1 \end{smallmatrix}\right] = (-\boldsymbol{VUV})^{-1} \boldsymbol{U} (-\boldsymbol{VUV}).$$

We also have

$$\boldsymbol{W} = \boldsymbol{UVU} = -\boldsymbol{Q}^0 \boldsymbol{V} \boldsymbol{Q}^1,$$

see [7, Lemma 5.2 (3), and p. 58], and

$$\boldsymbol{Z} = \boldsymbol{V}\boldsymbol{U}^{-1}\boldsymbol{VUVUV} = -\boldsymbol{Q}^0 \boldsymbol{V} \boldsymbol{Q}^1 \boldsymbol{V} \boldsymbol{Q}^1 \boldsymbol{V} \boldsymbol{Q}^0.$$

It is evident that $\boldsymbol{Z} = \boldsymbol{W} \left[\begin{smallmatrix} 0 & 1 \\ -1 & 1 \end{smallmatrix}\right]$ or, in other words,

$$\mathrm{SL}(2,\mathbb{Z}) \ni \boldsymbol{Z} = \boldsymbol{W}(-\boldsymbol{UVUV}),$$

but, in view of the order-reversing and bijective mappings $\frac{h}{k} \mapsto \frac{k-h}{k}$ in (1.8), and $\frac{h}{k} \mapsto \frac{k}{h+k}$ in (2.8), it is more natural, in the context of the book, to interpret the matrix $\boldsymbol{Z}$ as the following product in the general linear group:

$$\left[\begin{smallmatrix} 0 & 1 \\ 1 & 1 \end{smallmatrix}\right] \left[\begin{smallmatrix} -1 & 1 \\ 0 & 1 \end{smallmatrix}\right] = \boldsymbol{Z} \in \mathrm{GL}(2,\mathbb{Z}).$$

The group $\mathrm{SL}(2,\mathbb{R})$ of 2×2 matrices $\boldsymbol{R}$, with real entries, such that $\det \boldsymbol{R} = 1$ (see [92, 102]) is one of the two most important (together with the *Heisenberg group*) *Lie groups* in analysis [92, p. 1].

See [15, 78] on the connection between the *space of unimodular lattices in* $\mathbb{R}^2$ and the *BCZ map* (1.130).

The action of linear fractional transformations (or *Möbius transformations*) on the *Farey diagram* is discussed in [75, Chapter 3].

The Farey sequence $\mathcal{F}_m$ has a wide area of applications in mathematics, computer science and physics [45]. The Farey subsequence $\mathcal{F}(\mathbb{B}(2m), m)$, which is closely related to $\mathcal{F}_m$ via the Farey duality, as explained in Section 2.1.3, can be useful for an analysis of *collective decision-making* procedures. One such procedure is *pattern recognition* by means of *committee decision rules*; see the introductory chapter.

The groundwork for the modern committee theory was laid by the short notes [1, 2]; see also, e.g., [62, 90, 112, 114, 115].

As mentioned in the introductory chapter, the study of the structure of the family of all *committees of regions* for the *arrangements of oriented linear hyperplanes* might involve the Farey subsequences $\mathcal{F}^{\geq\frac{1}{2}}(\mathbb{B}(2m), m)$. Such an attempt is made in [112, Chapter 7] within the bounds of *oriented matroid* theory [21, 31].

Recall that well-structured neighborhoods of the critical value $\frac{1}{2}$ in the Farey subsequences $\mathcal{F}(\mathbb{B}(2m), m)$ are described in Remark 2.55 (ii) and (iii).

The neighbors of the central fraction $\frac{1}{2}$ in the Farey sequence $\mathcal{F}_m$ (see Remark 2.10) were found in [152, p. 272, Corollary 1]; see also, e.g., [22, Theorem II], [124, p. 300, Problem 1], the solution to [133, p. 85, Problem 6.31] and [165].

The sum $\sum_{d=1}^{m}\mu(d)\lfloor\frac{m}{d}\rfloor^2$ in (2.98) is characterized, e.g., in [119, p. 32 and p. 52]; see also page 66.

We conclude these notes by mentioning the unreduced Farey sequences. The (*positive*) *unreduced Farey sequence* $\mathcal{U}_n^*$ is the nondecreasing sequence

$$\left(\tfrac{\eta_1}{\kappa_1}, \tfrac{\eta_2}{\kappa_2}, \ldots, \tfrac{\eta_{n(n+1)/2}}{\kappa_{n(n+1)/2}}\right),$$

i.e.,

$$\mathcal{U}_n^* := \left(\tfrac{\eta}{\kappa} : 1 \leq \eta \leq \kappa \leq n\right),$$

of all such reduced and unreduced fractions. By convention, equal fractions are ordered by increasing denominator. For instance,

$$\mathcal{F}_4 - \{\tfrac{0}{1}\} =: \mathcal{F}_4^* := \left(\tfrac{1}{4} < \tfrac{1}{3} < \tfrac{1}{2} < \tfrac{2}{3} < \tfrac{3}{4} < \tfrac{1}{1}\right),$$

while

$$\mathcal{U}_4^* := \left(\tfrac{1}{4} < \tfrac{1}{3} < \tfrac{1}{2} = \tfrac{2}{4} < \tfrac{2}{3} < \tfrac{3}{4} < \tfrac{1}{1} = \tfrac{2}{2} = \tfrac{3}{3} = \tfrac{4}{4}\right).$$

The product of fractions in the sequence $\mathcal{U}_n^*$ (see [98, Theorem 2.1] and [99]) is the reciprocal of the product of binomial coefficients $\binom{n}{j}$:

$$\prod_{\frac{\eta}{\kappa}\in\mathcal{U}_n^*} \frac{\eta}{\kappa} = \prod_{j=1}^{n} \binom{n}{j}^{-1}.$$

As a consequence, the product $\prod_{\frac{\eta}{\kappa}\in\mathcal{U}_n^*} \frac{\kappa}{\eta}$ is always an integer. For the positive Farey sequences $\mathcal{F}_n^* := \mathcal{F}_n - \{\frac{0}{1}\}$, the product $\prod_{\frac{h}{k}\in\mathcal{F}_n^*} \frac{k}{h}$ is usually not an integer [98, Remark 2.3].

It was implicitly shown in [107, Corollary 8 (i)], via the Farey duality, that

$$\sum_{\frac{\eta}{\kappa}\in\mathcal{U}_n^*} \binom{n}{\eta}\binom{n}{\kappa} = \frac{1}{2}\binom{2n}{n} + 2^{2n-1} - 2^n.$$

Returning to the standard Farey sequence $\mathcal{F}_n$, note that it is an easy observation (see the solution to [133, p. 85, Exercise 6.32]), thanks to the order-reversing and bijective mapping $\frac{h}{k} \mapsto \frac{k-h}{k}$ in (1.8), that

$$\sum_{\frac{h}{k}\in\mathcal{F}_n^*} \frac{h}{k} = \tfrac{1}{2}|\mathcal{F}_n|.$$

In view of Corollary 1.30, we have

$$\sum_{\frac{h}{k}\in\mathcal{F}_n^*} \frac{h}{k} = \frac{3}{4} + \frac{1}{4}\sum_{d=1}^{n} \mu(d)\left\lfloor \frac{n}{d} \right\rfloor^2. \tag{2.100}$$

3 Monotone maps between Farey subsequences

Farey duality, the main subject in Chapter 2, characterizes monotone bijections between the Farey (sub)sequences $\mathcal{F}_m$ and $\mathcal{F}_m^{\ell}$ of a purely number-theoretic nature, on the one hand, and the half sequences $\mathcal{F}^{\leq\frac{1}{2}}(\mathbb{B}(n), m)$ and $\mathcal{F}^{\geq\frac{1}{2}}(\mathbb{B}(n), m)$ of the Farey subsequences $\mathcal{F}(\mathbb{B}(n), m)$ of a combinatorial nature, on the other hand.

In this chapter, Farey duality serves as an auxiliary technique for establishing *monotone bijective maps* $\mathcal{C}' \leftrightarrow \mathcal{C}''$ between subsequences $\mathcal{C}'$ and $\mathcal{C}''$ of the same Farey sequence $\mathcal{F}_n$ of large order n.

If $\mathcal{F}_m$ is a Farey sequence of small order m, then easily computed *matrix products* $\mathbf{M}$ and $\mathbf{N}$, of equal length and with the factors $\left[\begin{smallmatrix}1 & 0\\ 1 & 1\end{smallmatrix}\right]$ or $\left[\begin{smallmatrix}0 & 1\\ -1 & 2\end{smallmatrix}\right]$, enable us to give concise descriptions of the *bijections*

$$\begin{aligned}&\mathcal{F}_n \supset \mathbf{M}\cdot\mathcal{F}_m =: \mathcal{C}' \to \mathcal{C}'' := \mathbf{N}\cdot\mathcal{F}_m \subset \mathcal{F}_n,\\ &\left[\begin{smallmatrix}h\\ k\end{smallmatrix}\right] \mapsto \mathbf{N}\mathbf{M}^{-1}\cdot\left[\begin{smallmatrix}h\\ k\end{smallmatrix}\right] \quad \text{(order-preserving)}\end{aligned}$$

and

$$\begin{aligned}&\mathcal{F}_n \supset \mathcal{C}' \to \mathcal{C}'' \subset \mathcal{F}_n,\\ &\left[\begin{smallmatrix}h\\ k\end{smallmatrix}\right] \mapsto \mathbf{N}\left[\begin{smallmatrix}-1 & 1\\ 0 & 1\end{smallmatrix}\right]\mathbf{M}^{-1}\cdot\left[\begin{smallmatrix}h\\ k\end{smallmatrix}\right] \quad \text{(order-reversing)}\end{aligned}$$

between subsequences $\mathbf{M}\cdot\mathcal{F}_m := (\mathbf{M}\cdot\left[\begin{smallmatrix}h\\ k\end{smallmatrix}\right] : \left[\begin{smallmatrix}h\\ k\end{smallmatrix}\right] \in \mathcal{F}_m)$ and $\mathbf{N}\cdot\mathcal{F}_m$ of the sequence $\mathcal{F}_n$, whose fractions are written in vector form.

3.1 Monotone bijective maps between the sequences $\mathcal{F}^{\leq\frac{1}{2}}(\mathbb{B}(2m), m)$ and $\mathcal{F}^{\geq\frac{1}{2}}(\mathbb{B}(2m), m)$

One monotone bijection between the left half sequence $\mathcal{F}^{\leq\frac{1}{2}}(\mathbb{B}(2m), m)$ and the right half sequence $\mathcal{F}^{\geq\frac{1}{2}}(\mathbb{B}(2m), m)$ of any sequence $\mathcal{F}(\mathbb{B}(2m), m)$ is provided by the mapping $\left[\begin{smallmatrix}h\\ k\end{smallmatrix}\right] \mapsto \left[\begin{smallmatrix}-1 & 1\\ 0 & 1\end{smallmatrix}\right]\cdot\left[\begin{smallmatrix}h\\ k\end{smallmatrix}\right]$, defined in (1.8). We know that this bijection is order-reversing, and a question arises naturally: What bijections between half sequences of $\mathcal{F}(\mathbb{B}(2m), m)$ are order-preserving? Let us answer this question, and describe the order-reversing bijections on the half sequences themselves. Following the notes of Chapter 2, we will denote by $\boldsymbol{W}$ and $\boldsymbol{Z}$ the matrices that determine the order-preserving bijections in (2.13) and (2.15), respectively, namely,

$$\boldsymbol{W} := \left[\begin{smallmatrix}1 & 0\\ 1 & 1\end{smallmatrix}\right], \quad \boldsymbol{Z} := \left[\begin{smallmatrix}0 & 1\\ -1 & 2\end{smallmatrix}\right].$$

Theorem 3.1. (i) *The maps*

$$\mathcal{F}^{\leq\frac{1}{2}}(\mathbb{B}(2m), m) \to \mathcal{F}^{\leq\frac{1}{2}}(\mathbb{B}(2m), m), \quad \tfrac{h}{k} \mapsto \tfrac{k-2h}{2k-3h}, \quad \left[\begin{smallmatrix}h\\ k\end{smallmatrix}\right] \mapsto \left[\begin{smallmatrix}-2 & 1\\ -3 & 2\end{smallmatrix}\right]\cdot\left[\begin{smallmatrix}h\\ k\end{smallmatrix}\right], \tag{3.1}$$

$$\mathcal{F}^{\geq\frac{1}{2}}(\mathbb{B}(2m), m) \to \mathcal{F}^{\geq\frac{1}{2}}(\mathbb{B}(2m), m), \quad \tfrac{h}{k} \mapsto \tfrac{h}{3h-k}, \quad \left[\begin{smallmatrix}h\\ k\end{smallmatrix}\right] \mapsto \left[\begin{smallmatrix}1 & 0\\ 3 & -1\end{smallmatrix}\right]\cdot\left[\begin{smallmatrix}h\\ k\end{smallmatrix}\right], \tag{3.2}$$

are order-reversing and bijective.

DOI 10.1515/9783110547665-004

(ii) *The maps*

$$\mathcal{F}^{\leq\frac{1}{2}}(\mathbb{B}(2m),m)\to\mathcal{F}^{\geq\frac{1}{2}}(\mathbb{B}(2m),m),\quad \tfrac{h}{k}\mapsto\tfrac{k-h}{2k-3h},\quad \left[\begin{smallmatrix}h\\k\end{smallmatrix}\right]\mapsto\left[\begin{smallmatrix}-1&1\\-3&2\end{smallmatrix}\right]\cdot\left[\begin{smallmatrix}h\\k\end{smallmatrix}\right], \tag{3.3}$$

$$\mathcal{F}^{\geq\frac{1}{2}}(\mathbb{B}(2m),m)\to\mathcal{F}^{\leq\frac{1}{2}}(\mathbb{B}(2m),m),\quad \tfrac{h}{k}\mapsto\tfrac{2h-k}{3h-k},\quad \left[\begin{smallmatrix}h\\k\end{smallmatrix}\right]\mapsto\left[\begin{smallmatrix}2&-1\\3&-1\end{smallmatrix}\right]\cdot\left[\begin{smallmatrix}h\\k\end{smallmatrix}\right], \tag{3.4}$$

are order-preserving and bijective.

Proof. (i) The maps in (3.1) and (3.2) can be regarded as the composite maps

$$\mathcal{F}^{\leq\frac{1}{2}}(\mathbb{B}(2m),m)\xrightarrow{(2.12)}\mathcal{F}_m\xrightarrow{(2.17)}\mathcal{F}^{\leq\frac{1}{2}}(\mathbb{B}(2m),m),\quad \left[\begin{smallmatrix}h\\k\end{smallmatrix}\right]\mapsto\left[\begin{smallmatrix}-1&1\\-1&2\end{smallmatrix}\right]\left[\begin{smallmatrix}1&0\\-1&1\end{smallmatrix}\right]\cdot\left[\begin{smallmatrix}h\\k\end{smallmatrix}\right],$$

and

$$\mathcal{F}^{\geq\frac{1}{2}}(\mathbb{B}(2m),m)\xrightarrow{(2.14)}\mathcal{F}_m\xrightarrow{(2.19)}\mathcal{F}^{\geq\frac{1}{2}}(\mathbb{B}(2m),m),\quad \left[\begin{smallmatrix}h\\k\end{smallmatrix}\right]\mapsto\left[\begin{smallmatrix}0&1\\1&1\end{smallmatrix}\right]\left[\begin{smallmatrix}2&-1\\1&0\end{smallmatrix}\right]\cdot\left[\begin{smallmatrix}h\\k\end{smallmatrix}\right],$$

respectively.

(ii) The maps in (3.3) and (3.4) can be regarded as the composite maps

$$\mathcal{F}^{\leq\frac{1}{2}}(\mathbb{B}(2m),m)\xrightarrow{(2.12)}\mathcal{F}_m\xrightarrow{(2.15)}\mathcal{F}^{\leq\frac{1}{2}}(\mathbb{B}(2m),m),\quad \left[\begin{smallmatrix}h\\k\end{smallmatrix}\right]\mapsto\boldsymbol{Z}\left[\begin{smallmatrix}1&0\\-1&1\end{smallmatrix}\right]\cdot\left[\begin{smallmatrix}h\\k\end{smallmatrix}\right],$$

and

$$\mathcal{F}^{\geq\frac{1}{2}}(\mathbb{B}(2m),m)\xrightarrow{(2.14)}\mathcal{F}_m\xrightarrow{(2.13)}\mathcal{F}^{\geq\frac{1}{2}}(\mathbb{B}(2m),m),\quad \left[\begin{smallmatrix}h\\k\end{smallmatrix}\right]\mapsto\boldsymbol{W}\left[\begin{smallmatrix}2&-1\\1&0\end{smallmatrix}\right]\cdot\left[\begin{smallmatrix}h\\k\end{smallmatrix}\right],$$

respectively. □

Note that the observations made in Theorem 3.1 on the properties of the sequences $\mathcal{F}^{\leq\frac{1}{2}}(\mathbb{B}(2m),m)$ and $\mathcal{F}^{\geq\frac{1}{2}}(\mathbb{B}(2m),m)$ are no longer valid for the entire half sequences of the Farey sequence $\mathcal{F}_{2m}$.

Example 3.2. Suppose $m := 3$. We have

$$\begin{aligned}\mathcal{F}_{2m}&=(\tfrac{0}{1}<\tfrac{1}{6}<\tfrac{1}{5}<\tfrac{1}{4}<\tfrac{1}{3}<\tfrac{2}{5}<\tfrac{1}{2}<\tfrac{3}{5}<\tfrac{2}{3}<\tfrac{3}{4}<\tfrac{4}{5}<\tfrac{5}{6}<\tfrac{1}{1}),\\ \mathcal{F}(\mathbb{B}(2m),m)&=(\tfrac{0}{1}<\quad<\quad<\tfrac{1}{4}<\tfrac{1}{3}<\tfrac{2}{5}<\tfrac{1}{2}<\tfrac{3}{5}<\tfrac{2}{3}<\tfrac{3}{4}<\quad<\quad<\tfrac{1}{1}).\end{aligned}$$

Also,

$$\mathcal{F}^{\geq\frac{1}{2}}(\mathbb{B}(2m),m)\ni\tfrac{3}{4}\xmapsto{(3.2)}\tfrac{3}{3\cdot3-4}=\tfrac{3}{5}\in\mathcal{F}^{\geq\frac{1}{2}}(\mathbb{B}(2m),m),$$

but

$$([\![\tfrac{1}{2},1]\!]\cap\mathcal{F}_{2m})-\mathcal{F}^{\geq\frac{1}{2}}(\mathbb{B}(2m),m)\ni\tfrac{4}{5}\xmapsto{(3.2)}\tfrac{4}{3\cdot4-5}=\tfrac{4}{7}\notin\mathcal{F}_{2m}.$$

Further,

$$\mathcal{F}^{\leq\frac{1}{2}}(\mathbb{B}(2m),m)\ni\tfrac{1}{4}\xmapsto{(3.3)}\tfrac{4-1}{2\cdot4-3\cdot1}=\tfrac{3}{5}\in\mathcal{F}^{\geq\frac{1}{2}}(\mathbb{B}(2m),m),$$

but

$$([\![0,\tfrac{1}{2}]\!]\cap\mathcal{F}_{2m})-\mathcal{F}^{\leq\frac{1}{2}}(\mathbb{B}(2m),m)\ni\tfrac{1}{6}\xmapsto{(3.3)}\tfrac{6-1}{2\cdot6-3\cdot1}=\tfrac{5}{9}\notin\mathcal{F}_{2m}.$$

3.2 Monotone maps between the sequences $\mathcal{F}^{\leq\frac{1}{2}}(\mathbb{B}(n), m)$ and $\mathcal{F}^{\geq\frac{1}{2}}(\mathbb{B}(n), m)$, $n \neq 2m$

In this section we again use the Farey duality, this time for describing monotone maps between the half sequences of the sequences $\mathcal{F}(\mathbb{B}(n), m)$, with $n \neq 2m$.

Theorem 3.3. *Consider a Farey subsequence $\mathcal{F}(\mathbb{B}(n), m)$, where $n \neq 2m$.*

(i) *Suppose that $n < 2m$. Then the map*

$$\mathcal{F}^{\leq\frac{1}{2}}(\mathbb{B}(n), m) \to \mathcal{F}^{\leq\frac{1}{2}}(\mathbb{B}(n), m), \quad \tfrac{h}{k} \mapsto \tfrac{k-2h}{2k-3h}, \quad \left[\begin{smallmatrix} h \\ k \end{smallmatrix}\right] \mapsto \left[\begin{smallmatrix} -2 & 1 \\ -3 & 2 \end{smallmatrix}\right] \cdot \left[\begin{smallmatrix} h \\ k \end{smallmatrix}\right], \tag{3.5}$$

is order-reversing and bijective, and the map

$$\mathcal{F}^{\leq\frac{1}{2}}(\mathbb{B}(n), m) \to \mathcal{F}^{\geq\frac{1}{2}}(\mathbb{B}(n), m), \quad \tfrac{h}{k} \mapsto \tfrac{k-h}{2k-3h}, \quad \left[\begin{smallmatrix} h \\ k \end{smallmatrix}\right] \mapsto \left[\begin{smallmatrix} -1 & 1 \\ -3 & 2 \end{smallmatrix}\right] \cdot \left[\begin{smallmatrix} h \\ k \end{smallmatrix}\right], \tag{3.6}$$

is order-preserving and injective.

(ii) *Suppose that $n > 2m$. Then the map*

$$\mathcal{F}^{\geq\frac{1}{2}}(\mathbb{B}(n), m) \to \mathcal{F}^{\geq\frac{1}{2}}(\mathbb{B}(n), m), \quad \tfrac{h}{k} \mapsto \tfrac{h}{3h-k}, \quad \left[\begin{smallmatrix} h \\ k \end{smallmatrix}\right] \mapsto \left[\begin{smallmatrix} 1 & 0 \\ 3 & -1 \end{smallmatrix}\right] \cdot \left[\begin{smallmatrix} h \\ k \end{smallmatrix}\right], \tag{3.7}$$

is order-reversing and bijective, and the map

$$\mathcal{F}^{\geq\frac{1}{2}}(\mathbb{B}(n), m) \to \mathcal{F}^{\leq\frac{1}{2}}(\mathbb{B}(n), m), \quad \tfrac{h}{k} \mapsto \tfrac{2h-k}{3h-k}, \quad \left[\begin{smallmatrix} h \\ k \end{smallmatrix}\right] \mapsto \left[\begin{smallmatrix} 2 & -1 \\ 3 & -1 \end{smallmatrix}\right] \cdot \left[\begin{smallmatrix} h \\ k \end{smallmatrix}\right], \tag{3.8}$$

is order-preserving and injective.

Proof. (i) The map in (3.5) can be regarded as the composite map

$$\mathcal{F}^{\leq\frac{1}{2}}(\mathbb{B}(n), m) \xrightarrow{(2.20)} \mathcal{F}_{n-m} \xrightarrow{(2.22)} \mathcal{F}^{\leq\frac{1}{2}}(\mathbb{B}(n), m), \quad \left[\begin{smallmatrix} h \\ k \end{smallmatrix}\right] \mapsto \left[\begin{smallmatrix} -1 & 1 \\ -1 & 2 \end{smallmatrix}\right]\left[\begin{smallmatrix} 1 & 0 \\ -1 & 1 \end{smallmatrix}\right] \cdot \left[\begin{smallmatrix} h \\ k \end{smallmatrix}\right].$$

The map in (3.6) can be regarded as the composite map

$$\mathcal{F}^{\leq\frac{1}{2}}(\mathbb{B}(n), m) \xrightarrow{(2.20)} \mathcal{F}_{n-m} \xrightarrow{(2.15)} \mathcal{F}^{\geq\frac{1}{2}}(\mathbb{B}(2(n-m)), n-m) \subset \mathcal{F}^{\geq\frac{1}{2}}(\mathbb{B}(n), m),$$
$$\left[\begin{smallmatrix} h \\ k \end{smallmatrix}\right] \mapsto \boldsymbol{Z}\left[\begin{smallmatrix} 1 & 0 \\ -1 & 1 \end{smallmatrix}\right] \cdot \left[\begin{smallmatrix} h \\ k \end{smallmatrix}\right].$$

(ii) The map in (3.7) can be regarded as the composite map

$$\mathcal{F}^{\geq\frac{1}{2}}(\mathbb{B}(n), m) \xrightarrow{(2.27)} \mathcal{F}_{m} \xrightarrow{(2.30)} \mathcal{F}^{\geq\frac{1}{2}}(\mathbb{B}(n), m), \quad \left[\begin{smallmatrix} h \\ k \end{smallmatrix}\right] \mapsto \left[\begin{smallmatrix} 0 & 1 \\ 1 & 1 \end{smallmatrix}\right]\left[\begin{smallmatrix} 2 & -1 \\ 1 & 0 \end{smallmatrix}\right] \cdot \left[\begin{smallmatrix} h \\ k \end{smallmatrix}\right].$$

The map in (3.8) can be regarded as the composite map

$$\mathcal{F}^{\geq\frac{1}{2}}(\mathbb{B}(n), m) \xrightarrow{(2.27)} \mathcal{F}_{m} \xrightarrow{(2.13)} \mathcal{F}^{\leq\frac{1}{2}}(\mathbb{B}(2m), m) \subset \mathcal{F}^{\leq\frac{1}{2}}(\mathbb{B}(n), m),$$
$$\left[\begin{smallmatrix} h \\ k \end{smallmatrix}\right] \mapsto \boldsymbol{W}\left[\begin{smallmatrix} 2 & -1 \\ 1 & 0 \end{smallmatrix}\right] \cdot \left[\begin{smallmatrix} h \\ k \end{smallmatrix}\right].$$

□

Note that the map $\mathcal{F}^{\leq\frac{1}{2}}(\mathbb{B}(n), m) \to \mathcal{F}^{\geq\frac{1}{2}}(\mathbb{B}(n), m)$, where $n < 2m$, and the map $\mathcal{F}^{\geq\frac{1}{2}}(\mathbb{B}(n), m) \to \mathcal{F}^{\leq\frac{1}{2}}(\mathbb{B}(n), m)$, where $n > 2m$, determined by the mapping (1.8), are order-reversing and injective.

3.3 The sequences $\mathcal{F}_{2^s m}$, $\mathcal{F}(\mathbb{B}(2^{s+1}m), 2^s m)$, $\mathcal{F}(\mathbb{B}(2^{s+2}m), 2^{s+1}m)$, and monotone maps

In this section we consider monotone maps between (sub)sequences of the sequences $\mathcal{F}_{2^s m}$, $\mathcal{F}(\mathbb{B}(2^{s+1}m), 2^s m)$ and $\mathcal{F}(\mathbb{B}(2^{s+2}m), 2^{s+1}m)$, where $m \in \mathbb{P}$, $s \in \mathbb{N}$. The observations made for these maps can serve as instructive induction steps in an analysis of composite maps between Farey (sub)sequences that will be presented in Section 3.4.

Remark 3.4. Let $m \in \mathbb{P}$, $s \in \mathbb{N}$.

(i) The properties of the maps (2.13) and (2.15) imply that the following maps are order-preserving and injective:

$$\mathcal{F}_{2^s m} \to \mathcal{F}^{\leq\frac{1}{2}}(\mathbb{B}(2^{s+2}m), 2^{s+1}m),$$
$$\begin{bmatrix} h \\ k \end{bmatrix} \mapsto \boldsymbol{W}^2 \cdot \begin{bmatrix} h \\ k \end{bmatrix} = \begin{bmatrix} 1 & 0 \\ 2 & 1 \end{bmatrix} \cdot \begin{bmatrix} h \\ k \end{bmatrix}, \quad \begin{bmatrix} 0 \\ 1 \end{bmatrix} \mapsto \begin{bmatrix} 0 \\ 1 \end{bmatrix}, \quad \begin{bmatrix} 1 \\ 1 \end{bmatrix} \mapsto \begin{bmatrix} 1 \\ 3 \end{bmatrix};$$

$$\mathcal{F}_{2^s m} \to \mathcal{F}^{\leq\frac{1}{2}}(\mathbb{B}(2^{s+2}m), 2^{s+1}m),$$
$$\begin{bmatrix} h \\ k \end{bmatrix} \mapsto \boldsymbol{WZ} \cdot \begin{bmatrix} h \\ k \end{bmatrix} = \begin{bmatrix} 0 & 1 \\ -1 & 3 \end{bmatrix} \cdot \begin{bmatrix} h \\ k \end{bmatrix}, \quad \begin{bmatrix} 0 \\ 1 \end{bmatrix} \mapsto \begin{bmatrix} 1 \\ 3 \end{bmatrix}, \quad \begin{bmatrix} 1 \\ 1 \end{bmatrix} \mapsto \begin{bmatrix} 1 \\ 2 \end{bmatrix};$$

$$\mathcal{F}_{2^s m} \to \mathcal{F}^{\geq\frac{1}{2}}(\mathbb{B}(2^{s+2}m), 2^{s+1}m),$$
$$\begin{bmatrix} h \\ k \end{bmatrix} \mapsto \boldsymbol{ZW} \cdot \begin{bmatrix} h \\ k \end{bmatrix} = \begin{bmatrix} 1 & 1 \\ 1 & 2 \end{bmatrix} \cdot \begin{bmatrix} h \\ k \end{bmatrix}, \quad \begin{bmatrix} 0 \\ 1 \end{bmatrix} \mapsto \begin{bmatrix} 1 \\ 2 \end{bmatrix}, \quad \begin{bmatrix} 1 \\ 1 \end{bmatrix} \mapsto \begin{bmatrix} 2 \\ 3 \end{bmatrix};$$

$$\mathcal{F}_{2^s m} \to \mathcal{F}^{\geq\frac{1}{2}}(\mathbb{B}(2^{s+2}m), 2^{s+1}m),$$
$$\begin{bmatrix} h \\ k \end{bmatrix} \mapsto \boldsymbol{Z}^2 \cdot \begin{bmatrix} h \\ k \end{bmatrix} = \begin{bmatrix} -1 & 2 \\ -2 & 3 \end{bmatrix} \cdot \begin{bmatrix} h \\ k \end{bmatrix}, \quad \begin{bmatrix} 0 \\ 1 \end{bmatrix} \mapsto \begin{bmatrix} 2 \\ 3 \end{bmatrix}, \quad \begin{bmatrix} 1 \\ 1 \end{bmatrix} \mapsto \begin{bmatrix} 1 \\ 1 \end{bmatrix}.$$

(ii) As a consequence, the following maps are order-reversing and injective:

$$\mathcal{F}_{2^s m} \to \mathcal{F}^{\leq\frac{1}{2}}(\mathbb{B}(2^{s+2}m), 2^{s+1}m),$$
$$\begin{bmatrix} h \\ k \end{bmatrix} \mapsto \boldsymbol{W}^2 \begin{bmatrix} -1 & 1 \\ 0 & 1 \end{bmatrix} \cdot \begin{bmatrix} h \\ k \end{bmatrix} = \begin{bmatrix} -1 & 1 \\ -2 & 3 \end{bmatrix} \cdot \begin{bmatrix} h \\ k \end{bmatrix}, \quad \begin{bmatrix} 0 \\ 1 \end{bmatrix} \mapsto \begin{bmatrix} 1 \\ 3 \end{bmatrix}, \quad \begin{bmatrix} 1 \\ 1 \end{bmatrix} \mapsto \begin{bmatrix} 0 \\ 1 \end{bmatrix};$$

$$\mathcal{F}_{2^s m} \to \mathcal{F}^{\leq\frac{1}{2}}(\mathbb{B}(2^{s+2}m), 2^{s+1}m),$$
$$\begin{bmatrix} h \\ k \end{bmatrix} \mapsto \boldsymbol{WZ} \begin{bmatrix} -1 & 1 \\ 0 & 1 \end{bmatrix} \cdot \begin{bmatrix} h \\ k \end{bmatrix} = \begin{bmatrix} 0 & 1 \\ 1 & 2 \end{bmatrix} \cdot \begin{bmatrix} h \\ k \end{bmatrix}, \quad \begin{bmatrix} 0 \\ 1 \end{bmatrix} \mapsto \begin{bmatrix} 1 \\ 2 \end{bmatrix}, \quad \begin{bmatrix} 1 \\ 1 \end{bmatrix} \mapsto \begin{bmatrix} 1 \\ 3 \end{bmatrix};$$

$$\mathcal{F}_{2^s m} \to \mathcal{F}^{\geq\frac{1}{2}}(\mathbb{B}(2^{s+2}m), 2^{s+1}m),$$
$$\begin{bmatrix} h \\ k \end{bmatrix} \mapsto \boldsymbol{ZW} \begin{bmatrix} -1 & 1 \\ 0 & 1 \end{bmatrix} \cdot \begin{bmatrix} h \\ k \end{bmatrix} = \begin{bmatrix} -1 & 2 \\ -1 & 3 \end{bmatrix} \cdot \begin{bmatrix} h \\ k \end{bmatrix}, \quad \begin{bmatrix} 0 \\ 1 \end{bmatrix} \mapsto \begin{bmatrix} 2 \\ 3 \end{bmatrix}, \quad \begin{bmatrix} 1 \\ 1 \end{bmatrix} \mapsto \begin{bmatrix} 1 \\ 2 \end{bmatrix};$$

$$\mathcal{F}_{2^s m} \to \mathcal{F}^{\geq\frac{1}{2}}(\mathbb{B}(2^{s+2}m), 2^{s+1}m),$$
$$\begin{bmatrix} h \\ k \end{bmatrix} \mapsto \boldsymbol{Z}^2 \begin{bmatrix} -1 & 1 \\ 0 & 1 \end{bmatrix} \cdot \begin{bmatrix} h \\ k \end{bmatrix} = \begin{bmatrix} 1 & 1 \\ 2 & 1 \end{bmatrix} \cdot \begin{bmatrix} h \\ k \end{bmatrix}, \quad \begin{bmatrix} 0 \\ 1 \end{bmatrix} \mapsto \begin{bmatrix} 1 \\ 1 \end{bmatrix}, \quad \begin{bmatrix} 1 \\ 1 \end{bmatrix} \mapsto \begin{bmatrix} 2 \\ 3 \end{bmatrix}.$$

Proposition 3.5. *Let $m \in \mathbb{P}$, $s \in \mathbb{N}$.*

(i) *The following maps are order-preserving and injective:*

$$\mathcal{F}^{\leq\frac{1}{2}}(\mathbb{B}(2^{s+1}m), 2^s m) \to \mathcal{F}^{\leq\frac{1}{2}}(\mathbb{B}(2^{s+2}m), 2^{s+1}m),$$
$$\begin{bmatrix} h \\ k \end{bmatrix} \mapsto \boldsymbol{W} \cdot \begin{bmatrix} h \\ k \end{bmatrix}, \qquad \begin{bmatrix} 0 \\ 1 \end{bmatrix} \mapsto \begin{bmatrix} 0 \\ 1 \end{bmatrix}, \quad \begin{bmatrix} 1 \\ 2 \end{bmatrix} \mapsto \begin{bmatrix} 1 \\ 3 \end{bmatrix}; \tag{3.9}$$

$$\mathcal{F}^{\geq\frac{1}{2}}(\mathbb{B}(2^{s+1}m), 2^s m) \to \mathcal{F}^{\leq\frac{1}{2}}(\mathbb{B}(2^{s+2}m), 2^{s+1}m),$$
$$\begin{bmatrix} h \\ k \end{bmatrix} \mapsto \boldsymbol{W} \begin{bmatrix} 2 & -1 \\ 3 & -1 \end{bmatrix} \cdot \begin{bmatrix} h \\ k \end{bmatrix} = \begin{bmatrix} 2 & -1 \\ 5 & -2 \end{bmatrix} \cdot \begin{bmatrix} h \\ k \end{bmatrix}, \qquad \begin{bmatrix} 1 \\ 2 \end{bmatrix} \mapsto \begin{bmatrix} 0 \\ 1 \end{bmatrix}, \quad \begin{bmatrix} 1 \\ 1 \end{bmatrix} \mapsto \begin{bmatrix} 1 \\ 3 \end{bmatrix}; \tag{3.10}$$

$$\begin{aligned}
&\mathcal{F}^{\le\frac{1}{2}}(\mathbb{B}(2^{s+1}m), 2^s m) \to \mathcal{F}^{\le\frac{1}{2}}(\mathbb{B}(2^{s+2}m), 2^{s+1}m),\\
&\left[\begin{smallmatrix}h\\k\end{smallmatrix}\right] \mapsto \boldsymbol{W}\left[\begin{smallmatrix}-1&1\\-3&2\end{smallmatrix}\right]\cdot\left[\begin{smallmatrix}h\\k\end{smallmatrix}\right] = \left[\begin{smallmatrix}-1&1\\-4&3\end{smallmatrix}\right]\cdot\left[\begin{smallmatrix}h\\k\end{smallmatrix}\right], \qquad \left[\begin{smallmatrix}0\\1\end{smallmatrix}\right]\mapsto\left[\begin{smallmatrix}1\\3\end{smallmatrix}\right],\quad \left[\begin{smallmatrix}1\\2\end{smallmatrix}\right]\mapsto\left[\begin{smallmatrix}1\\2\end{smallmatrix}\right]; \quad (3.11)\\
&\mathcal{F}^{\ge\frac{1}{2}}(\mathbb{B}(2^{s+1}m), 2^s m) \to \mathcal{F}^{\le\frac{1}{2}}(\mathbb{B}(2^{s+2}m), 2^{s+1}m),\\
&\left[\begin{smallmatrix}h\\k\end{smallmatrix}\right] \mapsto \boldsymbol{W}\cdot\left[\begin{smallmatrix}h\\k\end{smallmatrix}\right], \qquad \left[\begin{smallmatrix}1\\2\end{smallmatrix}\right]\mapsto\left[\begin{smallmatrix}1\\3\end{smallmatrix}\right],\quad \left[\begin{smallmatrix}1\\1\end{smallmatrix}\right]\mapsto\left[\begin{smallmatrix}1\\2\end{smallmatrix}\right]; \quad (3.12)\\
&\mathcal{F}^{\le\frac{1}{2}}(\mathbb{B}(2^{s+1}m), 2^s m) \to \mathcal{F}^{\ge\frac{1}{2}}(\mathbb{B}(2^{s+2}m), 2^{s+1}m),\\
&\left[\begin{smallmatrix}h\\k\end{smallmatrix}\right] \mapsto \boldsymbol{Z}\cdot\left[\begin{smallmatrix}h\\k\end{smallmatrix}\right], \qquad \left[\begin{smallmatrix}0\\1\end{smallmatrix}\right]\mapsto\left[\begin{smallmatrix}1\\2\end{smallmatrix}\right],\quad \left[\begin{smallmatrix}1\\2\end{smallmatrix}\right]\mapsto\left[\begin{smallmatrix}2\\3\end{smallmatrix}\right]; \quad (3.13)\\
&\mathcal{F}^{\ge\frac{1}{2}}(\mathbb{B}(2^{s+1}m), 2^s m) \to \mathcal{F}^{\ge\frac{1}{2}}(\mathbb{B}(2^{s+2}m), 2^{s+1}m),\\
&\left[\begin{smallmatrix}h\\k\end{smallmatrix}\right] \mapsto \boldsymbol{Z}\left[\begin{smallmatrix}2&-1\\3&-1\end{smallmatrix}\right]\cdot\left[\begin{smallmatrix}h\\k\end{smallmatrix}\right] = \left[\begin{smallmatrix}3&-1\\4&-1\end{smallmatrix}\right]\cdot\left[\begin{smallmatrix}h\\k\end{smallmatrix}\right], \qquad \left[\begin{smallmatrix}1\\2\end{smallmatrix}\right]\mapsto\left[\begin{smallmatrix}1\\2\end{smallmatrix}\right],\quad \left[\begin{smallmatrix}1\\1\end{smallmatrix}\right]\mapsto\left[\begin{smallmatrix}2\\3\end{smallmatrix}\right]; \quad (3.14)\\
&\mathcal{F}^{\le\frac{1}{2}}(\mathbb{B}(2^{s+1}m), 2^s m) \to \mathcal{F}^{\ge\frac{1}{2}}(\mathbb{B}(2^{s+2}m), 2^{s+1}m),\\
&\left[\begin{smallmatrix}h\\k\end{smallmatrix}\right] \mapsto \boldsymbol{Z}\left[\begin{smallmatrix}-1&1\\-3&2\end{smallmatrix}\right]\cdot\left[\begin{smallmatrix}h\\k\end{smallmatrix}\right] = \left[\begin{smallmatrix}-3&2\\-5&3\end{smallmatrix}\right]\cdot\left[\begin{smallmatrix}h\\k\end{smallmatrix}\right], \qquad \left[\begin{smallmatrix}0\\1\end{smallmatrix}\right]\mapsto\left[\begin{smallmatrix}2\\3\end{smallmatrix}\right],\quad \left[\begin{smallmatrix}1\\2\end{smallmatrix}\right]\mapsto\left[\begin{smallmatrix}1\\1\end{smallmatrix}\right]; \quad (3.15)\\
&\mathcal{F}^{\ge\frac{1}{2}}(\mathbb{B}(2^{s+1}m), 2^s m) \to \mathcal{F}^{\ge\frac{1}{2}}(\mathbb{B}(2^{s+2}m), 2^{s+1}m),\\
&\left[\begin{smallmatrix}h\\k\end{smallmatrix}\right] \mapsto \boldsymbol{Z}\cdot\left[\begin{smallmatrix}h\\k\end{smallmatrix}\right], \qquad \left[\begin{smallmatrix}1\\2\end{smallmatrix}\right]\mapsto\left[\begin{smallmatrix}2\\3\end{smallmatrix}\right],\quad \left[\begin{smallmatrix}1\\1\end{smallmatrix}\right]\mapsto\left[\begin{smallmatrix}1\\1\end{smallmatrix}\right]. \quad (3.16)
\end{aligned}$$

(ii) *The following maps are order-reversing and injective:*

$$\begin{aligned}
&\mathcal{F}^{\le\frac{1}{2}}(\mathbb{B}(2^{s+1}m), 2^s m) \to \mathcal{F}^{\le\frac{1}{2}}(\mathbb{B}(2^{s+2}m), 2^{s+1}m),\\
&\left[\begin{smallmatrix}h\\k\end{smallmatrix}\right] \mapsto \boldsymbol{W}\left[\begin{smallmatrix}-2&1\\-3&2\end{smallmatrix}\right]\cdot\left[\begin{smallmatrix}h\\k\end{smallmatrix}\right] = \left[\begin{smallmatrix}-2&1\\-5&3\end{smallmatrix}\right]\cdot\left[\begin{smallmatrix}h\\k\end{smallmatrix}\right], \qquad \left[\begin{smallmatrix}1\\2\end{smallmatrix}\right]\mapsto\left[\begin{smallmatrix}0\\1\end{smallmatrix}\right],\quad \left[\begin{smallmatrix}0\\1\end{smallmatrix}\right]\mapsto\left[\begin{smallmatrix}1\\3\end{smallmatrix}\right]; \quad (3.17)\\
&\mathcal{F}^{\ge\frac{1}{2}}(\mathbb{B}(2^{s+1}m), 2^s m) \to \mathcal{F}^{\le\frac{1}{2}}(\mathbb{B}(2^{s+2}m), 2^{s+1}m),\\
&\left[\begin{smallmatrix}h\\k\end{smallmatrix}\right] \mapsto \boldsymbol{W}\left[\begin{smallmatrix}2&-1\\3&-1\end{smallmatrix}\right]\left[\begin{smallmatrix}1&0\\3&-1\end{smallmatrix}\right]\cdot\left[\begin{smallmatrix}h\\k\end{smallmatrix}\right] = \left[\begin{smallmatrix}-1&1\\-1&2\end{smallmatrix}\right]\cdot\left[\begin{smallmatrix}h\\k\end{smallmatrix}\right], \qquad \left[\begin{smallmatrix}1\\1\end{smallmatrix}\right]\mapsto\left[\begin{smallmatrix}0\\1\end{smallmatrix}\right],\quad \left[\begin{smallmatrix}1\\2\end{smallmatrix}\right]\mapsto\left[\begin{smallmatrix}1\\3\end{smallmatrix}\right]; \quad (3.18)\\
&\mathcal{F}^{\le\frac{1}{2}}(\mathbb{B}(2^{s+1}m), 2^s m) \to \mathcal{F}^{\le\frac{1}{2}}(\mathbb{B}(2^{s+2}m), 2^{s+1}m),\\
&\left[\begin{smallmatrix}h\\k\end{smallmatrix}\right] \mapsto \boldsymbol{W}\left[\begin{smallmatrix}-1&1\\-3&2\end{smallmatrix}\right]\left[\begin{smallmatrix}-2&1\\-3&2\end{smallmatrix}\right]\cdot\left[\begin{smallmatrix}h\\k\end{smallmatrix}\right] = \left[\begin{smallmatrix}-1&1\\-1&2\end{smallmatrix}\right]\cdot\left[\begin{smallmatrix}h\\k\end{smallmatrix}\right], \qquad \left[\begin{smallmatrix}1\\2\end{smallmatrix}\right]\mapsto\left[\begin{smallmatrix}1\\3\end{smallmatrix}\right],\quad \left[\begin{smallmatrix}0\\1\end{smallmatrix}\right]\mapsto\left[\begin{smallmatrix}1\\2\end{smallmatrix}\right]; \quad (3.19)\\
&\mathcal{F}^{\ge\frac{1}{2}}(\mathbb{B}(2^{s+1}m), 2^s m) \to \mathcal{F}^{\le\frac{1}{2}}(\mathbb{B}(2^{s+2}m), 2^{s+1}m),\\
&\left[\begin{smallmatrix}h\\k\end{smallmatrix}\right] \mapsto \boldsymbol{W}\left[\begin{smallmatrix}1&0\\3&-1\end{smallmatrix}\right]\cdot\left[\begin{smallmatrix}h\\k\end{smallmatrix}\right] = \left[\begin{smallmatrix}1&0\\4&-1\end{smallmatrix}\right]\cdot\left[\begin{smallmatrix}h\\k\end{smallmatrix}\right], \qquad \left[\begin{smallmatrix}1\\1\end{smallmatrix}\right]\mapsto\left[\begin{smallmatrix}1\\3\end{smallmatrix}\right],\quad \left[\begin{smallmatrix}1\\2\end{smallmatrix}\right]\mapsto\left[\begin{smallmatrix}1\\2\end{smallmatrix}\right]; \quad (3.20)\\
&\mathcal{F}^{\le\frac{1}{2}}(\mathbb{B}(2^{s+1}m), 2^s m) \to \mathcal{F}^{\ge\frac{1}{2}}(\mathbb{B}(2^{s+2}m), 2^{s+1}m),\\
&\left[\begin{smallmatrix}h\\k\end{smallmatrix}\right] \mapsto \boldsymbol{Z}\left[\begin{smallmatrix}-2&1\\-3&2\end{smallmatrix}\right]\cdot\left[\begin{smallmatrix}h\\k\end{smallmatrix}\right] = \left[\begin{smallmatrix}-3&2\\-4&3\end{smallmatrix}\right]\cdot\left[\begin{smallmatrix}h\\k\end{smallmatrix}\right], \qquad \left[\begin{smallmatrix}1\\2\end{smallmatrix}\right]\mapsto\left[\begin{smallmatrix}1\\2\end{smallmatrix}\right],\quad \left[\begin{smallmatrix}0\\1\end{smallmatrix}\right]\mapsto\left[\begin{smallmatrix}2\\3\end{smallmatrix}\right]; \quad (3.21)\\
&\mathcal{F}^{\ge\frac{1}{2}}(\mathbb{B}(2^{s+1}m), 2^s m) \to \mathcal{F}^{\ge\frac{1}{2}}(\mathbb{B}(2^{s+2}m), 2^{s+1}m),\\
&\left[\begin{smallmatrix}h\\k\end{smallmatrix}\right] \mapsto \boldsymbol{Z}\left[\begin{smallmatrix}2&-1\\3&-1\end{smallmatrix}\right]\left[\begin{smallmatrix}1&0\\3&-1\end{smallmatrix}\right]\cdot\left[\begin{smallmatrix}h\\k\end{smallmatrix}\right] = \left[\begin{smallmatrix}0&1\\1&1\end{smallmatrix}\right]\cdot\left[\begin{smallmatrix}h\\k\end{smallmatrix}\right], \qquad \left[\begin{smallmatrix}1\\1\end{smallmatrix}\right]\mapsto\left[\begin{smallmatrix}1\\2\end{smallmatrix}\right],\quad \left[\begin{smallmatrix}1\\2\end{smallmatrix}\right]\mapsto\left[\begin{smallmatrix}2\\3\end{smallmatrix}\right]; \quad (3.22)\\
&\mathcal{F}^{\le\frac{1}{2}}(\mathbb{B}(2^{s+1}m), 2^s m) \to \mathcal{F}^{\ge\frac{1}{2}}(\mathbb{B}(2^{s+2}m), 2^{s+1}m),\\
&\left[\begin{smallmatrix}h\\k\end{smallmatrix}\right] \mapsto \boldsymbol{Z}\left[\begin{smallmatrix}-1&1\\-3&2\end{smallmatrix}\right]\left[\begin{smallmatrix}-2&1\\-3&2\end{smallmatrix}\right]\cdot\left[\begin{smallmatrix}h\\k\end{smallmatrix}\right] = \left[\begin{smallmatrix}0&1\\1&1\end{smallmatrix}\right]\cdot\left[\begin{smallmatrix}h\\k\end{smallmatrix}\right], \qquad \left[\begin{smallmatrix}1\\2\end{smallmatrix}\right]\mapsto\left[\begin{smallmatrix}2\\3\end{smallmatrix}\right],\quad \left[\begin{smallmatrix}0\\1\end{smallmatrix}\right]\mapsto\left[\begin{smallmatrix}1\\1\end{smallmatrix}\right]; \quad (3.23)\\
&\mathcal{F}^{\ge\frac{1}{2}}(\mathbb{B}(2^{s+1}m), 2^s m) \to \mathcal{F}^{\ge\frac{1}{2}}(\mathbb{B}(2^{s+2}m), 2^{s+1}m),\\
&\left[\begin{smallmatrix}h\\k\end{smallmatrix}\right] \mapsto \boldsymbol{Z}\left[\begin{smallmatrix}1&0\\3&-1\end{smallmatrix}\right]\cdot\left[\begin{smallmatrix}h\\k\end{smallmatrix}\right] = \left[\begin{smallmatrix}3&-1\\5&-2\end{smallmatrix}\right]\cdot\left[\begin{smallmatrix}h\\k\end{smallmatrix}\right], \qquad \left[\begin{smallmatrix}1\\1\end{smallmatrix}\right]\mapsto\left[\begin{smallmatrix}2\\3\end{smallmatrix}\right],\quad \left[\begin{smallmatrix}1\\2\end{smallmatrix}\right]\mapsto\left[\begin{smallmatrix}1\\1\end{smallmatrix}\right]. \quad (3.24)
\end{aligned}$$

Proof. (i) (3.9) $\Leftarrow$ (2.13); (3.10) $\Leftarrow$ (3.4) and (2.13); (3.11) $\Leftarrow$ (3.3) and (2.13); (3.12) $\Leftarrow$ (2.13); (3.13) $\Leftarrow$ (2.15); (3.14) $\Leftarrow$ (3.4) and (2.15); (3.15) $\Leftarrow$ (3.3) and (2.15); (3.16) $\Leftarrow$ (2.15).

(ii) (3.17) $\Leftarrow$ (3.1) and (3.9); (3.18) $\Leftarrow$ (3.2) and (3.18); (3.19) $\Leftarrow$ (3.1) and (3.11); (3.20) $\Leftarrow$ (3.2) and (3.12); (3.21) $\Leftarrow$ (3.1) and (3.13); (3.22) $\Leftarrow$ (3.2) and (3.14); (3.23) $\Leftarrow$ (3.1) and (3.15); (3.24) $\Leftarrow$ (3.2) and (3.16). □

Let $\mathcal{C}$ be an increasing sequence of irreducible fractions $\frac{h}{k}$, written in vector form $\left[\begin{smallmatrix}h\\k\end{smallmatrix}\right]$. If $\boldsymbol{S}$ is a 2×2 matrix with integer entries and with determinant 1, then we denote by $\boldsymbol{S} \cdot \mathcal{C}$ the sequence of vectors $\boldsymbol{S} \cdot \left[\begin{smallmatrix}h\\k\end{smallmatrix}\right]$, that is,

$$\boldsymbol{S} \cdot \mathcal{C} := \left(\boldsymbol{S} \cdot \left[\begin{smallmatrix}h\\k\end{smallmatrix}\right] : \left[\begin{smallmatrix}h\\k\end{smallmatrix}\right] \in \mathcal{C}\right).$$

Proposition 3.6. *Let $m \in \mathbb{P}$, $s \in \mathbb{N}$.*
(i) *The following maps between subsequences of the sequence $\mathcal{F}(\mathbb{B}(2^{s+2}m), 2^{s+1}m)$ are order-preserving and bijective:*

$$\begin{aligned}&\left[\begin{smallmatrix}1&0\\2&1\end{smallmatrix}\right] \cdot \mathcal{F}_{2^s m} \to \left[\begin{smallmatrix}0&1\\-1&3\end{smallmatrix}\right] \cdot \mathcal{F}_{2^s m},\\ &\left[\begin{smallmatrix}h\\k\end{smallmatrix}\right] \mapsto \left[\begin{smallmatrix}-2&1\\-7&3\end{smallmatrix}\right] \cdot \left[\begin{smallmatrix}h\\k\end{smallmatrix}\right], \qquad \left[\begin{smallmatrix}0\\1\end{smallmatrix}\right] \mapsto \left[\begin{smallmatrix}1\\3\end{smallmatrix}\right], \quad \left[\begin{smallmatrix}1\\3\end{smallmatrix}\right] \mapsto \left[\begin{smallmatrix}1\\2\end{smallmatrix}\right];\end{aligned} \tag{3.25}$$

$$\begin{aligned}&\left[\begin{smallmatrix}1&0\\2&1\end{smallmatrix}\right] \cdot \mathcal{F}_{2^s m} \to \left[\begin{smallmatrix}1&1\\1&2\end{smallmatrix}\right] \cdot \mathcal{F}_{2^s m},\\ &\left[\begin{smallmatrix}h\\k\end{smallmatrix}\right] \mapsto \left[\begin{smallmatrix}-1&1\\-3&2\end{smallmatrix}\right] \cdot \left[\begin{smallmatrix}h\\k\end{smallmatrix}\right], \qquad \left[\begin{smallmatrix}0\\1\end{smallmatrix}\right] \mapsto \left[\begin{smallmatrix}1\\2\end{smallmatrix}\right], \quad \left[\begin{smallmatrix}1\\3\end{smallmatrix}\right] \mapsto \left[\begin{smallmatrix}2\\3\end{smallmatrix}\right];\end{aligned} \tag{3.26}$$

$$\begin{aligned}&\left[\begin{smallmatrix}1&0\\2&1\end{smallmatrix}\right] \cdot \mathcal{F}_{2^s m} \to \left[\begin{smallmatrix}-1&2\\-2&3\end{smallmatrix}\right] \cdot \mathcal{F}_{2^s m},\\ &\left[\begin{smallmatrix}h\\k\end{smallmatrix}\right] \mapsto \left[\begin{smallmatrix}-5&2\\-8&3\end{smallmatrix}\right] \cdot \left[\begin{smallmatrix}h\\k\end{smallmatrix}\right], \qquad \left[\begin{smallmatrix}0\\1\end{smallmatrix}\right] \mapsto \left[\begin{smallmatrix}2\\3\end{smallmatrix}\right], \quad \left[\begin{smallmatrix}1\\3\end{smallmatrix}\right] \mapsto \left[\begin{smallmatrix}1\\1\end{smallmatrix}\right];\end{aligned} \tag{3.27}$$

$$\begin{aligned}&\left[\begin{smallmatrix}0&1\\-1&3\end{smallmatrix}\right] \cdot \mathcal{F}_{2^s m} \to \left[\begin{smallmatrix}1&0\\2&1\end{smallmatrix}\right] \cdot \mathcal{F}_{2^s m},\\ &\left[\begin{smallmatrix}h\\k\end{smallmatrix}\right] \mapsto \left[\begin{smallmatrix}3&-1\\7&-2\end{smallmatrix}\right] \cdot \left[\begin{smallmatrix}h\\k\end{smallmatrix}\right], \qquad \left[\begin{smallmatrix}1\\3\end{smallmatrix}\right] \mapsto \left[\begin{smallmatrix}0\\1\end{smallmatrix}\right], \quad \left[\begin{smallmatrix}1\\2\end{smallmatrix}\right] \mapsto \left[\begin{smallmatrix}1\\3\end{smallmatrix}\right];\end{aligned} \tag{3.28}$$

$$\begin{aligned}&\left[\begin{smallmatrix}0&1\\-1&3\end{smallmatrix}\right] \cdot \mathcal{F}_{2^s m} \to \left[\begin{smallmatrix}1&1\\1&2\end{smallmatrix}\right] \cdot \mathcal{F}_{2^s m},\\ &\left[\begin{smallmatrix}h\\k\end{smallmatrix}\right] \mapsto \left[\begin{smallmatrix}4&-1\\5&-1\end{smallmatrix}\right] \cdot \left[\begin{smallmatrix}h\\k\end{smallmatrix}\right], \qquad \left[\begin{smallmatrix}1\\3\end{smallmatrix}\right] \mapsto \left[\begin{smallmatrix}1\\2\end{smallmatrix}\right], \quad \left[\begin{smallmatrix}1\\2\end{smallmatrix}\right] \mapsto \left[\begin{smallmatrix}2\\3\end{smallmatrix}\right];\end{aligned} \tag{3.29}$$

$$\begin{aligned}&\left[\begin{smallmatrix}0&1\\-1&3\end{smallmatrix}\right] \cdot \mathcal{F}_{2^s m} \to \left[\begin{smallmatrix}-1&2\\-2&3\end{smallmatrix}\right] \cdot \mathcal{F}_{2^s m},\\ &\left[\begin{smallmatrix}h\\k\end{smallmatrix}\right] \mapsto \left[\begin{smallmatrix}-1&1\\-3&2\end{smallmatrix}\right] \cdot \left[\begin{smallmatrix}h\\k\end{smallmatrix}\right], \qquad \left[\begin{smallmatrix}1\\3\end{smallmatrix}\right] \mapsto \left[\begin{smallmatrix}2\\3\end{smallmatrix}\right], \quad \left[\begin{smallmatrix}1\\2\end{smallmatrix}\right] \mapsto \left[\begin{smallmatrix}1\\1\end{smallmatrix}\right];\end{aligned} \tag{3.30}$$

$$\begin{aligned}&\left[\begin{smallmatrix}1&1\\1&2\end{smallmatrix}\right] \cdot \mathcal{F}_{2^s m} \to \left[\begin{smallmatrix}1&0\\2&1\end{smallmatrix}\right] \cdot \mathcal{F}_{2^s m},\\ &\left[\begin{smallmatrix}h\\k\end{smallmatrix}\right] \mapsto \left[\begin{smallmatrix}2&-1\\3&-1\end{smallmatrix}\right] \cdot \left[\begin{smallmatrix}h\\k\end{smallmatrix}\right], \qquad \left[\begin{smallmatrix}1\\2\end{smallmatrix}\right] \mapsto \left[\begin{smallmatrix}0\\1\end{smallmatrix}\right], \quad \left[\begin{smallmatrix}2\\3\end{smallmatrix}\right] \mapsto \left[\begin{smallmatrix}1\\3\end{smallmatrix}\right];\end{aligned} \tag{3.31}$$

$$\begin{aligned}&\left[\begin{smallmatrix}1&1\\1&2\end{smallmatrix}\right] \cdot \mathcal{F}_{2^s m} \to \left[\begin{smallmatrix}0&1\\-1&3\end{smallmatrix}\right] \cdot \mathcal{F}_{2^s m},\\ &\left[\begin{smallmatrix}h\\k\end{smallmatrix}\right] \mapsto \left[\begin{smallmatrix}-1&1\\-5&4\end{smallmatrix}\right] \cdot \left[\begin{smallmatrix}h\\k\end{smallmatrix}\right], \qquad \left[\begin{smallmatrix}1\\2\end{smallmatrix}\right] \mapsto \left[\begin{smallmatrix}1\\3\end{smallmatrix}\right], \quad \left[\begin{smallmatrix}2\\3\end{smallmatrix}\right] \mapsto \left[\begin{smallmatrix}1\\2\end{smallmatrix}\right];\end{aligned} \tag{3.32}$$

$$\begin{aligned}&\left[\begin{smallmatrix}1&1\\1&2\end{smallmatrix}\right] \cdot \mathcal{F}_{2^s m} \to \left[\begin{smallmatrix}-1&2\\-2&3\end{smallmatrix}\right] \cdot \mathcal{F}_{2^s m},\\ &\left[\begin{smallmatrix}h\\k\end{smallmatrix}\right] \mapsto \left[\begin{smallmatrix}-4&3\\-7&5\end{smallmatrix}\right] \cdot \left[\begin{smallmatrix}h\\k\end{smallmatrix}\right], \qquad \left[\begin{smallmatrix}1\\2\end{smallmatrix}\right] \mapsto \left[\begin{smallmatrix}2\\3\end{smallmatrix}\right], \quad \left[\begin{smallmatrix}2\\3\end{smallmatrix}\right] \mapsto \left[\begin{smallmatrix}1\\1\end{smallmatrix}\right];\end{aligned} \tag{3.33}$$

$$\begin{aligned}&\left[\begin{smallmatrix}-1&2\\-2&3\end{smallmatrix}\right] \cdot \mathcal{F}_{2^s m} \to \left[\begin{smallmatrix}1&0\\2&1\end{smallmatrix}\right] \cdot \mathcal{F}_{2^s m},\\ &\left[\begin{smallmatrix}h\\k\end{smallmatrix}\right] \mapsto \left[\begin{smallmatrix}3&-2\\8&-5\end{smallmatrix}\right] \cdot \left[\begin{smallmatrix}h\\k\end{smallmatrix}\right], \qquad \left[\begin{smallmatrix}2\\3\end{smallmatrix}\right] \mapsto \left[\begin{smallmatrix}0\\1\end{smallmatrix}\right], \quad \left[\begin{smallmatrix}1\\1\end{smallmatrix}\right] \mapsto \left[\begin{smallmatrix}1\\3\end{smallmatrix}\right];\end{aligned} \tag{3.34}$$

$$\begin{aligned}&\begin{bmatrix}-1&2\\-2&3\end{bmatrix}\cdot\mathcal{F}_{2^s m}\to\begin{bmatrix}0&1\\-1&3\end{bmatrix}\cdot\mathcal{F}_{2^s m},\\&\begin{bmatrix}h\\k\end{bmatrix}\mapsto\begin{bmatrix}2&-1\\3&-1\end{bmatrix}\cdot\begin{bmatrix}h\\k\end{bmatrix},\qquad\begin{bmatrix}2\\3\end{bmatrix}\mapsto\begin{bmatrix}1\\3\end{bmatrix},\quad\begin{bmatrix}1\\1\end{bmatrix}\mapsto\begin{bmatrix}1\\2\end{bmatrix};\end{aligned}\tag{3.35}$$

$$\begin{aligned}&\begin{bmatrix}-1&2\\-2&3\end{bmatrix}\cdot\mathcal{F}_{2^s m}\to\begin{bmatrix}1&1\\1&2\end{bmatrix}\cdot\mathcal{F}_{2^s m},\\&\begin{bmatrix}h\\k\end{bmatrix}\mapsto\begin{bmatrix}5&-3\\7&-4\end{bmatrix}\cdot\begin{bmatrix}h\\k\end{bmatrix},\qquad\begin{bmatrix}2\\3\end{bmatrix}\mapsto\begin{bmatrix}1\\2\end{bmatrix},\quad\begin{bmatrix}1\\1\end{bmatrix}\mapsto\begin{bmatrix}2\\3\end{bmatrix}.\end{aligned}\tag{3.36}$$

(ii) (a) *The following maps between subsequences of the sequence $\mathcal{F}(\mathbb{B}(2^{s+2}m), 2^{s+1}m)$ are order-reversing and bijective:*

$$\begin{aligned}&\begin{bmatrix}1&0\\2&1\end{bmatrix}\cdot\mathcal{F}_{2^s m}\to\begin{bmatrix}0&1\\-1&3\end{bmatrix}\cdot\mathcal{F}_{2^s m},\\&\begin{bmatrix}h\\k\end{bmatrix}\mapsto\begin{bmatrix}-2&1\\-3&2\end{bmatrix}\cdot\begin{bmatrix}h\\k\end{bmatrix},\qquad\begin{bmatrix}0\\1\end{bmatrix}\mapsto\begin{bmatrix}1\\2\end{bmatrix},\quad\begin{bmatrix}1\\3\end{bmatrix}\mapsto\begin{bmatrix}1\\3\end{bmatrix};\end{aligned}\tag{3.37}$$

$$\begin{aligned}&\begin{bmatrix}1&0\\2&1\end{bmatrix}\cdot\mathcal{F}_{2^s m}\to\begin{bmatrix}1&1\\1&2\end{bmatrix}\cdot\mathcal{F}_{2^s m},\\&\begin{bmatrix}h\\k\end{bmatrix}\mapsto\begin{bmatrix}-5&2\\-7&3\end{bmatrix}\cdot\begin{bmatrix}h\\k\end{bmatrix},\qquad\begin{bmatrix}0\\1\end{bmatrix}\mapsto\begin{bmatrix}2\\3\end{bmatrix},\quad\begin{bmatrix}1\\3\end{bmatrix}\mapsto\begin{bmatrix}1\\2\end{bmatrix};\end{aligned}\tag{3.38}$$

$$\begin{aligned}&\begin{bmatrix}1&0\\2&1\end{bmatrix}\cdot\mathcal{F}_{2^s m}\to\begin{bmatrix}-1&2\\-2&3\end{bmatrix}\cdot\mathcal{F}_{2^s m},\\&\begin{bmatrix}h\\k\end{bmatrix}\mapsto\begin{bmatrix}-1&1\\0&1\end{bmatrix}\cdot\begin{bmatrix}h\\k\end{bmatrix},\qquad\begin{bmatrix}0\\1\end{bmatrix}\mapsto\begin{bmatrix}1\\1\end{bmatrix},\quad\begin{bmatrix}1\\3\end{bmatrix}\mapsto\begin{bmatrix}2\\3\end{bmatrix};\end{aligned}\tag{3.39}$$

$$\begin{aligned}&\begin{bmatrix}0&1\\-1&3\end{bmatrix}\cdot\mathcal{F}_{2^s m}\to\begin{bmatrix}1&0\\2&1\end{bmatrix}\cdot\mathcal{F}_{2^s m},\\&\begin{bmatrix}h\\k\end{bmatrix}\mapsto\begin{bmatrix}-2&1\\-3&2\end{bmatrix}\cdot\begin{bmatrix}h\\k\end{bmatrix},\qquad\begin{bmatrix}1\\3\end{bmatrix}\mapsto\begin{bmatrix}1\\3\end{bmatrix},\quad\begin{bmatrix}1\\2\end{bmatrix}\mapsto\begin{bmatrix}0\\1\end{bmatrix};\end{aligned}\tag{3.40}$$

$$\begin{aligned}&\begin{bmatrix}0&1\\-1&3\end{bmatrix}\cdot\mathcal{F}_{2^s m}\to\begin{bmatrix}1&1\\1&2\end{bmatrix}\cdot\mathcal{F}_{2^s m},\\&\begin{bmatrix}h\\k\end{bmatrix}\mapsto\begin{bmatrix}-1&1\\0&1\end{bmatrix}\cdot\begin{bmatrix}h\\k\end{bmatrix},\qquad\begin{bmatrix}1\\3\end{bmatrix}\mapsto\begin{bmatrix}2\\3\end{bmatrix},\quad\begin{bmatrix}1\\2\end{bmatrix}\mapsto\begin{bmatrix}1\\2\end{bmatrix};\end{aligned}\tag{3.41}$$

$$\begin{aligned}&\begin{bmatrix}0&1\\-1&3\end{bmatrix}\cdot\mathcal{F}_{2^s m}\to\begin{bmatrix}-1&2\\-2&3\end{bmatrix}\cdot\mathcal{F}_{2^s m},\\&\begin{bmatrix}h\\k\end{bmatrix}\mapsto\begin{bmatrix}4&-1\\7&-2\end{bmatrix}\cdot\begin{bmatrix}h\\k\end{bmatrix},\qquad\begin{bmatrix}1\\3\end{bmatrix}\mapsto\begin{bmatrix}1\\1\end{bmatrix},\quad\begin{bmatrix}1\\2\end{bmatrix}\mapsto\begin{bmatrix}2\\3\end{bmatrix};\end{aligned}\tag{3.42}$$

$$\begin{aligned}&\begin{bmatrix}1&1\\1&2\end{bmatrix}\cdot\mathcal{F}_{2^s m}\to\begin{bmatrix}1&0\\2&1\end{bmatrix}\cdot\mathcal{F}_{2^s m},\\&\begin{bmatrix}h\\k\end{bmatrix}\mapsto\begin{bmatrix}-3&2\\-7&5\end{bmatrix}\cdot\begin{bmatrix}h\\k\end{bmatrix},\qquad\begin{bmatrix}1\\2\end{bmatrix}\mapsto\begin{bmatrix}1\\3\end{bmatrix},\quad\begin{bmatrix}2\\3\end{bmatrix}\mapsto\begin{bmatrix}0\\1\end{bmatrix};\end{aligned}\tag{3.43}$$

$$\begin{aligned}&\begin{bmatrix}1&1\\1&2\end{bmatrix}\cdot\mathcal{F}_{2^s m}\to\begin{bmatrix}0&1\\-1&3\end{bmatrix}\cdot\mathcal{F}_{2^s m},\\&\begin{bmatrix}h\\k\end{bmatrix}\mapsto\begin{bmatrix}-1&1\\0&1\end{bmatrix}\cdot\begin{bmatrix}h\\k\end{bmatrix},\qquad\begin{bmatrix}1\\2\end{bmatrix}\mapsto\begin{bmatrix}1\\2\end{bmatrix},\quad\begin{bmatrix}2\\3\end{bmatrix}\mapsto\begin{bmatrix}1\\3\end{bmatrix};\end{aligned}\tag{3.44}$$

$$\begin{aligned}&\begin{bmatrix}1&1\\1&2\end{bmatrix}\cdot\mathcal{F}_{2^s m}\to\begin{bmatrix}-1&2\\-2&3\end{bmatrix}\cdot\mathcal{F}_{2^s m},\\&\begin{bmatrix}h\\k\end{bmatrix}\mapsto\begin{bmatrix}1&0\\3&-1\end{bmatrix}\cdot\begin{bmatrix}h\\k\end{bmatrix},\qquad\begin{bmatrix}1\\2\end{bmatrix}\mapsto\begin{bmatrix}1\\1\end{bmatrix},\quad\begin{bmatrix}2\\3\end{bmatrix}\mapsto\begin{bmatrix}2\\3\end{bmatrix};\end{aligned}\tag{3.45}$$

$$\begin{aligned}&\begin{bmatrix}-1&2\\-2&3\end{bmatrix}\cdot\mathcal{F}_{2^s m}\to\begin{bmatrix}1&0\\2&1\end{bmatrix}\cdot\mathcal{F}_{2^s m},\\&\begin{bmatrix}h\\k\end{bmatrix}\mapsto\begin{bmatrix}-1&1\\0&1\end{bmatrix}\cdot\begin{bmatrix}h\\k\end{bmatrix},\qquad\begin{bmatrix}2\\3\end{bmatrix}\mapsto\begin{bmatrix}1\\3\end{bmatrix},\quad\begin{bmatrix}1\\1\end{bmatrix}\mapsto\begin{bmatrix}0\\1\end{bmatrix};\end{aligned}\tag{3.46}$$

$$\begin{aligned}&\begin{bmatrix}-1&2\\-2&3\end{bmatrix}\cdot\mathcal{F}_{2^s m}\to\begin{bmatrix}0&1\\-1&3\end{bmatrix}\cdot\mathcal{F}_{2^s m},\\&\begin{bmatrix}h\\k\end{bmatrix}\mapsto\begin{bmatrix}2&-1\\7&-4\end{bmatrix}\cdot\begin{bmatrix}h\\k\end{bmatrix},\qquad\begin{bmatrix}2\\3\end{bmatrix}\mapsto\begin{bmatrix}1\\2\end{bmatrix},\quad\begin{bmatrix}1\\1\end{bmatrix}\mapsto\begin{bmatrix}1\\3\end{bmatrix};\end{aligned}\tag{3.47}$$

$$\begin{aligned}&\begin{bmatrix}-1&2\\-2&3\end{bmatrix}\cdot\mathcal{F}_{2^s m}\to\begin{bmatrix}1&1\\1&2\end{bmatrix}\cdot\mathcal{F}_{2^s m},\\&\begin{bmatrix}h\\k\end{bmatrix}\mapsto\begin{bmatrix}1&0\\3&-1\end{bmatrix}\cdot\begin{bmatrix}h\\k\end{bmatrix},\qquad\begin{bmatrix}2\\3\end{bmatrix}\mapsto\begin{bmatrix}2\\3\end{bmatrix},\quad\begin{bmatrix}1\\1\end{bmatrix}\mapsto\begin{bmatrix}1\\2\end{bmatrix}.\end{aligned}\tag{3.48}$$

(b) *The following maps of subsequences of the sequence* $\mathcal{F}(\mathbb{B}(2^{s+2}m), 2^{s+1}m)$ *onto themselves are order-reversing and bijective:*

$$\begin{aligned}
&\left[\begin{smallmatrix} 1 & 0 \\ 2 & 1 \end{smallmatrix}\right]\cdot\mathcal{F}_{2^s m} \to \left[\begin{smallmatrix} 1 & 0 \\ 2 & 1 \end{smallmatrix}\right]\cdot\mathcal{F}_{2^s m},\\
&\left[\begin{smallmatrix} h \\ k \end{smallmatrix}\right] \mapsto \left[\begin{smallmatrix} -3 & 1 \\ -8 & 3 \end{smallmatrix}\right]\cdot\left[\begin{smallmatrix} h \\ k \end{smallmatrix}\right], \qquad \left[\begin{smallmatrix} 0 \\ 1 \end{smallmatrix}\right]\mapsto\left[\begin{smallmatrix} 1 \\ 3 \end{smallmatrix}\right],\quad \left[\begin{smallmatrix} 1 \\ 4 \end{smallmatrix}\right]\mapsto\left[\begin{smallmatrix} 1 \\ 4 \end{smallmatrix}\right],\quad \left[\begin{smallmatrix} 1 \\ 3 \end{smallmatrix}\right]\mapsto\left[\begin{smallmatrix} 0 \\ 1 \end{smallmatrix}\right]; \qquad (3.49)\\
&\left[\begin{smallmatrix} 0 & 1 \\ -1 & 3 \end{smallmatrix}\right]\cdot\mathcal{F}_{2^s m} \to \left[\begin{smallmatrix} 0 & 1 \\ -1 & 3 \end{smallmatrix}\right]\cdot\mathcal{F}_{2^s m},\\
&\left[\begin{smallmatrix} h \\ k \end{smallmatrix}\right] \mapsto \left[\begin{smallmatrix} 1 & 0 \\ 5 & -1 \end{smallmatrix}\right]\cdot\left[\begin{smallmatrix} h \\ k \end{smallmatrix}\right], \qquad \left[\begin{smallmatrix} 1 \\ 3 \end{smallmatrix}\right]\mapsto\left[\begin{smallmatrix} 1 \\ 2 \end{smallmatrix}\right],\quad \left[\begin{smallmatrix} 2 \\ 5 \end{smallmatrix}\right]\mapsto\left[\begin{smallmatrix} 2 \\ 5 \end{smallmatrix}\right],\quad \left[\begin{smallmatrix} 1 \\ 2 \end{smallmatrix}\right]\mapsto\left[\begin{smallmatrix} 1 \\ 3 \end{smallmatrix}\right]; \qquad (3.50)\\
&\left[\begin{smallmatrix} 1 & 1 \\ 1 & 2 \end{smallmatrix}\right]\cdot\mathcal{F}_{2^s m} \to \left[\begin{smallmatrix} 1 & 1 \\ 1 & 2 \end{smallmatrix}\right]\cdot\mathcal{F}_{2^s m},\\
&\left[\begin{smallmatrix} h \\ k \end{smallmatrix}\right] \mapsto \left[\begin{smallmatrix} -4 & 3 \\ -5 & 4 \end{smallmatrix}\right]\cdot\left[\begin{smallmatrix} h \\ k \end{smallmatrix}\right], \qquad \left[\begin{smallmatrix} 1 \\ 2 \end{smallmatrix}\right]\mapsto\left[\begin{smallmatrix} 2 \\ 3 \end{smallmatrix}\right],\quad \left[\begin{smallmatrix} 3 \\ 5 \end{smallmatrix}\right]\mapsto\left[\begin{smallmatrix} 3 \\ 5 \end{smallmatrix}\right],\quad \left[\begin{smallmatrix} 2 \\ 3 \end{smallmatrix}\right]\mapsto\left[\begin{smallmatrix} 1 \\ 2 \end{smallmatrix}\right]; \qquad (3.51)\\
&\left[\begin{smallmatrix} -1 & 2 \\ -2 & 3 \end{smallmatrix}\right]\cdot\mathcal{F}_{2^s m} \to \left[\begin{smallmatrix} -1 & 2 \\ -2 & 3 \end{smallmatrix}\right]\cdot\mathcal{F}_{2^s m},\\
&\left[\begin{smallmatrix} h \\ k \end{smallmatrix}\right] \mapsto \left[\begin{smallmatrix} 5 & -3 \\ 8 & -5 \end{smallmatrix}\right]\cdot\left[\begin{smallmatrix} h \\ k \end{smallmatrix}\right], \qquad \left[\begin{smallmatrix} 2 \\ 3 \end{smallmatrix}\right]\mapsto\left[\begin{smallmatrix} 1 \\ 1 \end{smallmatrix}\right],\quad \left[\begin{smallmatrix} 3 \\ 4 \end{smallmatrix}\right]\mapsto\left[\begin{smallmatrix} 3 \\ 4 \end{smallmatrix}\right],\quad \left[\begin{smallmatrix} 1 \\ 1 \end{smallmatrix}\right]\mapsto\left[\begin{smallmatrix} 2 \\ 3 \end{smallmatrix}\right]. \qquad (3.52)
\end{aligned}$$

Proof. (i) We have

$$\begin{aligned}
&(3.25):\ \boldsymbol{W}^2\cdot\mathcal{F}_{2^s m}\to\boldsymbol{WZ}\cdot\mathcal{F}_{2^s m}, && \left[\begin{smallmatrix} h \\ k \end{smallmatrix}\right]\mapsto\boldsymbol{WZW}^{-2}\cdot\left[\begin{smallmatrix} h \\ k \end{smallmatrix}\right];\\
&(3.26):\ \boldsymbol{W}^2\cdot\mathcal{F}_{2^s m}\to\boldsymbol{ZW}\cdot\mathcal{F}_{2^s m}, && \left[\begin{smallmatrix} h \\ k \end{smallmatrix}\right]\mapsto\boldsymbol{ZWW}^{-2}\cdot\left[\begin{smallmatrix} h \\ k \end{smallmatrix}\right];\\
&(3.27):\ \boldsymbol{W}^2\cdot\mathcal{F}_{2^s m}\to\boldsymbol{Z}^2\cdot\mathcal{F}_{2^s m}, && \left[\begin{smallmatrix} h \\ k \end{smallmatrix}\right]\mapsto\boldsymbol{Z}^2\boldsymbol{W}^{-2}\cdot\left[\begin{smallmatrix} h \\ k \end{smallmatrix}\right];\\
&(3.28):\ \boldsymbol{WZ}\cdot\mathcal{F}_{2^s m}\to\boldsymbol{W}^2\cdot\mathcal{F}_{2^s m}, && \left[\begin{smallmatrix} h \\ k \end{smallmatrix}\right]\mapsto\boldsymbol{W}^2\boldsymbol{Z}^{-1}\boldsymbol{W}^{-1}\cdot\left[\begin{smallmatrix} h \\ k \end{smallmatrix}\right];\\
&(3.29):\ \boldsymbol{WZ}\cdot\mathcal{F}_{2^s m}\to\boldsymbol{ZW}\cdot\mathcal{F}_{2^s m}, && \left[\begin{smallmatrix} h \\ k \end{smallmatrix}\right]\mapsto\boldsymbol{ZWZ}^{-1}\boldsymbol{W}^{-1}\cdot\left[\begin{smallmatrix} h \\ k \end{smallmatrix}\right];\\
&(3.30):\ \boldsymbol{WZ}\cdot\mathcal{F}_{2^s m}\to\boldsymbol{Z}^2\cdot\mathcal{F}_{2^s m}, && \left[\begin{smallmatrix} h \\ k \end{smallmatrix}\right]\mapsto\boldsymbol{Z}^2\boldsymbol{Z}^{-1}\boldsymbol{W}^{-1}\cdot\left[\begin{smallmatrix} h \\ k \end{smallmatrix}\right];\\
&(3.31):\ \boldsymbol{ZW}\cdot\mathcal{F}_{2^s m}\to\boldsymbol{W}^2\cdot\mathcal{F}_{2^s m}, && \left[\begin{smallmatrix} h \\ k \end{smallmatrix}\right]\mapsto\boldsymbol{W}^2\boldsymbol{W}^{-1}\boldsymbol{Z}^{-1}\cdot\left[\begin{smallmatrix} h \\ k \end{smallmatrix}\right];\\
&(3.32):\ \boldsymbol{ZW}\cdot\mathcal{F}_{2^s m}\to\boldsymbol{WZ}\cdot\mathcal{F}_{2^s m}, && \left[\begin{smallmatrix} h \\ k \end{smallmatrix}\right]\mapsto\boldsymbol{WZW}^{-1}\boldsymbol{Z}^{-1}\cdot\left[\begin{smallmatrix} h \\ k \end{smallmatrix}\right];\\
&(3.33):\ \boldsymbol{ZW}\cdot\mathcal{F}_{2^s m}\to\boldsymbol{Z}^2\cdot\mathcal{F}_{2^s m}, && \left[\begin{smallmatrix} h \\ k \end{smallmatrix}\right]\mapsto\boldsymbol{Z}^2\boldsymbol{W}^{-1}\boldsymbol{Z}^{-1}\cdot\left[\begin{smallmatrix} h \\ k \end{smallmatrix}\right];\\
&(3.34):\ \boldsymbol{Z}^2\cdot\mathcal{F}_{2^s m}\to\boldsymbol{W}^2\cdot\mathcal{F}_{2^s m}, && \left[\begin{smallmatrix} h \\ k \end{smallmatrix}\right]\mapsto\boldsymbol{W}^2\boldsymbol{Z}^{-2}\cdot\left[\begin{smallmatrix} h \\ k \end{smallmatrix}\right];\\
&(3.35):\ \boldsymbol{Z}^2\cdot\mathcal{F}_{2^s m}\to\boldsymbol{WZ}\cdot\mathcal{F}_{2^s m}, && \left[\begin{smallmatrix} h \\ k \end{smallmatrix}\right]\mapsto\boldsymbol{WZZ}^{-2}\cdot\left[\begin{smallmatrix} h \\ k \end{smallmatrix}\right];\\
&(3.36):\ \boldsymbol{Z}^2\cdot\mathcal{F}_{2^s m}\to\boldsymbol{ZW}\cdot\mathcal{F}_{2^s m}, && \left[\begin{smallmatrix} h \\ k \end{smallmatrix}\right]\mapsto\boldsymbol{ZWZ}^{-2}\cdot\left[\begin{smallmatrix} h \\ k \end{smallmatrix}\right].
\end{aligned}$$

(ii) (a) We have

$$\begin{aligned}
&(3.37):\ \boldsymbol{W}^2\cdot\mathcal{F}_{2^s m}\to\boldsymbol{WZ}\cdot\mathcal{F}_{2^s m}, && \left[\begin{smallmatrix} h \\ k \end{smallmatrix}\right]\mapsto\boldsymbol{WZ}\left[\begin{smallmatrix} -1 & 1 \\ 0 & 1 \end{smallmatrix}\right]\boldsymbol{W}^{-2}\cdot\left[\begin{smallmatrix} h \\ k \end{smallmatrix}\right];\\
&(3.38):\ \boldsymbol{W}^2\cdot\mathcal{F}_{2^s m}\to\boldsymbol{ZW}\cdot\mathcal{F}_{2^s m}, && \left[\begin{smallmatrix} h \\ k \end{smallmatrix}\right]\mapsto\boldsymbol{ZW}\left[\begin{smallmatrix} -1 & 1 \\ 0 & 1 \end{smallmatrix}\right]\boldsymbol{W}^{-2}\cdot\left[\begin{smallmatrix} h \\ k \end{smallmatrix}\right];\\
&(3.39):\ \boldsymbol{W}^2\cdot\mathcal{F}_{2^s m}\to\boldsymbol{Z}^2\cdot\mathcal{F}_{2^s m}, && \left[\begin{smallmatrix} h \\ k \end{smallmatrix}\right]\mapsto\boldsymbol{Z}^2\left[\begin{smallmatrix} -1 & 1 \\ 0 & 1 \end{smallmatrix}\right]\boldsymbol{W}^{-2}\cdot\left[\begin{smallmatrix} h \\ k \end{smallmatrix}\right];\\
&(3.40):\ \boldsymbol{WZ}\cdot\mathcal{F}_{2^s m}\to\boldsymbol{W}^2\cdot\mathcal{F}_{2^s m}, && \left[\begin{smallmatrix} h \\ k \end{smallmatrix}\right]\mapsto\boldsymbol{W}^2\left[\begin{smallmatrix} -1 & 1 \\ 0 & 1 \end{smallmatrix}\right]\boldsymbol{Z}^{-1}\boldsymbol{W}^{-1}\cdot\left[\begin{smallmatrix} h \\ k \end{smallmatrix}\right];\\
&(3.41):\ \boldsymbol{WZ}\cdot\mathcal{F}_{2^s m}\to\boldsymbol{ZW}\cdot\mathcal{F}_{2^s m}, && \left[\begin{smallmatrix} h \\ k \end{smallmatrix}\right]\mapsto\boldsymbol{ZW}\left[\begin{smallmatrix} -1 & 1 \\ 0 & 1 \end{smallmatrix}\right]\boldsymbol{Z}^{-1}\boldsymbol{W}^{-1}\cdot\left[\begin{smallmatrix} h \\ k \end{smallmatrix}\right];\\
&(3.42):\ \boldsymbol{WZ}\cdot\mathcal{F}_{2^s m}\to\boldsymbol{Z}^2\cdot\mathcal{F}_{2^s m}, && \left[\begin{smallmatrix} h \\ k \end{smallmatrix}\right]\mapsto\boldsymbol{Z}^2\left[\begin{smallmatrix} -1 & 1 \\ 0 & 1 \end{smallmatrix}\right]\boldsymbol{Z}^{-1}\boldsymbol{W}^{-1}\cdot\left[\begin{smallmatrix} h \\ k \end{smallmatrix}\right];\\
&(3.43):\ \boldsymbol{ZW}\cdot\mathcal{F}_{2^s m}\to\boldsymbol{W}^2\cdot\mathcal{F}_{2^s m}, && \left[\begin{smallmatrix} h \\ k \end{smallmatrix}\right]\mapsto\boldsymbol{W}^2\left[\begin{smallmatrix} -1 & 1 \\ 0 & 1 \end{smallmatrix}\right]\boldsymbol{W}^{-1}\boldsymbol{Z}^{-1}\cdot\left[\begin{smallmatrix} h \\ k \end{smallmatrix}\right];\\
&(3.44):\ \boldsymbol{ZW}\cdot\mathcal{F}_{2^s m}\to\boldsymbol{WZ}\cdot\mathcal{F}_{2^s m}, && \left[\begin{smallmatrix} h \\ k \end{smallmatrix}\right]\mapsto\boldsymbol{WZ}\left[\begin{smallmatrix} -1 & 1 \\ 0 & 1 \end{smallmatrix}\right]\boldsymbol{W}^{-1}\boldsymbol{Z}^{-1}\cdot\left[\begin{smallmatrix} h \\ k \end{smallmatrix}\right];
\end{aligned}$$

$$
\begin{aligned}
&(3.45): && \boldsymbol{ZW}\cdot\mathcal{F}_{2^s m}\to\boldsymbol{Z}^2\cdot\mathcal{F}_{2^s m}, && \left[\begin{smallmatrix}h\\k\end{smallmatrix}\right]\mapsto \boldsymbol{Z}^2\left[\begin{smallmatrix}-1&1\\0&1\end{smallmatrix}\right]\boldsymbol{W}^{-1}\boldsymbol{Z}^{-1}\cdot\left[\begin{smallmatrix}h\\k\end{smallmatrix}\right];\\
&(3.46): && \boldsymbol{Z}^2\cdot\mathcal{F}_{2^s m}\to\boldsymbol{W}^2\cdot\mathcal{F}_{2^s m}, && \left[\begin{smallmatrix}h\\k\end{smallmatrix}\right]\mapsto \boldsymbol{W}^2\left[\begin{smallmatrix}-1&1\\0&1\end{smallmatrix}\right]\boldsymbol{Z}^{-2}\cdot\left[\begin{smallmatrix}h\\k\end{smallmatrix}\right];\\
&(3.47): && \boldsymbol{Z}^2\cdot\mathcal{F}_{2^s m}\to\boldsymbol{WZ}\cdot\mathcal{F}_{2^s m}, && \left[\begin{smallmatrix}h\\k\end{smallmatrix}\right]\mapsto \boldsymbol{WZ}\left[\begin{smallmatrix}-1&1\\0&1\end{smallmatrix}\right]\boldsymbol{Z}^{-2}\cdot\left[\begin{smallmatrix}h\\k\end{smallmatrix}\right];\\
&(3.48): && \boldsymbol{Z}^2\cdot\mathcal{F}_{2^s m}\to\boldsymbol{ZW}\cdot\mathcal{F}_{2^s m}, && \left[\begin{smallmatrix}h\\k\end{smallmatrix}\right]\mapsto \boldsymbol{ZW}\left[\begin{smallmatrix}-1&1\\0&1\end{smallmatrix}\right]\boldsymbol{Z}^{-2}\cdot\left[\begin{smallmatrix}h\\k\end{smallmatrix}\right].
\end{aligned}
$$

(ii) (b) We have

$$
\begin{aligned}
&(3.49): && \boldsymbol{W}^2\cdot\mathcal{F}_{2^s m}\to\boldsymbol{W}^2\cdot\mathcal{F}_{2^s m}, && \left[\begin{smallmatrix}h\\k\end{smallmatrix}\right]\mapsto \boldsymbol{W}^2\left[\begin{smallmatrix}-1&1\\0&1\end{smallmatrix}\right]\boldsymbol{W}^{-2}\cdot\left[\begin{smallmatrix}h\\k\end{smallmatrix}\right];\\
&(3.50): && \boldsymbol{WZ}\cdot\mathcal{F}_{2^s m}\to\boldsymbol{WZ}\cdot\mathcal{F}_{2^s m}, && \left[\begin{smallmatrix}h\\k\end{smallmatrix}\right]\mapsto \boldsymbol{WZ}\left[\begin{smallmatrix}-1&1\\0&1\end{smallmatrix}\right]\boldsymbol{Z}^{-1}\boldsymbol{W}^{-1}\cdot\left[\begin{smallmatrix}h\\k\end{smallmatrix}\right];\\
&(3.51): && \boldsymbol{ZW}\cdot\mathcal{F}_{2^s m}\to\boldsymbol{ZW}\cdot\mathcal{F}_{2^s m}, && \left[\begin{smallmatrix}h\\k\end{smallmatrix}\right]\mapsto \boldsymbol{ZW}\left[\begin{smallmatrix}-1&1\\0&1\end{smallmatrix}\right]\boldsymbol{W}^{-1}\boldsymbol{Z}^{-1}\cdot\left[\begin{smallmatrix}h\\k\end{smallmatrix}\right];\\
&(3.52): && \boldsymbol{Z}^2\cdot\mathcal{F}_{2^s m}\to\boldsymbol{Z}^2\cdot\mathcal{F}_{2^s m}, && \left[\begin{smallmatrix}h\\k\end{smallmatrix}\right]\mapsto \boldsymbol{Z}^2\left[\begin{smallmatrix}-1&1\\0&1\end{smallmatrix}\right]\boldsymbol{Z}^{-2}\cdot\left[\begin{smallmatrix}h\\k\end{smallmatrix}\right]. \quad \square
\end{aligned}
$$

Example 3.7. Suppose $m := 1$, $s := 1$, and consider the sequences

$$
\begin{aligned}
\mathcal{F}_{2^{s+2}m} &= \left(\tfrac{0}{1} < \tfrac{1}{8} < \tfrac{1}{7} < \tfrac{1}{6} < \tfrac{1}{5} < \tfrac{1}{4} < \tfrac{2}{7} < \tfrac{1}{3} < \tfrac{3}{8} < \tfrac{2}{5} < \tfrac{3}{7} < \tfrac{1}{2} < \tfrac{4}{7}\right.\\
&\qquad \left. < \tfrac{3}{5} < \tfrac{5}{8} < \tfrac{2}{3} < \tfrac{5}{7} < \tfrac{3}{4} < \tfrac{4}{5} < \tfrac{5}{6} < \tfrac{6}{7} < \tfrac{7}{8} < \tfrac{1}{1}\right),\\
\mathcal{F}(\mathbb{B}(2^{s+2}m), 2^{s+1}m) &= \left(\tfrac{0}{1} < \;\; < \;\; < \;\; < \tfrac{1}{5} < \tfrac{1}{4} < \;\; < \tfrac{1}{3} < \;\; < \tfrac{2}{5} < \tfrac{3}{7} < \tfrac{1}{2} < \tfrac{4}{7}\right.\\
&\qquad \left. < \tfrac{3}{5} < \;\; < \tfrac{2}{3} < \;\; < \tfrac{3}{4} < \tfrac{4}{5} < \;\; < \;\; < \;\; < \tfrac{1}{1}\right),\\
\mathcal{F}_{2^{s+1}m} &= \left(\tfrac{0}{1} < \;\; < \;\; < \;\; < \;\; < \tfrac{1}{4} < \;\; < \tfrac{1}{3} < \;\; < \;\; < \;\; < \tfrac{1}{2} <\right.\\
&\qquad \left. < \;\; < \;\; < \tfrac{2}{3} < \;\; < \tfrac{3}{4} < \;\; < \;\; < \;\; < \;\; < \tfrac{1}{1}\right),\\
\mathcal{F}(\mathbb{B}(2^{s+1}m), 2^{s}m) &= \left(\tfrac{0}{1} < \;\; < \;\; < \;\; < \;\; < \;\; < \;\; < \tfrac{1}{3} < \;\; < \;\; < \;\; < \tfrac{1}{2} <\right.\\
&\qquad \left. < \;\; < \;\; < \tfrac{2}{3} < \;\; < \;\; < \;\; < \;\; < \;\; < \;\; < \tfrac{1}{1}\right),\\
\mathcal{F}_{2^s m} = \mathcal{F}(\mathbb{B}(2^{s}m), 2^{s-1}m) &= \left(\tfrac{0}{1} < \;\; < \;\; < \;\; < \;\; < \;\; < \;\; < \;\; < \;\; < \;\; < \;\; < \tfrac{1}{2} <\right.\\
&\qquad \left. < \;\; < \;\; < \;\; < \;\; < \;\; < \;\; < \;\; < \;\; < \;\; < \tfrac{1}{1}\right),\\
\mathcal{F}_{2^{s-1}m} &= \left(\tfrac{0}{1} < \;\; < \;\; < \;\; < \;\; < \;\; < \;\; < \;\; < \;\; < \;\; < \;\; < \;\; <\right.\\
&\qquad \left. < \;\; < \;\; < \;\; < \;\; < \;\; < \;\; < \;\; < \;\; < \;\; < \tfrac{1}{1}\right).
\end{aligned}
$$

It follows from (3.25) that the map

$$
\begin{aligned}
&\mathcal{F}_{2^{s+2}m} \supset \left[\begin{smallmatrix}1&0\\2&1\end{smallmatrix}\right]\cdot\mathcal{F}_{2^s m} = \left[\begin{smallmatrix}1&0\\2&1\end{smallmatrix}\right]\cdot\left(\left[\begin{smallmatrix}0\\1\end{smallmatrix}\right],\left[\begin{smallmatrix}1\\2\end{smallmatrix}\right],\left[\begin{smallmatrix}1\\1\end{smallmatrix}\right]\right) = \underbrace{\left(\left[\begin{smallmatrix}0\\1\end{smallmatrix}\right],\left[\begin{smallmatrix}1\\4\end{smallmatrix}\right],\left[\begin{smallmatrix}1\\3\end{smallmatrix}\right]\right)}_{\downarrow(3.25)}\\
&\to \left[\begin{smallmatrix}0&1\\-1&3\end{smallmatrix}\right]\cdot\mathcal{F}_{2^s m} = \left[\begin{smallmatrix}0&1\\-1&3\end{smallmatrix}\right]\cdot\left(\left[\begin{smallmatrix}0\\1\end{smallmatrix}\right],\left[\begin{smallmatrix}1\\2\end{smallmatrix}\right],\left[\begin{smallmatrix}1\\1\end{smallmatrix}\right]\right) = \overbrace{\left(\left[\begin{smallmatrix}1\\3\end{smallmatrix}\right],\left[\begin{smallmatrix}2\\5\end{smallmatrix}\right],\left[\begin{smallmatrix}1\\2\end{smallmatrix}\right]\right)} \subset \mathcal{F}_{2^{s+2}m}
\end{aligned}
$$

given by

$$
\left[\begin{smallmatrix}h\\k\end{smallmatrix}\right]\mapsto\left[\begin{smallmatrix}-2&1\\-7&3\end{smallmatrix}\right]\cdot\left[\begin{smallmatrix}h\\k\end{smallmatrix}\right]
$$

is order-preserving and bijective. (3.37) and (3.50) imply that the maps

$$\mathcal{F}_{2^{s+2}m} \supset \left([\begin{smallmatrix}0\\1\end{smallmatrix}],[\begin{smallmatrix}1\\4\end{smallmatrix}],[\begin{smallmatrix}1\\3\end{smallmatrix}]\right) \xrightarrow{(3.37)} \left([\begin{smallmatrix}1\\3\end{smallmatrix}],[\begin{smallmatrix}2\\5\end{smallmatrix}],[\begin{smallmatrix}1\\2\end{smallmatrix}]\right) \subset \mathcal{F}_{2^{s+2}m},$$

$$[\begin{smallmatrix}h\\k\end{smallmatrix}] \longmapsto [\begin{smallmatrix}-2 & 1\\-3 & 2\end{smallmatrix}]\cdot[\begin{smallmatrix}h\\k\end{smallmatrix}],$$

$$\mathcal{F}_{2^{s+2}m} \supset \left([\begin{smallmatrix}1\\3\end{smallmatrix}],[\begin{smallmatrix}2\\5\end{smallmatrix}],[\begin{smallmatrix}1\\2\end{smallmatrix}]\right) \xrightarrow{(3.50)} \left([\begin{smallmatrix}1\\3\end{smallmatrix}],[\begin{smallmatrix}2\\5\end{smallmatrix}],[\begin{smallmatrix}1\\2\end{smallmatrix}]\right),$$

$$[\begin{smallmatrix}h\\k\end{smallmatrix}] \longmapsto \underset{\substack{\uparrow\\ \text{involutory}}}{[\begin{smallmatrix}1 & 0\\5 & -1\end{smallmatrix}]}\cdot[\begin{smallmatrix}h\\k\end{smallmatrix}], \quad [\begin{smallmatrix}2\\5\end{smallmatrix}] \mapsto [\begin{smallmatrix}2\\5\end{smallmatrix}],$$

are order-reversing and bijective.

3.4 Monotone bijective maps between subsequences of the Farey sequence $\mathcal{F}_n$

In this section we will describe monotone bijections between certain distinguished subsequences of the Farey sequences.

The following auxiliary statement that uses the approach sketched in Remark 3.4 (i) is based on the properties of the order-preserving bijections (2.13) and (2.15).

Lemma 3.8. *Let m and n be positive integers, with $n \geq 2m$. Suppose $s := \lfloor \log_2(\frac{n}{m}) \rfloor$. Then, for any ordered collection $(\boldsymbol{M}_1, \dots, \boldsymbol{M}_s) \in \{\boldsymbol{W}, \boldsymbol{Z}\}^s$, of length s, whose entries are matrices $\boldsymbol{W} := [\begin{smallmatrix}1 & 0\\1 & 1\end{smallmatrix}]$ or $\boldsymbol{Z} := [\begin{smallmatrix}0 & 1\\-1 & 2\end{smallmatrix}]$, the map*

$$\mathcal{F}_m \to \mathcal{F}_n, \quad [\begin{smallmatrix}h\\k\end{smallmatrix}] \mapsto \prod_{i=1}^{s} \boldsymbol{M}_{s-i+1}\cdot[\begin{smallmatrix}h\\k\end{smallmatrix}] \begin{cases} \leq \frac{1}{2} & \textit{if } \boldsymbol{M}_s = \boldsymbol{W}, \\ \geq \frac{1}{2} & \textit{if } \boldsymbol{M}_s = \boldsymbol{Z} \end{cases} \tag{3.53}$$

is order-preserving and injective.

The image of any Farey (sub)sequence retains its two generic properties.

Remark 3.9. Let $\mathcal{C}$ be any Farey (sub)sequence of the form $\mathcal{F}_m$, $\mathcal{F}_m^\ell$, $\mathcal{G}_m^\ell$ or $\mathcal{F}(\mathbb{B}(m), \ell)$. If $\frac{h_j}{k_j} < \frac{h_{j+1}}{k_{j+1}} < \frac{h_{j+2}}{k_{j+2}}$ are three consecutive fractions in the subsequence $\prod_{i=1}^{s} \boldsymbol{M}_{s-i+1}\cdot\mathcal{C}$ of the Farey sequence $\mathcal{F}_n$ considered in Lemma 3.8, then

$$k_j h_{j+1} - h_j k_{j+1} = 1$$

and

$$\frac{h_{j+1}}{k_{j+1}} = \frac{h_j + h_{j+2}}{\gcd(h_j + h_{j+2}, k_j + k_{j+2})} \Big/ \frac{k_j + k_{j+2}}{\gcd(h_j + h_{j+2}, k_j + k_{j+2})};$$

cf. Propositions 1.23 and 1.24. Therefore, the sequence

$$\prod_{i=1}^{s} \boldsymbol{M}_{s-i+1}\cdot\mathcal{C}$$

can be regarded as a true Farey subsequence.

We conclude this section with a straightforward extension of Proposition 3.6, see also Example 3.7.

Theorem 3.10. *Let m and n be positive integers, with $n \geq 2m$. Suppose $s := \lfloor \log_2(\frac{n}{m}) \rfloor$. Let $(\boldsymbol{M}_1, \ldots, \boldsymbol{M}_s), (\boldsymbol{N}_1, \ldots, \boldsymbol{N}_s) \in \{\boldsymbol{W}, \boldsymbol{Z}\}^s$ be two ordered collections of matrices, and*

$$\mathbf{M} := \prod_{i=1}^{s} \boldsymbol{M}_{s-i+1}, \quad \mathbf{N} := \prod_{i=1}^{s} \boldsymbol{N}_{s-i+1}.$$

The map

$$\mathcal{F}_n \supset \mathbf{M} \cdot \mathcal{F}_m \to \mathbf{N} \cdot \mathcal{F}_m \subset \mathcal{F}_n, \quad \left[\begin{smallmatrix} h \\ k \end{smallmatrix}\right] \mapsto \mathbf{N}\mathbf{M}^{-1} \cdot \left[\begin{smallmatrix} h \\ k \end{smallmatrix}\right],$$

between subsequences $\mathbf{M} \cdot \mathcal{F}_m$ and $\mathbf{N} \cdot \mathcal{F}_m$ of the Farey sequence $\mathcal{F}_n$ is order-preserving and bijective.

The map

$$\mathcal{F}_n \supset \mathbf{M} \cdot \mathcal{F}_m \to \mathbf{N} \cdot \mathcal{F}_m \subset \mathcal{F}_n, \quad \left[\begin{smallmatrix} h \\ k \end{smallmatrix}\right] \mapsto \mathbf{N}\left[\begin{smallmatrix} -1 & 1 \\ 0 & 1 \end{smallmatrix}\right]\mathbf{M}^{-1} \cdot \left[\begin{smallmatrix} h \\ k \end{smallmatrix}\right],$$

is order-reversing and bijective; in particular, the map

$$\mathcal{F}_n \supset \mathbf{M} \cdot \mathcal{F}_m \to \mathbf{M} \cdot \mathcal{F}_m, \quad \left[\begin{smallmatrix} h \\ k \end{smallmatrix}\right] \mapsto \underbrace{\mathbf{M}\left[\begin{smallmatrix} -1 & 1 \\ 0 & 1 \end{smallmatrix}\right]\mathbf{M}^{-1}}_{\text{involutory}} \cdot \left[\begin{smallmatrix} h \\ k \end{smallmatrix}\right],$$

with the fixed point $\mathbf{M} \cdot \left[\begin{smallmatrix} 1 \\ 2 \end{smallmatrix}\right]$ in the case $m > 1$, is order-reversing and bijective.

3.5 Useful matrix products

Let us survey several subproducts of the matrix products mentioned in Remark 3.4, Lemma 3.8, Theorem 3.10, and in the proof of Proposition 3.6.

If $j \in \mathbb{Z}$, then

$$\boldsymbol{W}^j := \left[\begin{smallmatrix} 1 & 0 \\ 1 & 1 \end{smallmatrix}\right]^j = \left[\begin{smallmatrix} 1 & 0 \\ j & 1 \end{smallmatrix}\right], \quad \boldsymbol{Z}^j := \left[\begin{smallmatrix} 0 & 1 \\ -1 & 2 \end{smallmatrix}\right]^j = \left[\begin{smallmatrix} 1-j & j \\ -j & 1+j \end{smallmatrix}\right];$$

in particular,

$$\boldsymbol{W}^{-1} := \left[\begin{smallmatrix} 1 & 0 \\ 1 & 1 \end{smallmatrix}\right]^{-1} = \left[\begin{smallmatrix} 1 & 0 \\ -1 & 1 \end{smallmatrix}\right], \quad \boldsymbol{Z}^{-1} := \left[\begin{smallmatrix} 0 & 1 \\ -1 & 2 \end{smallmatrix}\right]^{-1} = \left[\begin{smallmatrix} 2 & -1 \\ 1 & 0 \end{smallmatrix}\right].$$

If $i \in \mathbb{Z}$, then

$$\boldsymbol{W}^i\boldsymbol{Z}^j := \left[\begin{smallmatrix} 1 & 0 \\ 1 & 1 \end{smallmatrix}\right]^i\left[\begin{smallmatrix} 0 & 1 \\ -1 & 2 \end{smallmatrix}\right]^j = \left[\begin{smallmatrix} 1-j & j \\ i-j-ij & 1+j+ij \end{smallmatrix}\right],$$
$$\boldsymbol{Z}^i\boldsymbol{W}^j := \left[\begin{smallmatrix} 0 & 1 \\ -1 & 2 \end{smallmatrix}\right]^i\left[\begin{smallmatrix} 1 & 0 \\ 1 & 1 \end{smallmatrix}\right]^j = \left[\begin{smallmatrix} 1-i+ij & i \\ -i+j+ij & 1+i \end{smallmatrix}\right].$$

We also have

$$\boldsymbol{W}^i\left[\begin{smallmatrix} -1 & 1 \\ 0 & 1 \end{smallmatrix}\right] := \left[\begin{smallmatrix} 1 & 0 \\ 1 & 1 \end{smallmatrix}\right]^i\left[\begin{smallmatrix} -1 & 1 \\ 0 & 1 \end{smallmatrix}\right] = \left[\begin{smallmatrix} -1 & 1 \\ -i & 1+i \end{smallmatrix}\right],$$
$$\boldsymbol{Z}^i\left[\begin{smallmatrix} -1 & 1 \\ 0 & 1 \end{smallmatrix}\right] := \left[\begin{smallmatrix} 0 & 1 \\ -1 & 2 \end{smallmatrix}\right]^i\left[\begin{smallmatrix} -1 & 1 \\ 0 & 1 \end{smallmatrix}\right] = \left[\begin{smallmatrix} -1+i & 1 \\ i & 1 \end{smallmatrix}\right],$$
$$\left[\begin{smallmatrix} -1 & 1 \\ 0 & 1 \end{smallmatrix}\right]\boldsymbol{W}^j := \left[\begin{smallmatrix} -1 & 1 \\ 0 & 1 \end{smallmatrix}\right]\left[\begin{smallmatrix} 1 & 0 \\ 1 & 1 \end{smallmatrix}\right]^j = \left[\begin{smallmatrix} -1+j & 1 \\ j & 1 \end{smallmatrix}\right],$$
$$\left[\begin{smallmatrix} -1 & 1 \\ 0 & 1 \end{smallmatrix}\right]\boldsymbol{Z}^j := \left[\begin{smallmatrix} -1 & 1 \\ 0 & 1 \end{smallmatrix}\right]\left[\begin{smallmatrix} 0 & 1 \\ -1 & 2 \end{smallmatrix}\right]^j = \left[\begin{smallmatrix} -1 & 1 \\ -j & 1+j \end{smallmatrix}\right],$$

and

$$\begin{aligned}
\boldsymbol{W}^i\left[\begin{smallmatrix}-1 & 1\\ 0 & 1\end{smallmatrix}\right]\boldsymbol{Z}^j &:= \left[\begin{smallmatrix}1 & 0\\ 1 & 1\end{smallmatrix}\right]^i\left[\begin{smallmatrix}-1 & 1\\ 0 & 1\end{smallmatrix}\right]\left[\begin{smallmatrix}0 & 1\\ -1 & 2\end{smallmatrix}\right]^j = \left[\begin{smallmatrix}-1 & 1\\ -i-j & 1+i+j\end{smallmatrix}\right],\\
\boldsymbol{Z}^i\left[\begin{smallmatrix}-1 & 1\\ 0 & 1\end{smallmatrix}\right]\boldsymbol{W}^j &:= \left[\begin{smallmatrix}0 & 1\\ -1 & 2\end{smallmatrix}\right]^i\left[\begin{smallmatrix}-1 & 1\\ 0 & 1\end{smallmatrix}\right]\left[\begin{smallmatrix}1 & 0\\ 1 & 1\end{smallmatrix}\right]^j = \left[\begin{smallmatrix}-1+i+j & 1\\ i+j & 1\end{smallmatrix}\right],\\
\boldsymbol{W}^i\left[\begin{smallmatrix}-1 & 1\\ 0 & 1\end{smallmatrix}\right]\boldsymbol{W}^j &:= \left[\begin{smallmatrix}1 & 0\\ 1 & 1\end{smallmatrix}\right]^i\left[\begin{smallmatrix}-1 & 1\\ 0 & 1\end{smallmatrix}\right]\left[\begin{smallmatrix}1 & 0\\ 1 & 1\end{smallmatrix}\right]^j = \left[\begin{smallmatrix}-1+j & 1\\ -i+j+ij & 1+i\end{smallmatrix}\right],\\
\boldsymbol{Z}^i\left[\begin{smallmatrix}-1 & 1\\ 0 & 1\end{smallmatrix}\right]\boldsymbol{Z}^j &:= \left[\begin{smallmatrix}0 & 1\\ -1 & 2\end{smallmatrix}\right]^i\left[\begin{smallmatrix}-1 & 1\\ 0 & 1\end{smallmatrix}\right]\left[\begin{smallmatrix}0 & 1\\ -1 & 2\end{smallmatrix}\right]^j = \left[\begin{smallmatrix}-1+i-ij & 1+ij\\ i-j-ij & 1+j+ij\end{smallmatrix}\right].
\end{aligned}$$

3.6 Order-reversing and bijective mapping $\frac{h}{k} \mapsto \frac{k-h}{k}$. II

Let us cast a parting look at Farey (sub)sequences and the mapping $\left[\begin{smallmatrix}h\\ k\end{smallmatrix}\right] \mapsto \left[\begin{smallmatrix}-1 & 1\\ 0 & 1\end{smallmatrix}\right] \cdot \left[\begin{smallmatrix}h\\ k\end{smallmatrix}\right]$, defined in (1.8) and surveyed in Table 1.2.

Let us assign to an ordered collection of matrices $(\boldsymbol{M}_1, \ldots, \boldsymbol{M}_t) \in \{\boldsymbol{W}, \boldsymbol{Z}\}^t$ a collection $(\boldsymbol{P}_1, \ldots, \boldsymbol{P}_t)$ such that

$$\boldsymbol{P}_i := \begin{cases} \boldsymbol{Z} & \text{if } \boldsymbol{M}_i = \boldsymbol{W},\\ \boldsymbol{W} & \text{if } \boldsymbol{M}_i = \boldsymbol{Z}, \end{cases}$$

$1 \le i \le t$. For any order m, the map

$$\begin{gathered}
\prod_{i=1}^{t} \boldsymbol{M}_{t-i+1} \cdot \mathcal{F}_m \to \prod_{i=1}^{t} \boldsymbol{P}_{t-i+1} \cdot \mathcal{F}_m,\\
\left[\begin{smallmatrix}h\\ k\end{smallmatrix}\right] \mapsto \left[\begin{smallmatrix}-1 & 1\\ 0 & 1\end{smallmatrix}\right] \cdot \left[\begin{smallmatrix}h\\ k\end{smallmatrix}\right], \quad \tfrac{h}{k} \mapsto \tfrac{k-h}{k},
\end{gathered}$$

is order-reversing and bijective.

Notes

Throughout Chapter 3, we manipulate elements of the general linear group $\mathrm{GL}(2, \mathbb{Z})$; see also the notes of Chapter 2. Here we merely find the inverses, and compute products of 2×2 matrices with integer entries and with determinant -1 or 1.

It is a basic fact in matrix analysis that the inverse $\boldsymbol{A}^{-1}$ of a matrix $\boldsymbol{A} := \left[\begin{smallmatrix}a & b\\ c & d\end{smallmatrix}\right]$, with $a, b, c, d \in \mathbb{Z}$ and $\det \boldsymbol{A} := ad - bc \in \{-1, 1\}$, is

$$\boldsymbol{A}^{-1} := \left[\begin{smallmatrix}a & b\\ c & d\end{smallmatrix}\right]^{-1} = \begin{cases} \left[\begin{smallmatrix}-d & b\\ c & -a\end{smallmatrix}\right] & \text{if } \det \boldsymbol{A} = -1,\\ \left[\begin{smallmatrix}d & -b\\ -c & a\end{smallmatrix}\right] & \text{if } \det \boldsymbol{A} = 1. \end{cases}$$

Monotone and bijective maps between Farey subsequences were interpreted in Section 2.1.2 as the restrictions of *linear fractional transformations* to finite subsets of real intervals, see Proposition 2.4. In particular, the order-preserving and bijective maps (2.13) and (2.15), that is,

$$\mathcal{F}_m \to \mathcal{F}^{\leq\frac{1}{2}}(\mathbb{B}(2m), m), \quad \left[\begin{smallmatrix} h \\ k \end{smallmatrix}\right] \mapsto \boldsymbol{W} \cdot \left[\begin{smallmatrix} h \\ k \end{smallmatrix}\right] := \left[\begin{smallmatrix} 1 & 0 \\ 1 & 1 \end{smallmatrix}\right] \cdot \left[\begin{smallmatrix} h \\ k \end{smallmatrix}\right],$$
$$\mathcal{F}_m \to \mathcal{F}^{\geq\frac{1}{2}}(\mathbb{B}(2m), m), \quad \left[\begin{smallmatrix} h \\ k \end{smallmatrix}\right] \mapsto \boldsymbol{Z} \cdot \left[\begin{smallmatrix} h \\ k \end{smallmatrix}\right] := \left[\begin{smallmatrix} 0 & 1 \\ -1 & 2 \end{smallmatrix}\right] \cdot \left[\begin{smallmatrix} h \\ k \end{smallmatrix}\right],$$

can be regarded as the strictly increasing maps

$$[\![0,1]\!] \supset \mathcal{F}_m \to \mathcal{F}^{\leq\frac{1}{2}}(\mathbb{B}(n), m) \subset [\![0, \tfrac{1}{2}]\!], \quad x \mapsto F_0^{-1}(x) := \frac{x}{1+x},$$
$$[\![0,1]\!] \supset \mathcal{F}_m \to \mathcal{F}^{\geq\frac{1}{2}}(\mathbb{B}(n), m) \subset [\![\tfrac{1}{2}, 1]\!], \quad x \mapsto F_1^{-1}(x) := \frac{1}{2-x},$$

respectively; see the definition (2.11) of the *modified Farey map* $F\colon [\![0,1]\!] \to [\![0,1]\!]$. From this point of view, the order-preserving and injective composite maps

$$\mathcal{F}_m \to \mathcal{F}_n, \quad \left[\begin{smallmatrix} h \\ k \end{smallmatrix}\right] \mapsto \boldsymbol{W}^s \cdot \left[\begin{smallmatrix} h \\ k \end{smallmatrix}\right],$$
$$\mathcal{F}_m \to \mathcal{F}_n, \quad \left[\begin{smallmatrix} h \\ k \end{smallmatrix}\right] \mapsto \boldsymbol{Z}^s \cdot \left[\begin{smallmatrix} h \\ k \end{smallmatrix}\right],$$

in Lemma 3.8, are the strictly increasing *iterated* maps

$$[\![0,1]\!] \supset \mathcal{F}_m \to \mathcal{F}_n \cap [\![0, \tfrac{1}{2}]\!], \quad x \mapsto F_0^{-s}(x),$$
$$[\![0,1]\!] \supset \mathcal{F}_m \to \mathcal{F}_n \cap [\![\tfrac{1}{2}, 1]\!], \quad x \mapsto F_1^{-s}(x),$$

respectively. Finally, the general order-preserving and injective maps

$$\mathcal{F}_m \to \mathcal{F}_n, \quad \left[\begin{smallmatrix} h \\ k \end{smallmatrix}\right] \mapsto \prod_{i=1}^{s} \boldsymbol{M}_{s-i+1} \cdot \left[\begin{smallmatrix} h \\ k \end{smallmatrix}\right],$$

in (3.53), are induced by length s compositions of the linear fractional transformations F_0^{-1} and F_1^{-1}.

The matrices $\left[\begin{smallmatrix} 0 & 1 \\ -1 & p \end{smallmatrix}\right] = \boldsymbol{W}^{p-2}\boldsymbol{Z}$, with $p \in \mathbb{P}$, are closely related to the *BKZ map* $\Upsilon\colon \mathcal{T} \to \mathcal{T}$, defined by (1.130). This map acts on the region $\mathcal{T}_p := \{\left[\begin{smallmatrix} x \\ y \end{smallmatrix}\right] \in \mathcal{T} \colon \lfloor \frac{1+x}{y} \rfloor = p\}$ of the *Farey triangle* $\mathcal{T} = \bigcup_{p \in \mathbb{P}} \mathcal{T}_p$ via $\Upsilon(\left[\begin{smallmatrix} x \\ y \end{smallmatrix}\right]) = \left[\begin{smallmatrix} 0 & 1 \\ -1 & p \end{smallmatrix}\right] \cdot \left[\begin{smallmatrix} x \\ y \end{smallmatrix}\right]$; see [15] and [24–26, 45, 162].

Bibliography

[1] C. M. Ablow and D. J. Kaylor, A committee solution of the pattern recognition problem, *IEEE Trans. Inform. Theory* **11** (1965), no. 3, 453–455.

[2] C. M. Ablow and D. J. Kaylor, Inconsistent homogeneous linear inequalities, *Bull. Amer. Math. Soc.* **71** (1965), 724.

[3] D. M. Acketa and J. D. Žunić, On the number of linear partitions of the (m, n)-grid, *Inform. Process. Lett.* **38** (1991), no. 3, 163–168.

[4] D. M. Acketa and J. D. Žunić, An asymptotically optimal algorithm for construction of optimal digital star-shaped polygons, *Zb. Rad. Prirod.-Mat. Fak. Ser. Mat.* **24** (1994), no. 2, 101–116.

[5] M. Aigner, *Combinatorial Theory*, Classics Math., Springer, Berlin, 1997.

[6] M. Aigner, *A Course in Enumeration*, Grad. Texts in Math. 238, Springer, Berlin, 2007.

[7] M. Aigner, *Markov's Theorem and 100 Years of the Uniqueness Conjecture*, Springer, Cham, 2013.

[8] T. Andreescu and D. Andrica, *Number Theory. Structures, Examples, and Problems*, Birkhäuser, Boston, 2009.

[9] T. Andreescu, D. Andrica and I. Cucurezeanu, *An Introduction to Diophantine Equations. A Problem-Based Approach*, Birkhäuser, New York, 2010.

[10] W. S. Anglin, *The Queen of Mathematics. An Introduction to Number Theory*, Kluwer Texts Math. Sci. 8, Kluwer Academic Publishers, Dordrecht, 1995.

[11] T. M. Apostol, *Introduction to Analytic Number Theory*, Undergrad. Texts Math., Springer, New York, 1976.

[12] T. M. Apostol, *Modular Functions and Dirichlet Series in Number Theory*, 2nd ed., Grad. Texts in Math. 41, Springer, New York, 1990.

[13] R. Artuso, P. Cvitanović and G. Tanner, Cycle expansions for intermittent maps, *Progr. Theoret. Phys. Suppl.* (2003), no. 150, 1–21.

[14] A. Ash and R. Gross, *Fearless Symmetry*, Princeton University Press, Princeton, 2006.

[15] J. S. Athreya and Y. Cheung, A Poincaré section for the horocycle flow on the space of lattices, *Int. Math. Res. Not. IMRN* **2014** (2014), no. 10, 2643–2690.

[16] A. Baker, *A Comprehensive Course in Number Theory*, Cambridge University Press, Cambridge, 2012.

[17] M. W. Baldoni, C. Ciliberto and G. M. Piacentini Cattaneo, *Elementary Number Theory, Cryptography and Codes*, Universitext, Springer, Berlin, 2009.

[18] A. H. Beiler, *Recreations in the Theory of Numbers. The Queen of Mathematics Entertains*, 2nd ed., Dover Publications, New York, 1966.

[19] S. Ben Ammou, C. Bonanno, I. Chouari and S. Isola, On the spectrum of the transfer operators of a one-parameter family with intermittency transition, *Far East J. Dyn. Syst.* **27** (2015), no. 1, 13–25.

[20] A. Björner, Subspace arrangements, in: *First European Congress of Mathematics. Vol. I (Paris 1992)*, Progr. Math. 119, Birkhäuser, Basel (1994), 321–370.

[21] A. Björner, M. Las Vergnas, B. Sturmfels, N. White and G. M. Ziegler, *Oriented Matroids*, 2nd ed., Encyclopedia Math. Appl. 46, Cambridge University Press, Cambridge, 1999.

[22] J. A. Blake, Mathematical notes: Some characteristic properties of the Farey series, *Amer. Math. Monthly* **73** (1966), no. 1, 50–52.

[23] F. P. Boca, An AF algebra associated with the Farey tessellation, *Canad. J. Math.* **60** (2008), no. 5, 975–1000.

[24] F. P. Boca, C. Cobeli and A. Zaharescu, A conjecture of R. R. Hall on Farey points, *J. Reine Angew. Math.* **535** (2001), 207–236.

DOI 10.1515/9783110547665-005

[25] F. P. Boca and A. Zaharescu, Farey fractions and two-dimensional tori, in: *Noncommutative Geometry and Number Theory*, Aspects Math. E37, Friedr. Vieweg, Wiesbaden (2006), 57–77.

[26] F. P. Boca, A. Zaharescu and R. N. Gologan, On the index of Farey sequences, *Q. J. Math.* **53** (2002), no. 4, 377–391.

[27] J. C. A. Boeyens, *New Theories for Chemistry*, Elsevier, Amsterdam, 2005.

[28] J. C. A. Boeyens, *Chemistry from First Principles*, Springer, New York, 2008.

[29] J. C. A. Boeyens, *Chemical Cosmology*, Springer, Dordrecht, 2010.

[30] J. C. A. Boeyens and D. C. Levendis, *Number Theory and the Periodicity of Matter*, Springer, New York, 2008.

[31] J. G. Bokowski, *Computational Oriented Matroids. Equivalence Classes of Matrices Within a Natural Framework*, Cambridge University Press, Cambridge, 2006.

[32] M. Bóna, *A Walk Through Combinatorics. An Introduction to Enumeration and Graph Theory. With a Foreword by Richard Stanley*, 3rd ed., World Scientific, Hackensack, 2011.

[33] M. Bóna, *Handbook of Enumerative Combinatorics*, Discrete Math. Appl. (Boca Raton), CRC Press, Boca Raton, 2015.

[34] F. Bonahon, *Low-Dimensional Geometry. From Euclidean Surfaces to Hyperbolic Knots*, Stud. Math. Libr. 49, American Mathematical Society, Providence, 2009.

[35] C. Bonanno, S. Graffi and S. Isola, Spectral analysis of transfer operators associated to Farey fractions, *Atti Accad. Naz. Lincei Rend. Lincei Mat. Appl.* **19** (2008), no. 1, 1–23.

[36] C. Bonanno and S. Isola, Orderings of the rationals and dynamical systems, *Colloq. Math.* **116** (2009), no. 2, 165–189.

[37] O. Bordellès, *Arithmetic Tales*, Universitext, Springer, London, 2012.

[38] N. V. Bougaïeff, *The Theory of Numerical Derivatives* (in Russian), Mamontov, Moscow, 1876.

[39] R. A. Brualdi, *Introductory Combinatorics*, 5th ed., Pearson Prentice Hall, Upper Saddle River, 2010.

[40] M. Bruckheimer and A. Arcavi, Farey series and Pick's area theorem, *Math. Intelligencer* **17** (1995), no. 4, 64–67.

[41] A. A. Buhštab, *Number Theory* (in Russian), 2nd corr. ed., Izdat. "Prosveščenie", Moscow, 1966.

[42] D. M. Burton, *Elementary Number Theory*, 7th ed., McGraw–Hill, New York, 2007.

[43] E. Busche, Über eine Kroneckersche Beziehung zwischen Geometrie und Zahlentheorie, *Math. Ann.* **60** (1905), no. 2, 285–316.

[44] A.-L. Cauchy, Démonstration d'un théorême curieux sur les nombres, in: *Exercices de Mathématiques*, Paris (1826), 114–116.

[45] C. Cobeli and A. Zaharescu, The Haros–Farey sequence at two hundred years, *Acta Univ. Apulensis Math. Inform.* (2003), no. 5, 1–38.

[46] M. C. Crabb, Counting nilpotent endomorphisms, *Finite Fields Appl.* **12** (2006), no. 1, 151–154.

[47] P. Cvitanović, Circle maps: Irrationally winding, in: *From Number Theory to Physics (Les Houches 1989)*, Springer, Berlin (1992), 631–658.

[48] C. De Concini and C. Procesi, *Topics in Hyperplane Arrangements, Polytopes and Box-Splines*, Universitext, Springer, New York, 2011.

[49] M. Degli Esposti, S. Isola and A. Knauf, Generalized Farey trees, transfer operators and phase transitions, *Comm. Math. Phys.* **275** (2007), no. 2, 297–329.

[50] L. E. Dickson, *History of the Theory of Numbers. Vol. I: Divisibility and Primality*, Chelsea Publishing, New York, 1952.

[51] A. Dimca, *Hyperplane Arrangements. An Introduction*, Universitext, Springer, Cham, 2017.

[52] M. D'Ocagne, Sur certaines suites de fractions irréductibles, *Bull. Soc. Math. France* **14** (1886), 93–97.

[53] P. S. Dodds, *On the thermodynamic formalism for the Farey map*, Masters thesis, University of Melbourne, Victoria, 1995.

[54] U. Dudley, *A Guide to Elementary Number Theory*, Dolciani Math. Exp. 41, Mathematical Association of America, Washington, 2009.

[55] J. Farey, On a curious property of vulgar fractions, *Philos. Mag.* **47** (1816), 385–386.

[56] J. Farey, Propriété curieuse des fractions ordinaires, *Bull. Sci. Soc. Phil.* **3** (1816), 105–112.

[57] M. J. Feigenbaum, Presentation functions, fixed points, and a theory of scaling function dynamics, *J. Statist. Phys.* **52** (1988), no. 3-4, 527–569.

[58] M. J. Feigenbaum, I. Procaccia and T. Tél, Scaling properties of multifractals as an eigenvalue problem, *Phys. Rev. A (3)* **39** (1989), no. 10, 5359–5372.

[59] J. Fiala and P. Kleban, Generalized number theoretic spin chain-connections to dynamical systems and expectation values, *J. Stat. Phys.* **121** (2005), no. 3-4, 553–577.

[60] J. Fiala, P. Kleban and A. Özlük, The phase transition in statistical models defined on Farey fractions, *J. Statist. Phys.* **110** (2003), no. 1-2, 73–86.

[61] B. Fine and G. Rosenberger, *Number Theory. An Introduction via the Density of Primes*, 2nd ed., Birkhäuser/Springer, Cham, 2016.

[62] D. N. Gainanov, *Graphs for Pattern Recognition. Infeasible Systems of Linear Inequalities*, De Gruyter, Berlin, 2016.

[63] S. B. Gashkov and V. N. Chubarikov, *Arithmetic. Algorithms. Complexity of Computation* (in Russian), 3rd ed., Drofa, Moscow, 2005.

[64] J. W. L. Glaisher, On a property of vulgar fractions, *Philos. Mag. Ser. 5* **7** (1879), no. 44, 321–336.

[65] R. L. Graham, D. E. Knuth and O. Patashnik, *Concrete Mathematics: A Foundation for Computer Science*, 2nd ed., Addison-Wesley, Reading, 1994.

[66] H. Goodwyn, *The First Centenary of a Series of Concise and Useful Tables of all the Complete Decimal Quotients, Which Can Arise from Dividing a Unit, or Any Whole Number less than Each Divisor by all Integers from* 1 *to* 1024. *To which is now added a tabular series of complete decimal quotients for all the proper vulgar fractions of which when in their lowest terms, neither the numerator nor the denominator is greater than* 100: *With the equivalent vulgar fractions prefixed*, Payne & Foss, London, 1818.

[67] S. B. Guthery, *A Motif of Mathematics. History and Application of the Mediant and the Farey Sequence*, Docent Press, Boston, 2011.

[68] R. R. Hall, A note on Farey series, *J. Lond. Math. Soc. (2)* **2** (1970), 139–148.

[69] R. R. Hall, The parity of Farey denominators and the Farey index, *J. Number Theory* **115** (2005), no. 1, 71–86.

[70] R. R. Hall and P. Shiu, The index of a Farey sequence, *Michigan Math. J.* **51** (2003), no. 1, 209–223.

[71] R. R. Hall and G. Tenenbaum, On consecutive Farey arcs, *Acta Arith.* **44** (1984), no. 4, 397–405.

[72] G. H. Hardy and E. M. Wright, *An Introduction to the Theory of Numbers*, 6th ed., Oxford University Press, Oxford, 2008.

[73] C. Haros, Tables pour évaluer une fraction ordinaire avec autant de decimals qu'on voudra; et pour trouver la fraction ordinaire la plus simple, et qui a approche sensiblement d'une fraction décimale, *J. Éc. Polytechnique* **6** (1802), no. 11, 364–368.

[74] M. Hata, *Neurons—A Mathematical Ignition*, Ser. Number Theory Appl. 9, World Scientific, Hackensack, 2015.

[75] A. Hatcher, *Topology of Numbers*, in preparation, http://www.math.cornell.edu/~hatcher/TN/TNpage.html.

[76] A. Haynes, The distribution of special subsets of the Farey sequence, *J. Number Theory* **107** (2004), no. 1, 95–104.

[77] B. Heersink, An effective estimate for the Lebesgue measure of preimages of iterates of the Farey map, *Adv. Math.* **291** (2016), 621–634.

[78] B. Heersink, Poincaré sections for the horocycle flow in covers of $SL(2,\mathbb{R})/SL(2,\mathbb{Z})$ and applications to Farey fraction statistics, *Monatsh. Math.* **179** (2016), no. 3, 389–420.

[79] L. K. Hua, *Introduction to Number Theory*, Springer, Berlin, 1982.

[80] M. N. Huxley, *The Distribution of Prime Numbers. Large Sieves and Zero-Density Theorems*, Oxford Math. Monogr., Clarendon Press, Oxford, 1972.

[81] M. N. Huxley, *Area, Lattice Points, and Exponential Sums*, London Math. Soc. Monogr. (N. S.) 13, Clarendon Press, New York, 1996.

[82] S. Isola, On the spectrum of Farey and Gauss maps, *Nonlinearity* **15** (2002), no. 5, 1521–1539.

[83] S. Isola, From infinite ergodic theory to number theory (and possibly back), *Chaos Solitons Fractals* **44** (2011), no. 7, 467–479.

[84] S. Isola, Continued fractions and dynamics, *Appl. Math.* **5** (2014), 1067–1090.

[85] S. Ito, Algorithms with mediant convergents and their metrical theory, *Osaka J. Math.* **26** (1989), no. 3, 557–578.

[86] M. Jarnicki and P. Pflug, *Continuous Nowhere Differentiable Functions. The Monsters of Analysis*, Springer Monogr. Math., Springer, Cham, 2015.

[87] O. Karpenkov, *Geometry of Continued Fractions*, Algorithms Comput. Math. 26, Springer, Heidelberg, 2013.

[88] M. Kesseböhmer, S. Munday and B. O. Stratmann, *Infinite Ergodic Theory of Numbers*, De Gruyter Grad., De Gruyter, Berlin, 2016.

[89] M. Kesseböhmer and B. O. Stratmann, On the asymptotic behaviour of the Lebesgue measure of sum-level sets for continued fractions, *Discrete Contin. Dyn. Syst.* **32** (2012), no. 7, 2437–2451.

[90] M. Y. Khachai, V. D. Mazurov and A. I. Rybin, Committee constructions for solving problems of selection, diagnostics, and prediction, *Proc. Steklov Inst. Math.* (2002), no. S.1, S67–S101.

[91] S. Khrushchev, *Orthogonal Polynomials and Continued Fractions. From Euler's Point of View*, Encyclopedia Math. Appl. 122, Cambridge University Press, Cambridge, 2008.

[92] V. V. Kisil, *Geometry of Möbius Transformations. Elliptic, Parabolic and Hyperbolic Actions of* $SL_2(\mathbb{R})$, Imperial College Press, London, 2012.

[93] D. E. Knuth, *The Art of Computer Programming. Vol. 1: Fundamental Algorithms*, 2nd ed., Addison-Wesley, Reading, 1969.

[94] T. Koshy, *Elementary Number Theory with Applications*, 2nd ed., Academic Press, Burlington, 2007.

[95] J. S. Kraft and L. C. Washington, *An Introduction to Number Theory with Cryptography*, CRC Press, Boca Raton, 2014.

[96] J. P. S. Kung, G.-C. Rota and C. H. Yan, *Combinatorics: The Rota Way*, Cambridge Math. Lib., Cambridge University Press, Cambridge, 2009.

[97] J. C. Lagarias, Number theory and dynamical systems, in: *The Unreasonable Effectiveness of Number Theory (Orono 1991)*, Proc. Sympos. Appl. Math. 46, American Mathematical Society, Providence (1992), 35–72.

[98] J. C. Lagarias and H. Mehta, Products of binomial coefficients and unreduced Farey fractions, *Int. J. Number Theory* **12** (2016), no. 1, 57–91.

[99] J. C. Lagarias and H. Mehta, Products of Farey fractions, *Exp. Math.* **26** (2017), no. 1, 1–21.

[100] J. C. Lagarias and C. P. Tresser, A walk along the branches of the extended Farey tree, *IBM J. Res. Develop.* **39** (1995), no. 3, 283–294.

[101] E. Landau, *Elementary Number Theory*, Chelsea Publishing, New York, 1958.
[102] S. Lang, $SL_2(\mathbb{R})$, Grad. Texts in Math. 105, Springer, New York, 1985.
[103] M. Lewinter and J. Meyer, *Elementary Number Theory with Programming*, John Wiley & Sons, Hoboken, 2016.
[104] R. Lipschitz, Sur des séries relatives à la théorie des nombres, *C. R. Acad. Sci. Paris* **89** (1879), 948–950.
[105] N. A. Loehr, *Bijective Combinatorics*, Discrete Math. Appl. (Boca Raton), CRC Press, Boca Raton, 2011.
[106] S. Lozić, V. Mudrinski and J. D. Žunić, Some results in discrete mathematics and digital geometry by Professor Dragan Acketa, *Novi Sad J. Math.* **33** (2003), no. 1, 1–14.
[107] A. O. Matveev, A note on Boolean lattices and Farey sequences, *Integers* **7** (2007), Paper No. A20.
[108] A. O. Matveev, Relative blocking in posets, *J. Comb. Optim.* **13** (2007), no. 4, 379–403; corrigendum, http://arxiv.org/abs/math/0411026.
[109] A. O. Matveev, A note on Boolean lattices and Farey sequences. II, *Integers* **8** (2008), Paper No. A24.
[110] A. O. Matveev, Neighboring fractions in Farey subsequences, preprint (2008), http://arxiv.org/abs/0801.1981.
[111] A. O. Matveev, A note on Boolean lattices and Farey sequences. III, preprint (2015), http://arxiv.org/abs/1507.07236.
[112] A. O. Matveev, *Pattern Recognition on Oriented Matroids*, De Gruyter, Berlin, 2017.
[113] D. R. Mazur, *Combinatorics. A Guided Tour*, MAA Textbooks, Mathematical Association of America, Washington, 2010.
[114] V. D. Mazurov, *The Committee Method for Optimization and Classification Problems* (in Russian), "Nauka", Moscow, 1990.
[115] V. D. Mazurov and M. Y. Khachai, Committees of systems of linear inequalities, *Avtomat. i Telemekh.* (2004), no. 2, 43–54.
[116] E. Meissel, Observations quaedam in theoria numerorum, *J. Reine Angew. Math.* **48** (1854), 301–316.
[117] R. S. Millman, P. J. Shiue and E. B. Kahn, *Problems and Proofs in Numbers and Algebra*, Springer, Cham, 2015.
[118] R. A. Mollin, *Fundamental Number Theory with Applications*, 2nd ed., Discrete Math. Appl. (Boca Raton), Chapman & Hall/CRC, Boca Raton, 2008.
[119] L. Moser, *The Trillia Lectures on Mathematics. An Introduction to the Theory of Numbers*, Trillia Group, West Lafayette, 2009.
[120] N. Moshchevitin and A. Zhigljavsky, Entropies of the partitions of the unit interval generated by the Farey tree, *Acta Arith.* **115** (2004), no. 1, 47–58.
[121] M. R. Murty, *Problems in Analytic Number Theory. Readings in Mathematics*, 2nd ed., Grad. Texts in Math. 206, Springer, New York, 2008.
[122] W. Narkiewicz, *The Development of Prime Number Theory. From Euclid to Hardy and Littlewood*, Springer Monogr. Math., Springer, Berlin, 2000.
[123] M. B. Nathanson, *Elementary Methods in Number Theory*, Grad. Texts in Math. 195, Springer, New York, 2000.
[124] I. Niven, H. S. Zuckerman and H. L. Montgomery, *An Introduction to the Theory of Numbers*, 5th ed., John Wiley & Sons, New York, 1991.
[125] P. Orlik and H. Terao, *Arrangements of Hyperplanes*, Grundlehren Math. Wiss. 300, Springer, Berlin, 1992.
[126] P. Orlik and V. Welker, *Algebraic Combinatorics. Lectures from the Summer School held in Nordfjordeid, June 2003*, Universitext, Springer, Berlin, 2007.

[127] B. Paria, S. Pratihar and P. Bhowmick, On Farey table and its compression for space optimization with guaranteed error bounds, *Math. Appl. (Brno)* **5** (2016), no. 2, 123–145.

[128] C. E. Pătraşcu and M. Pătraşcu, Computing order statistics in the Farey sequence, in: *Algorithmic Number Theory*, Lecture Notes in Comput. Sci. 3076, Springer, Berlin (2004), 358–366.

[129] J. Pawlewicz and M. Pătraşcu, Order statistics in the Farey sequences in sublinear time and counting primitive lattice points in polygons, *Algorithmica* **55** (2009), no. 2, 271–282.

[130] P. Philippon, A Farey tail, *Notices Amer. Math. Soc.* **59** (2012), no. 6, 746–757.

[131] P. Pongsriiam, A remark on relatively prime sets, *Integers* **13** (2013), Paper No. A49.

[132] V. Pop and O. Furdui, *Square Matrices of Order 2. Theory, Applications, and Problems*, Springer, Cham, 2017.

[133] V. V. Prasolov, *Problems in Algebra, Arithmetic and Analysis* (in Russian), MCCME, Moscow, 2007.

[134] T. Prellberg, Towards a complete determination of the spectrum of a transfer operator associated with intermittency, *J. Phys. A* **36** (2003), no. 10, 2455–2461.

[135] T. Prellberg, J. Fiala and P. Kleban, Cluster approximation for the Farey fraction spin chain, *J. Stat. Phys.* **123** (2006), no. 2, 455–471.

[136] T. Prellberg and J. Slawny, Maps of intervals with indifferent fixed points: Thermodynamic formalism and phase transitions, *J. Statist. Phys.* **66** (1992), no. 1-2, 503–514.

[137] H. Rademacher, *Lectures on Elementary Number Theory*, Robert E. Krieger, Huntington, 1977.

[138] R. A. Rankin, *Modular Forms and Functions*, Cambridge University Press, Cambridge, 1977.

[139] M. T. Rassias, *Problem-Solving and Selected Topics in Number Theory*, Springer, New York, 2011.

[140] P. Ribenboim, *The Little Book of Bigger Primes*, 2nd ed., Springer, New York, 2004.

[141] H. Riesel, *Prime Numbers and Computer Methods for Factorization*, Mod. Birkhäuser Class., Birkhäuser/Springer, New York, 2012.

[142] K. H. Rosen, *Elementary Number Theory and Its Applications*, 5th ed., Addison-Wesley, Reading, 2005.

[143] N. Routledge, Computing Farey series, *Math. Gazette* **92** (2008), no. 523, 55–62.

[144] W. M. Schmidt, *Diophantine Approximation*, 2nd printing, Lecture Notes in Math. 785, Springer, Berlin, 1996.

[145] M. R. Schroeder, *Number Theory in Science and Communication. With Applications in Cryptography, Physics, Digital Information, Computing, and Self-Similarity*, 5th ed., Springer Ser. Inform. Sci. 7, Springer, Berlin, 2009.

[146] T. Shonhiwa, On relatively prime sets counting functions, *Integers* **10** (2010), no. A39, 465–476.

[147] W. Sierpiński, *Elementary Theory of Numbers*, 2nd ed., North-Holland Math. Libr. 31, North-Holland, Amsterdam, 1988.

[148] R. P. Stanley, An introduction to hyperplane arrangements, in: *Geometric Combinatorics*, IAS/Park City Math. Ser. 13, American Mathematical Society, Providence (2007), 389–496.

[149] R. P. Stanley, *Enumerative Combinatorics. Volume 1*, 2nd ed., Cambridge Stud. Adv. Math. 49, Cambridge University Press, Cambridge, 2012.

[150] H. M. Stark, *An Introduction to Number Theory*, MIT Press, Cambridge, 1978.

[151] J. Stillwell, *Elements of Number Theory*, Undergrad. Texts Math., Springer, New York, 2003.

[152] M. V. Stouvenel, Note sur une certaine suite des fractions ordinaires, *J. Math. Pures Appl.* **5** (1840), no. 5, 265–275.

[153] I. Svalbe and A. Kingston, On correcting the unevenness of angle distributions arising from integer ratios lying in restricted portions of the Farey plane, in: *Combinatorial Image Analysis*, Lecture Notes in Comput. Sci. 3322, Springer, Berlin, (2004), 110–121.

[154] J. J. Tattersall, *Elementary Number Theory in Nine Chapters*, 2nd ed., Cambridge University Press, Cambridge, 2005.

[155] G. Tenenbaum and M. Mendès France, *The Prime Numbers and Their Distribution*, Stud. Math. Libr. 6, American Mathematical Society, Providence, 2000.

[156] R. Tomás, Asymptotic behavior of a series of Euler's totient function $\varphi(k)$ times the index of $1/k$ in a Farey sequence, preprint (2014), http://arxiv.org/abs/1406.6991.

[157] L. Tóth, On the number of certain relatively prime subsets of $\{1, 2, \dots, n\}$, *Integers* **10** (2010), no. A35, 407–421.

[158] J. H. van Lint and R. M. Wilson, *A Course in Combinatorics*, 2nd ed., Cambridge University Press, Cambridge, 2001.

[159] B. A. Venkov, *Elementary Number Theory*, Wolters-Noordhoff, Groningen, 1970.

[160] H. S. Warren, Jr., *Hacker's Delight*, 2nd ed., Addison-Wesley, Upper Saddle River, 2012.

[161] J. J. Watkins, *Number Theory. A Historical Approach*, Princeton University Press, Princeton, 2014.

[162] M. Xiong and A. Zaharescu, A problem of Erdős–Szüsz–Turán on Diophantine approximation, *Acta Arith.* **125** (2006), no. 2, 163–177.

[163] S. Y. Yan, *Number Theory for Computing*, 2nd ed., Springer, Berlin, 2002.

[164] S. Y. Yan, *Computational Number Theory and Modern Cryptography*, John Wiley & Sons, Singapore, 2013.

[165] A. Zame, Mathematical Notes: A Comment on a Recent Note on the Farey Series, *Amer. Math. Monthly* **74** (1967), no. 8, 977.

[166] U. Zannier, *Lecture Notes on Diophantine Analysis*, Appunti. Sc. Norm. Super. Pisa (N. S.) 8, Edizioni della Normale, Pisa, 2009.

[167] OEIS, The on-line encyclopedia of integer sequences, (2017), http://oeis.org.

[168] Sage Math Cloud, Collaborative computational mathematics, (2017), http://cloud.sagemath.com.

List of notation

- $\mathbb{P}$ – positive integers
- $\mathbb{N}$ – nonnegative integers
- $\mathbb{Q}$ – rational numbers
- $\mathbb{R}$ – real numbers
- $\mathbb{C}$ – complex numbers
- $:=$ – equal by definition
- $|A|$ – the number of elements in a finite set A
- $\gcd(a, b)$ – greatest common divisor of integers a and b
- $\dim \mathbf{A}$ – dimension of a vector space $\mathbf{A}$
- $[a, b]$ – set $\{a, a+1, \dots, b\}$ of consecutive integers
- $[\![a, b]\!]$ – interval $\{x \in \mathbb{R} : a \le x \le b\}$ of real numbers
- $\mathbb{B}(n)$ – Boolean lattice of rank n
- $\rho(a)$ – rank function of a poset
- $\mathbb{V}(n)$ – n-dimensional vector space
- $\dot{\cup}$ – disjoint union
- $\lfloor x \rfloor$ – greatest integer less than or equal to x (floor function)
- $\lceil x \rceil$ – least integer greater than or equal to x (ceiling function)
- $\varphi(n; [s', s''])$ – quantity $|\{s \in [s', s''] : \gcd(n, s) = 1\}|$
- $\varphi(n)$ – Euler's totient function
- $d|n$ – d divides n
- $\mu(d)$ – Möbius function
- $M(x)$ – Mertens function
- $\binom{n}{j}$ – binomial coefficient $\frac{n!}{j!(n-j)!}$; the number of j-subsets of an n-set
- $\boldsymbol{W}$ – matrix $\left[\begin{smallmatrix} 1 & 0 \\ 1 & 1 \end{smallmatrix}\right]$
- $\boldsymbol{Z}$ – matrix $\left[\begin{smallmatrix} 0 & 1 \\ -1 & 2 \end{smallmatrix}\right]$

DOI 10.1515/9783110547665-006

Index

DOI 10.1515/9783110547665-007

Bei Fragen zur Produktsicherheit wenden Sie sich bitte an:
If you have any questions regarding product safety,
please contact:

Walter de Gruyter GmbH
Genthiner Straße 13
10785 Berlin
productsafety@degruyterbrill.com